# STEREOCHEMISTRY RESEARCH TRENDS

# Stereochemistry Research Trends

Monika A. Horvat and
Jana H. Golob
Editors

Nova Science Publishers, Inc.
*New York*

For permission to use material from this book please contact us:
Telephone 631-231-7269; Fax 631-231-8175
Web Site: http://www.novapublishers.com

**NOTICE TO THE READER**



**LIBRARY OF CONGRESS CATALOGING-IN-PUBLICATION DATA**

Stereochemistry research trends / Monika A. Horvat and Jana H. Golob, editors.
p. cm.
ISBN 978-1-60456-825-7 (hardcover)
1. Stereochemistry. I. Horvat, Monika A. II. Golob, Jana H.
QD481.S764 2008
541'.223--dc22
2008022895

*Published by Nova Science Publishers, Inc. ✦ New York*

# CONTENTS

# PREFACE

Stereochemistry is an important facet of chemistry. The study of stereochemical problems spans the entire range of organic, inorganic, biological, physical and supramolecular chemistries. Stereochemistry includes methods for determining and describing these relationships; the effect on the physical or biological properties these relationships impart upon the molecules in question, and the manner in which these relationships influence the reactivity of the molecules in question (dynamic stereochemistry). This book provides new research from around the globe on this field.

Chapter 1 – In this text is presented a summary of the synthetic strategies used to build up either D- or L-pyranose skeleton of sugar framework structures under stereoselective methodologies. Polihydroxylated piperidines were generated from simple starting material like amino-acids, tartaric acid or glyceraldehyde derivatives or even from achiral reagents. Introduction of chiral centres in achiral materials intermediates are generally achieved by catalytic methods, using different classes of ligands. In other cases, diastereoselective methods and chemoenzymatic synthesis have been developed.

Chapter 2 - Both enantiomers of a pheromone from the tea tussock moth were synthesized from (*S*)- and (*R*)-citronellol to examine the stereochemistry of the natural product. The absolute configurations of hydroxy-fatty acids isolated from the resinous secretion of lac insects were determined by NMR. The novel chiral resolving agents, MαNP acid, MαNP acid, and M9PP acid, were synthesized by asymmetric induction, using L-(–)-menthol. The absolute configurations of these agents and the conformations of their ester derivatives were determined by X-ray crystallography and NMR. The utilities of the single-enantiomer MαNP and M9PP acids were demonstrated by enantioresolution of insect pheromones. Both MαNP and M9PP acids exhibited superior resolution ability in HPLC and powerful shielding in $^1$H NMR.

Chapter 3 - Aflastatin A and blasticidin A are aflatoxin production inhibitors produced by *Streptomyces* sp. They have similar unique structures, which are tetramic acid derivatives with a highly oxygenated long alkyl chain containing a lot of chiral centers. In this article, our studies on stereochemistry of aflastatin A and blasticidin A are reviewed. Several fragment molecules were prepared from each of aflastatin A and blasticidin A. Relative or absolute configurations of the fragment molecules were assigned by a variety of methods including *J*-based, acetonide, MTPA and NMR database ones and by preparing model compounds. Basic concept and application of *J*-based method are also reviewed since the method was effectively used for assigning the relative stereochemistry of the acyclic moieties of aflastatin A and blasticidin A.

Chapter 4 – Nine new complexes of assemblies containing diastereomers of nickel(II) complexes incorporating 1,2-diaminocyclohexane ligands and tetracyanometalates, [Ni(*RR*-chxn)$_3$][M(CN)$_4$]·3H$_2$O (M = Ni (*1*), Pt (*4*), Pd (*7*)), [Ni(*trans*-chxn)$_3$][M(CN)$_4$]·3H$_2$O (M = Ni (*2*), Pt (*5*), Pd (*8*)), and [Ni(*cis*-chxn)$_3$][M(CN)$_4$]·3H$_2$O (M = Ni (*3*), Pt (*6*), Pd (*9*)) (*RR*-chxn = cyclohexane-(1*R*,2*R*)-diamine, *trans*-chxn = racemic *trans*-cyclohexane-(1,2)-diamine, and *cis*-chxn = racemic *cis*-cyclohexane-(1,2)-diamine) have been synthesized and characterized by SQUID and IR, XPS, and electronic spectra. Crystal structure of *1* (space group $P2_12_12_1$) revealed that both $\Delta$- and $\Lambda$-[Ni(*RR*-chxn)$_3$]$^{2+}$ cations and isolated square planar [Ni(CN)$_4$]$^{2-}$ anions without cyano-bridges were contained in the assembly. The emission spectra of *4*, *5*, *6* ($\lambda_{ex}$ = 360 nm) exhibited that different intensity of a peak around 470 nm which depended on interaction of quenching between paramagnetic [Ni(*RR*-chxn)$_3$]$^{2+}$ moieties and [Pt(CN)$_4$]$^{2-}$ moieties. The CD spectra for *1*, *4*, and *7* showed not only d-d bands around 20000 cm$^{-1}$ of [Ni(*RR*-chxn)$_3$]$^{2+}$ moieties by vicnal effect but also charge transfer bands of [M(CN)$_4$]$^{2-}$ moieties induced by chiral ligands.

Chapter 5 - (4*R**,5*R**)-4-Benzoylamino-5-phenyl-3-pyrazolidinone (*1*) is a useful building block, which is available in two steps from *N*-benzoylglycine *via* transformation into the azlactone followed by treatment with excess hydrazine hydrate. Since pyrazolidinone *1* is actually a cyclic $\alpha$-amino acid derivative, it can be successfully employed for the preparation of functionalized pyrazoles containing an $\alpha$-amino acid or a dipeptide structural motif. The most useful derivatives of *1* are stable (1*Z*,4*R**,5*R**)-1-alkylidene-4-benzoylamino-5-phenyl-3-pyrazolidinone-1-azomethine imines *6*, which are obtained in the reactions of *1* with aldehydes and ketones. Azomethine imines *6* derived from aromatic aldehydes are particularly stable and easily available by parallel synthesis. Reduction of azomethine imines *6* with sodium borohydride leads to 1-alkylated 3-pyrazolidinones, which were transformed into (2*R**,3*R**)-3-alkylamino-*N*-benzoyl-3-phenylalaninamides *via* reductive N–N bond cleavage. Similar N–N bond fission was observed also in base-catalyzed reactions of *6* with potassium cyanide. With aliphatic 1,3-diketones in methanol, pyrazolidinone *1* undergoes a 'ring switching' transformation into (2*R**,3*R**)-*N*-benzoyl-3-alkylamino-3-phenyl-3-(1*H*-pyrazol-1-yl)alanine methyl esters. Analogous 3-pyrazolylalanine esters were also obtained in two-steps *via* treatment of azomethine imines *6* with dimethyl acetylenedicarboxylate followed by oxidative opening of the pyrazolidinone ring. So far, [3+2] cycloadditions of (1*Z*,4*R**,5*R**)-1-alkylidene-4-benzoylamino-5-phenyl-3-pyrazolidinone-1-azomethine imines *6* were the most extensively studied reactions witihin the context of (4*R**,5*R**)-4-benzoylamino-5-phenyl-3-pyrazolidinone (*1*) chemistry. Azomethine imines *6* react with typical electron-deficient dipolarophiles, *i.e.* with esters of acrylic, methacrylic, maleic, fumaric, propiolic, and acetylenedicarboxylic acid, as well as with electron-rich $\beta$-keto esters. In addition to broad reactivity (connectivity) spectrum, most of their cycloaddition reactions were highly stereoselective. Stereocontrol depended on (a) stereodirecting phenyl group at position 5, (b) on *ortho*-substituents at the 1'-aryl group in dipole *6*, and (c) on the structure (shape) of the dipolarophile. High selectivity of cycloadditions and simplicity of isolation of the products enable combinatorial solution-phase synthesis of pyrazolo[1,2–*a*]pyrazolone derivatives. In this manner, cycloadditions of dipoles *6* to maleimides and $\beta$-keto esters were carried out to furnish the corresponding libraties of analytically pure cycloadducts.

Chapter 6 - Specific binding of carbohydrates by proteins has been the focus of increased study during the last decades because it plays a key role in important biological processes. The specificity of the carbohydrate-protein interactions is determined in many cases by stereochemical diversity of small sugars which are constituents of oligosaccharide chains of glycoproteins and glycolipids. Most monosaccharides are stereoisomers which only differ in the spatial configuration of specific OH groups. However it is enough for amazing diversity in their chemical and biological activity. As oligosaccharides and proteins are particular complex molecules, one useful approach that can be used to understand the nature and mechanism of protein-carbohydrate interactions is to study model compounds. Monosaccharides are monomer units of the oligosaccharide chains of glycoproteins and glycolipids. Amino acids are basic components of proteins and can be considered as their model compounds. The aim of this article is to report the data on interactions between some isomeric monosaccharides and L-α-amino acids in aqueous solutions. It has been found that stereochemistry of the monosaccharides has significant influence on their selective binding with the amino acids.

Chapter 7 - In the last two decades a new key role of carbohydrates in living organisms has emerged: they have demonstrated to be the major information carriers between cells and their surroundings, whether free or in the form of glycoconjugates with proteins or lipids. This finding has rapidly brought increasing interest in several aspects of carbohydrate chemistry, driven by their potential use in biochemical and pharmaceutical research. The need for specific saccharides in significant amounts, in order to thoroughly understand their biological functions, has inspired modern synthetic approaches. The demand for compounds which can mimic natural substances by way of similar structure and/or biological function has stimulated elaboration of innovative preparative procedures in carbohydrate synthesis. The search for analogues of mono- and oligosaccharides resistant to hydrolytic/enzymatic cleavage has been carried out as well. All these efforts in chemistry and biology have converged in establishing a well defined discipline, called glycoscience.

Especially with respect to investigations on mono- and oligosaccharides, an interest on mirror-image carbohydrates has emerged as an extremely intriguing topic, since they have been found to be key components of several bioactive compounds, such as oligosaccharides, antibiotics, glycopeptides, terpene glycosides, steroid glycosides and other clinically useful agents. However, while most D-sugars are available in large quantity from natural materials, their L-enantiomers are rather rare in nature and then not easily accessible from inexpensive sources. As demand for their synthesis in considerable amount and high purity is more and more pressing, in recent years intense efforts have been focused on the development of new and general methodologies for the construction of L-hexoses.

In this chapter the authors present a survey of the synthetic routes to L-sugars, mainly the ones coming from the new century, although some former and creative approaches will also be reviewed. These methodologies include specific or general approaches and range from carbon chain elongation (homologation), epimerization of D-sugars by chemical or enzymatic methods, up to the most recent amino acid-catalyzed stereoselective aldol reactions.

In: Stereochemistry Research Trends
Editor: M. A. Horvat and J. H. Golob

ISBN 978-1-60456-825-7


*Chapter 1*

# STEREOSELECTIVE METHODS TOWARDS THE SYNTHESIS OF IMINOSUGARS

***Maria José Alves*and Nuno G. Azoia***
Universidade do Minho, Departamento de Química, Campus de Gualtar
4710-057 Braga, Portugal

## ABSTRACT

In this text is presented a summary of the synthetic strategies used to build up either D- or L-pyranose skeleton of sugar framework structures under stereoselective methodologies. Polihydroxylated piperidines were generated from simple starting material like amino-acids, tartaric acid or glyceraldehyde derivatives or even from achiral reagents. Introduction of chiral centres in achiral materials intermediates are generally achieved by catalytic methods, using different classes of ligands. In other cases diastereoselective methods and chemoenzymatic synthesis have been developed.

## INTRODUCTION

Imino sugars, also known as azasugars or imino cyclitols, are a group of compounds that have received a lot of attention in recent years because they typically exhibit excellent inhibitory properties over a range of enzymes involved in carbohydrate handling and carbohydrate recognizing receptors, widely found in living organisms. For this reason many efforts have been made to develop simple synthesis of such compounds. The common approach is based on transformation of naturally occurring D-hexoses. Polihydroxylated piperidines can be made however from simpler starting materials like amino-acids, tartaric acid or glyceraldehyde derivatives or even from achiral reagents. Introduction of new chiral centres in intermediates are generally achieved by catalytic methods, using different classes

* mja@quimica.uminho.pt

of ligands such as proline, ephedrine, some phthalazines or their derivatives. Chemoenzymatic synthesis is another methodology used to develop chirality in achiral precursors of imino sugars giving generally excellent enantiomeric excesses.

The aim of this work is to give a summary of the synthetic strategies developed to build up sugar pyranose framework. The eight types of D-pyranose skeleton (*gluco*, *manno*, *galacto*, *gulo*, *altro*, *allo*, *ido* and *tallo*) and a number of L-pyranoses had been achieved by using known stereoselective methodologies under convenient strategies.

This review will cover recent advances, within the last 15 years, in the preparation of pyranoside iminosugars. Other authors have focused their attention in therapeutical applications and natural occurance of imino sugars [1-3]. Strategies based on carbohydrate manipulation will not be presented in this occasion, as they have been described elsewhere [2-5].

Subtitles were recorded in accordance with leading imino sugar compounds, Nojirimycin, Mannojirimycin, Adenophorine, Fagomine and Isofagomine (figure 1). Miscellaneous configuration types were presented together under "other imino sugars" subtitle.

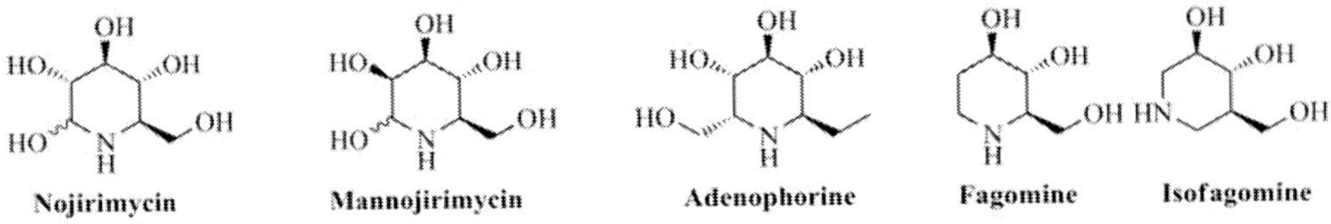


Figure 1. Principal types of imino sugars targets.

## Synthesis of Nojirimycin, Deoxynojirimycin, Homonojirimycin and Derivatives

The earlier synthesis of nojirimycin and deoxynojirimycin were reported in 1967 [6,7], and after then the amount of work in this area do not stop to grow. Vogel [8] reported a synthesis of deoxynojirimycin starting from furan and 2,2-diethoxyacetonitrile *N*-oxide as described in scheme 1. A 1,3-dipolar cycloaddition of the *N*-oxide compound to the furan [9] formed the primary intermediate *1*. Dihydroxylation with osmium tetroxide gave a mixture of anomeric diols *2* in a 3:1 ratio. Diol protection and transacetalation of compound *2* provide the racemic aldehyde *3*. To resolve the mixture, the aldehyde *3* was reacted with (1*S*,2*S*)-diphenylethanediamine [10] in ether. A mixture of diastereomers *4* was formed, being readily separated by crystallization [(+)*4*] and chromatographic purification [(-)*4*]. The diastereoisomeric purity of both compounds was determined by HPLC, and exceeds 98%. The enantiomeric pure aldehydes were obtained by selective acidic hydrolysis with recovery of the chiral auxiliary diamine. The absolute configuration of each aldehyde was established by chemical correlation, i.e. by conversion into 1-deoxy-D-nojirimycin and 1-deoxy-L-nojirimycin. The aldehyde *5* was reduced with sodium borohydride in methanol to afford the alcohol *5*. Deprotection of *5* with TFA/$H_2O$ gave the anomers *6* in 100% yield. Catalytic hydrogenation finally provided deoxy-D-nojirimycin *7* in 65% yield. The same sequence starting with (-)*4* furnished the enantiomeric deoxy-L-nojirimycin *8*.

Comins synthesized[11] 1-deoxynojirimycin *7* together with some 1-deoxymanojirimycin starting from 4-methoxypyridine, applying reported methods (scheme 2). This strategy allowed the preparation of *7* in eight steps from readily available 4-methoxypyridine in 13% overall yield. First 4-methoxypyridine was silylated at the position 3 forming *9*[12]. Compound *9* was then acylated with *trans*-(α-cumyl)cyclohexyl chloroformate[13] giving 1-acylpyridinium salt *10*. Addition of a (benzyloxy)methylcuprate to this salt furnished the *N*-acyldihydropyridone *11* with 90% *de*. In one-pot reaction the chiral auxiliary (95% recovery) and the TIPS group were removed providing enantiopure *12*. Protection of the amine function with benzyl chloroformate gave carbamate *13* with 99% yield. Acetoxylation with $Pb(OAc)_4$[14] in refluxing toluene afforded the dihydropyridone *14*. The *trans* stereochemistry was obtained by delivery of the acetate that attacks from the axial direction via a chair-like transition state[14,15]. Pyridone *14* was hydrolyzed in aqueous HCl in ethanol, and further reduced with tetramethylammonium triacetoxyborohydride[16] forming diol *15* with complete stereoselectivity. Treatment of *15* with osmium tetroxide and *N*-methylmorpholine *N*-oxide afforded the tetrahydroxypiperidines *16*, which prove to be unstable to silica gel chromatography. Hydrogenation of the crude mixture in the presence of $Pd(OH)_2$ resulted in a mixture of *7* and *17* in a ratio of 2.7:1. Chromatographic purification of this mixture provided 1-deoxynojirimycin *7* in 55% yield and 1-deoxymanojirimycin *17* in 21% yield.

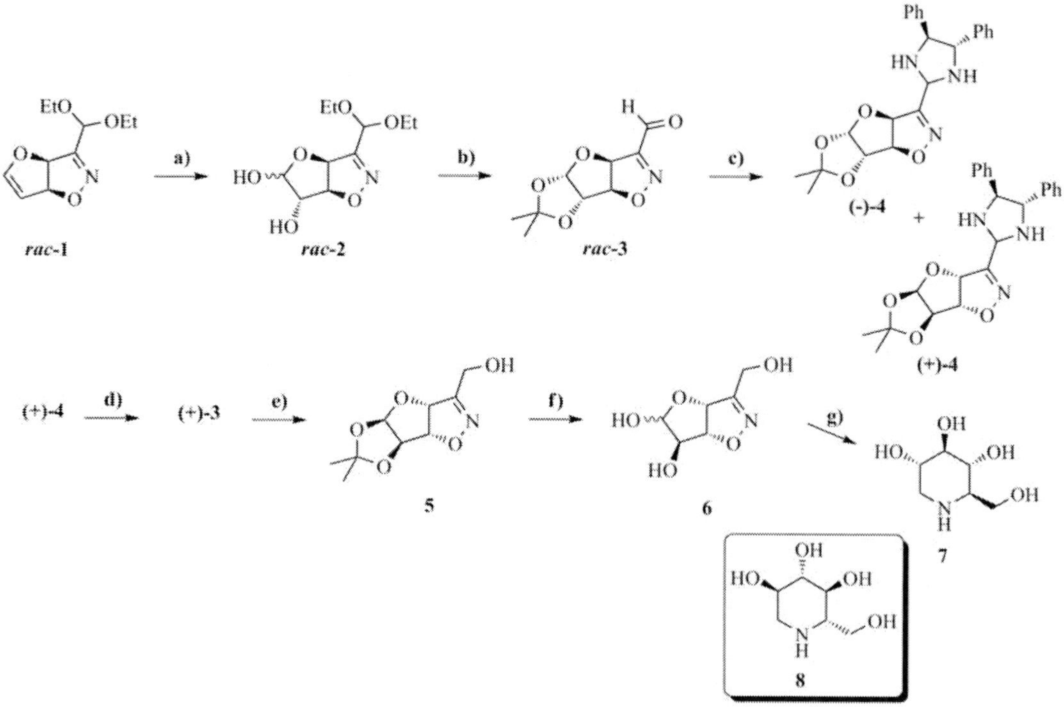


Scheme 1. Reagents and conditions [8]. a) $OsO_4$, NMO, acetone/$H_2O$ 4:1, 60°C, 88%. b) Conc. $H_2SO_4$, anhyd. acetone, $CaSO_4$, 25°C, 92%. c) (-)-(1S,2S)-diphenylethanediamine, $Et_2O$, Molecular Sieves 4 Å, 25°C, 45% for *(+)-4*; 40% for *(-)-4*. d) 1N $H_2SO_4$, $Et_2O$, 25°C, 97%. e) $NaBH_4$, MeOH, 5°C, 90%. f) TFA/$H_2O$ 8:1, 4°C, 100%. g) $H_2$ (0.1 MPa), Pd/C, MeOH, 25°C, 65%.

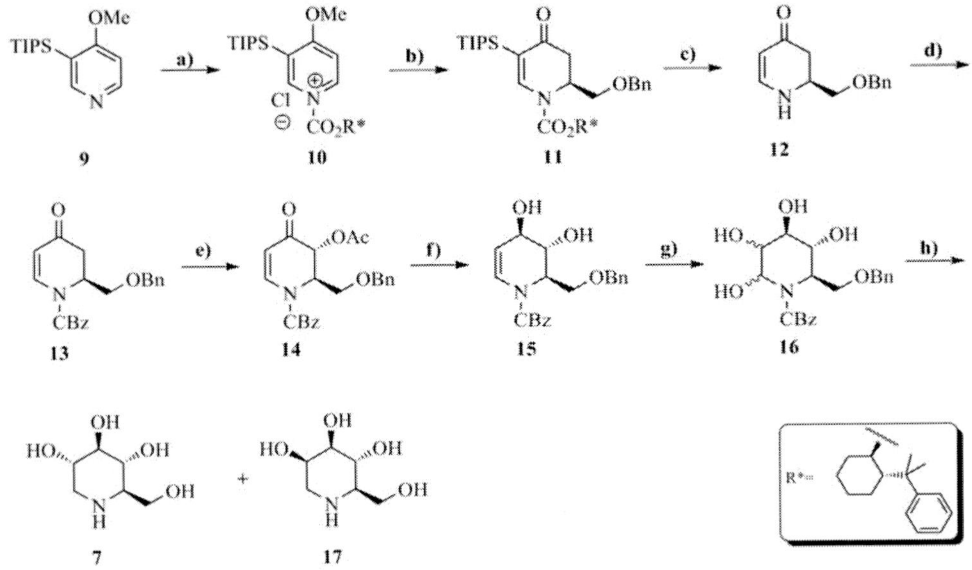

Scheme 2. Reagents and conditions [11] a) [12] R*OCOCl (generated *in situ* from R*OH e phosgene in the presence of quinoline), toluene. b) $BnOCH_2$(2-thyenil)Cu(CN)Li[13], THF, -78°C, 64% from *9*. c) NaOMe, MeOH, Δ; HCl, i-PrOH, 74%. d) BnOCOCl, *n*-BuLi, THF, -78°C, 99%. e) $Pb(OAc)_4$, toluene, Δ, 78%. f) i) HCl aq., EtOH, 75%; ii) $Me_4NBH(OAc)_3$, acetone/AcOH, 83%. g) $OsO_4$, NMO, acetone/$H_2O$. h) $H_2$, $Pd(OH)_2$, MeOH, HCl, neutralization, chromatography, *7* 55%, *17* 21%, from *15*.

Kibayashi reported[17] a synthesis of 1-deoxynojirimycin *7* in 11 steps and 26.5% yield from diol *19* (scheme 3). Diol *19* was easily obtained from diethyl L-tartrate in two steps[18]. Monosilylation of the diol was obtained by treating it with 1 equivalent of TBSCl and sodium hydride. The silylated product *20* was transformed into aldehyde *21* by Swern oxidation. Carbon chain elongation was obtained by Horner-Emmons reaction of *21* to give ester *22* as a single isomer in good yield. The reduction of the ester group with DIBAL afforded allylic alcohol *23*, and its epoxidation gave *24* under the Sharpless asymmetric epoxidation conditions[19]. Treatment of the epoxide with sodium azide furnished the azide *25* in 75% yield with total regioselectivity. The nucleophilic attack proceeded as expected with high selectivity on the basis of both steric and inductive effects of the acetal oxygen atoms[20]. The hydroxyl groups in compound *25* were protected as methoxymethyl ether. Desilylation with TBAF, followed by mesylation provide *28*. Catalytic hydrogenation of *28* led to *29* that after removal of protecting groups with hydrochloric acid form 1-deoxynojirimycin *7*.

The synthesis of nojirimycin *35* can be accomplished by catalytic hydrogenation of the azide function in *26* to give the amino alcohol *30* (scheme 4). The formation of a six-membered ring by intramolecular nucleophilic attack of the amine moity requires some previous transformations on *30*. The free amine group was protected as a carbamate and the silyl group was cleaved with TBAF to give alcohol *32*. Swern oxidation of this alcohol generated aldehyde *33* which in turn was treated with aqueous sulfurous acid to provide 1-deoxynojirimycin-1-sulfonic acid *34* via an open-chain salt. Nojirimycin was generated by treatment of sulfonic acid *34* with Dowex 1X2 resin ($OH^-$ form) in 90%. Nojirimycin *35* was obtained in 13 steps and 12.4% overall yield from diol *19*.

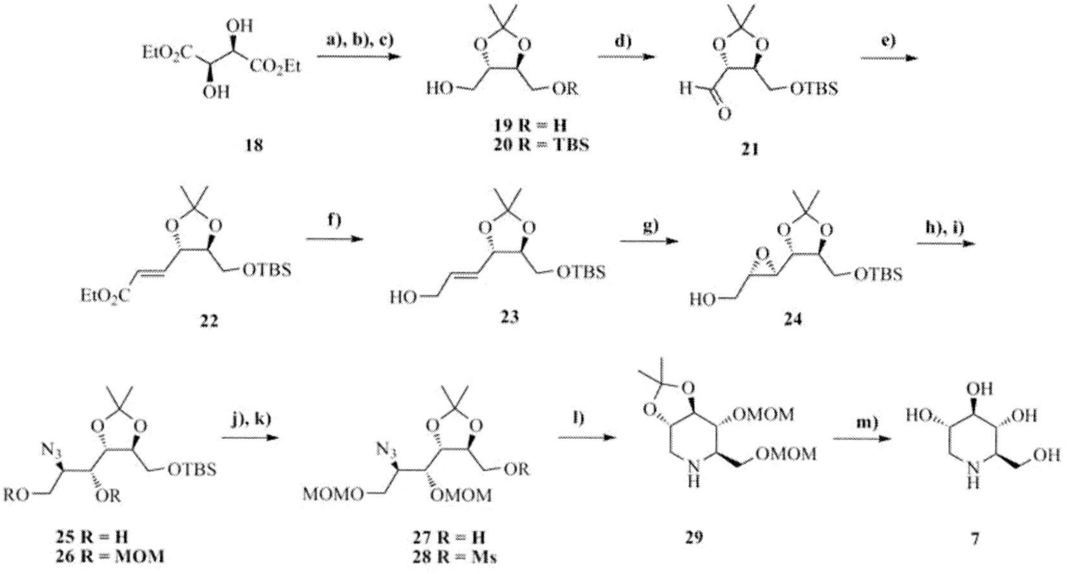

Scheme 3. Reagents and conditions [17] a) Acetone, conc. $H_2SO_4$, petroleum ether. b) $LiAlH_4$, $Et_2O$, Δ. c) NaH, TBSCl, glyme, 99.7%. d) $(ClCO)_2$, DMSO, DCM, $Et_3N$, 85%. e) NaH, $(MeO)_2POCH_2CO_2Me$, $C_6H_6$, 95%. f) DIBAL, DCM, 81%. g) (+)-Diisopropyl L-tartrate, Ti(i-PrO)$_4$, *tert*-butyl hydroperoxide, DCM, 78%. h) $NaN_3$, $NH_4Cl$, glyme/2-methoxyethanol/$H_2O$ 1:2:1, 75%. i) MOMCl, i-$Pr_2EtN$, $CHCl_3$, 91%. j) TBAF, THF, 98%. k) MsCl, $Et_3N$, DCM, 94%. l) $H_2$, Pd/C, MeOH; $Et_3N$, Δ, 92%. m) Conc. HCl, MeOH, 90%.

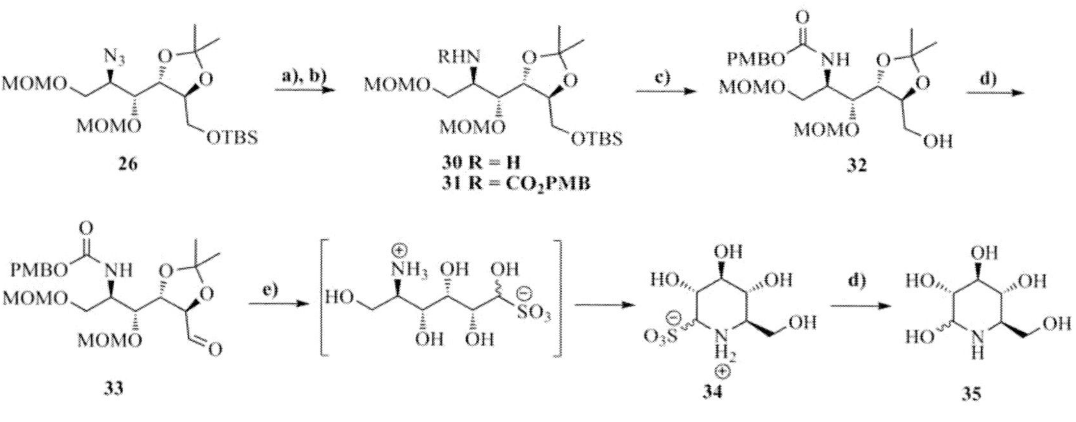

Scheme 4. Reagents and conditions[17] a) $H_2$, Pd/C, MeOH, 86%. b) *p*-Methoxybenzyl-(*S*)-(4,6-dimethylpyrimidin-2-yl)thiocarbonate[21], $Et_3N$, dioxane, 91%. c) TBAF, THF, 98%. d) $(ClCO)_2$, DMSO, DCM, $Et_3N$, 82%. e) $SO_2/H_2O$; MeOH, $SO_2$, 63%. f) Ionic exchange column, Dowex 1X2 ($OH^-$ form), 90%.

Some years later Kibayashi reported an extension of this research[22] to the synthesis of α-homonojirimycin *46* from epoxide *24*, as presented in scheme 5. The epoxide ring opening was achieved with dialkylaluminium amide, according to Overman's method[23], with total regio and stereoselectivity. The amino group was selectively protected to afford carbamate *37*, which was subjected to treatment with methoxymethyl chloride to protect the hydroxyl groups, followed by removal of the silyl group with TBAF. Swern oxidation of alcohol *38* gave the aldehyde *39*, which was transformed into alkene *40* by Wittig reaction with methyltriphenylphosphonium bromide. Hydroxylation of alkene *40* with osmium tetroxide afforded a 1:2.5 mixture of diastereoisomers *41* and *42*. After separation, diol *42* was converted into *43* by selective silylation of the primary alcohol function with TBSCl, followed by mesylation of the secondary alcohol under standard conditions. Deprotection of the amino group was carried out *via* hydrogenolysis over palladium hydroxide in methanol, followed by treatment of the reaction mixture with TEA in refluxing ethanol, furnishing *45*. All the protecting groups were removed at the same time with conc. HCl in methanol to provide α-homonojirimycin *46*. This methodology furnished α-homonojirimycin *46* in 14 steps with 6.9% overall yield from diol *19*.

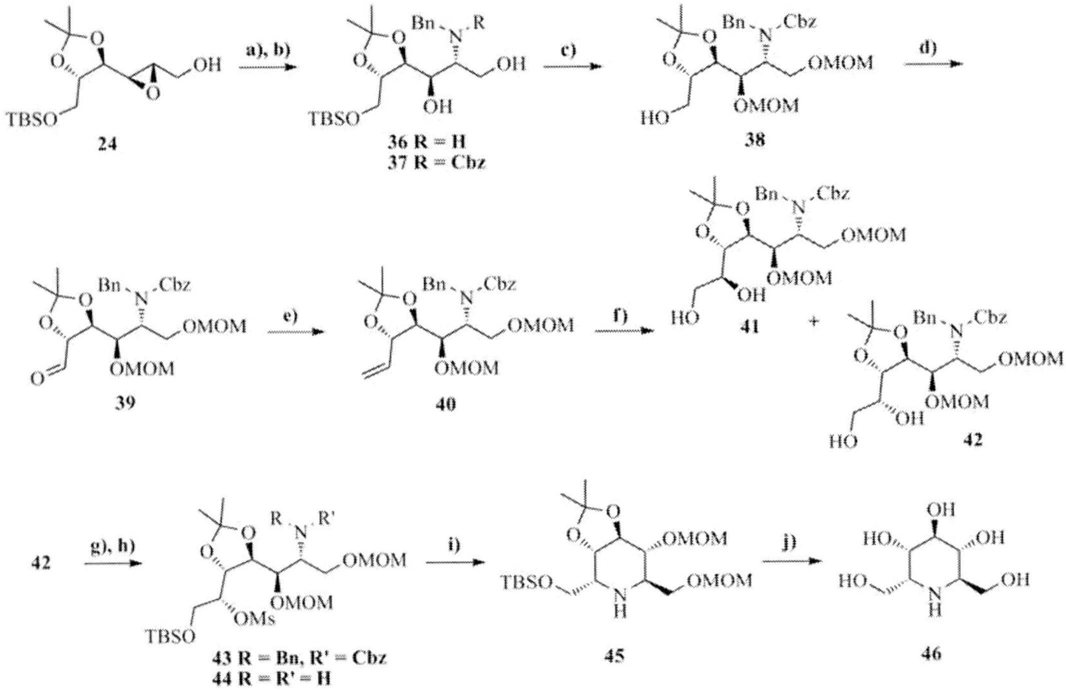


Scheme 5. Reagents and conditions[22] a) $Et_2AlNHCH_2Ph$, DCM, 70%. b) $PhCH_2OCOCl$, aq. $Na_2CO_3$, DCM, 98%. c) MOMCl, *i*-$Pr_2EtN$, $CHCl_3$; TBAF, THF, 88%. d) $(ClCO)_2$, DMSO, DCM, $Et_3N$, 98%. e) $Ph_3PCH_3Br$, *n*-BuLi, THF, 84%. f) $OsO_4$, NMO, acetone/$H_2O$, 90%, (*41*:*42*, 1:2.5). g) TBSCl, imidazole, DMF; MsCl, $Et_3N$, DCM, 77%. h) $H_2$, $Pd(OH)_2$, MeOH. i) $Et_3N$, MeOH, Δ, 81% from *43*. j) Conc. HCl, MeOH, Δ; Dowex 1X8, 68%.

Using a similar approach, Somfai reported[24] a synthesis of 1-deoxynojirimycin from diene *47* (scheme 6). The synthesis was focused on the known propensity of vinyl epoxides to be regio- and stereoselectively attacked at the allylic position by nucleophiles as amines. Opening of the epoxide moiety by a primary amine and subsequent ring closure was expected to give a piperidine derivative. Towards this end *p*-methoxybenzyl ether *47*[25,26] was subjected to the AD-mix-α reagent[27], producing the *cis* diol *48* in moderated yield but with excellent enantioselectivity (97% *ee*). The nature of the hydroxyl protecting group was shown to be crucial for the enantioselectivity of the asymmetric dihydroxylation[28]. Protection of the diol moiety in *48* with 2-methoxypropene, followed by reduction of the ester function with DIBAL gave allylic alcohol *49*. Epoxidation of this compound was only possible using the stoichiometric Sharpless protocol[29], affording *50* in good yield and 90% *de*. Swern oxidation of *50* followed by Wittig olefination of the resulting aldehyde gave vinyl epoxide *51*. Standard deprotection of the ether group gave the corresponding alcohol, that was mesylated to give *52*. Treating of *52* with benzyl amine in DMSO and in the presence of a catalytic amount of *p*-toluenesulfonic acid resulted in the epoxide ring opening and concomitant intramolecular displacement of the mesyl group, furnishing piranoside *53* in 76% yield. Initial protection of the primary alcohol with mesyl chloride is important to activate the corresponding carbon towards nucleophilic attack with promotion of the six-membered ring closure in *53* formation. The vinyl group in *53* was then dihydroxylated with $OsO_4$/NMO, and the diol submitted to oxidative cleavage with $NaIO_4$. Reduction of the aldehyde obtained with $LiAlH_4$ furnished the protected deoxynojirimycin derivative *54*. Sequential removal of the isopropylidene and benzyl protecting groups by standard procedures gave 1deoxynojirimycin *7* in 14 steps and 12.5% overall yield from *47*.

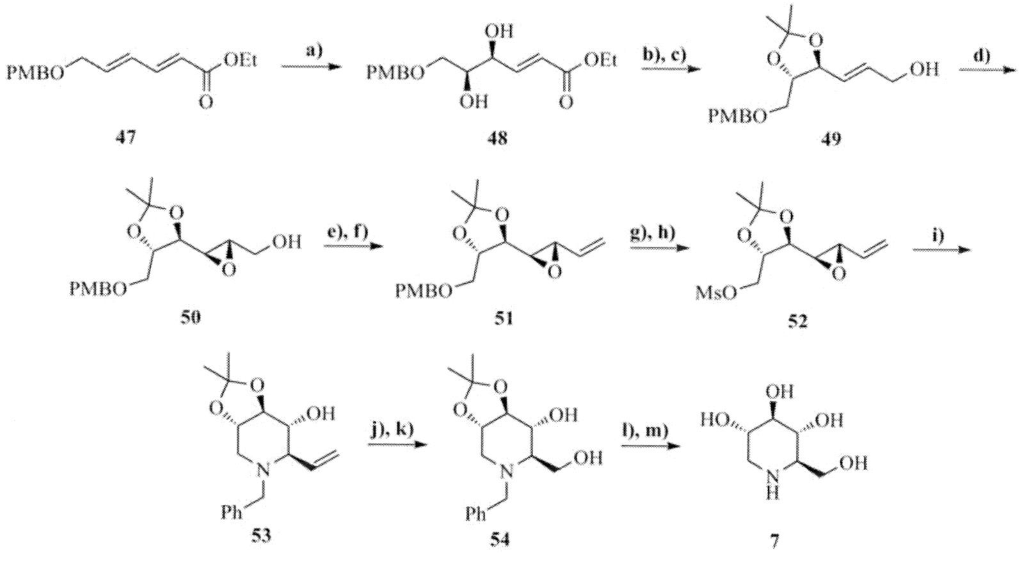


Scheme 6. Reagents and conditions [24] a) AD-mix-α, *t-BuOH*, $H_2O$, 62%, 97% ee. b) 2-methoxypropene, DMF, 97%. c) DIBAL, DCM, -78°C, 93%. d) (+)-Diisopropyl L-tartrate, Ti(i-PrO)$_4$, *tert*-butyl hydroperoxide, DCM, -20°C, 80%, >90% de. e) $(COCl_2)_2$, DMSO, $Et_3N$, DCM, -78°C. f) $Ph_3PCH_3Br$, KHMDS, THF, toluene, 73%. g) DDQ, DCM, $H_2O$, 88%. h) MsCl, i-$Pr_2$NEt, DCM, 0°C, 88%. i) $BnNH_2$, TsOH, DMSO, 120°C, 76%. j) $OsO_4$, NMO,*t*-BuOH, THF, $H_2O$. k) $LiAlH_4$, THF, 0°C, 84% from *52*. l) TFA, MeOH, 87%. m) $H_2$, Pd/C, EtOH, 89%.

Some years later, Somfai published[30] a modified procedure to obtain 1-deoxynojirimycin, scheme 7, in just 11 steps with 36% overall yield. The main difference between the two methods relies on the cyclization reaction. Compound *50* was obtained from diene *47* by the procedure described in scheme 6. The alcohol *50* was protected with TBDPSCl to give *55*. Selective removal of the PMB group afforded the corresponding primary alcohol. Activation of the alcohol towards nucleophilic attack was obtained by its conversion into mesylate *56*. Although this mesylated compound proved to be inert to benzylamine, it could be converted into azide *57* without opening the epoxide. The azide was reduced to amine *58* using Staudinger[31] conditions, which was heated in refluxing ethanol to smoothly give the piperidine derivative *59* in quantitative yield. A proof that the epoxide opening was selective is that only one product was detected. The nucleophilic nitrogen attack to the vicinal carbon is not a favored process since it would produce a *trans*-fused bicyclo[3.3.0]-nonane type of system. Concomitant removal of both the silyl and acetonide protecting groups in *59* by aq. HCl in refluxing EtOH gave 1deoxynojirimycin *7* in quantitative yield.

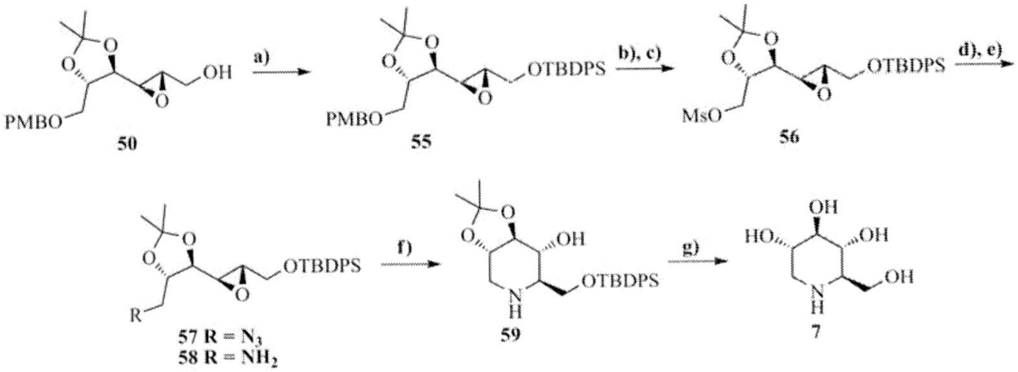


Scheme 7. Reagents and conditions [30] a)*t*-$BuPh_2SiCl$, $Et_3N$, DMAP, DCM, 97%. b) DDQ, DCM, $H_2O$, 92%. c) MsCl, i-$Pr_2EtN$, DCM, 100%. d) $NaN_3$, DMF, 70°C, 91%. e) $Ph_3P$, THF, $H_2O$, 83%. f) EtOH, Δ, 100%. g) HCl (37%), MeOH, 100%.

## Synthesis of Mannojirimycin, Deoxymannojirimycin, Homomannojirimycin and Derivatives

Hudlicky reported[32] a synthesis of mannojirimycin from (1*S*-*cis*)-3-chloro-3,5-cyclohexadiene-1,2-diol *60* (scheme 8), obtained by fermentation of chlorobenzene with *Pseudomonas putida*[33]. Protection of the diol moiety with dimethoxypropane, followed by a regioselective double bond epoxidation with *m*-CPBA formed epoxide *61*. Opening of the epoxide with lithium chloride (Bajwa procedure[34]) followed by displacement of the chloride by the azide furnished *62*. Direct treatment of epoxide *61* with sodium azide would originate an unwanted inverse configuration of the carbon attached to the azide group. Ozonolysis of *62* followed by sodium borohydride workup gave *63*. Reduction of azide group in *63* with $Me_3P$ in wet THF followed by acid-catalyzed deprotection gave mannojirimycin *64* in a total of 5 steps from chlorobenzene.

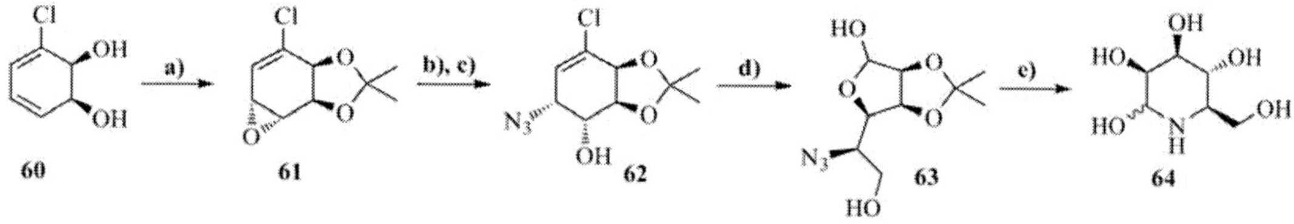

Scheme 8. Reagents and conditions [32] a) i) 2,2-Dimethoxypropane, *p*-TsOH, DCM, 61%; ii) *m*-CPBA, DCM, 80%. b) Ethyl acetoacetate, LiCl, THF, 91%. c) $NaN_3$, DMF, 91%. d) $O_3$, $NaHCO_3$, MeOH; $NaBH_4$, 58%. e) $Ph_3P$, THF, $H_2O$, Δ; TFA.

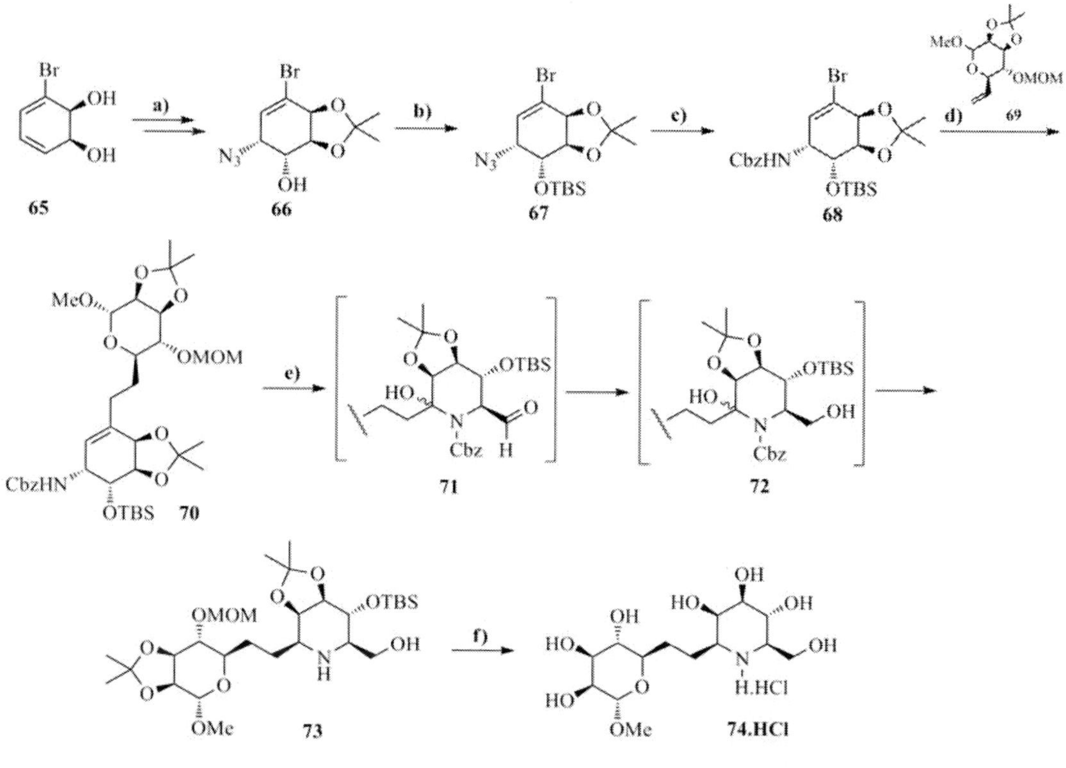

Scheme 9. Reagents and conditions [35] a) i) 2,2-Dimethoxypropane, *p*-TsOH, DCM, 61%; ii) *m*-CPBA, DCM, 88%; iii) LiCl, $MeNO_2$, THF, 88%; iv) $NaN_3$, DMF, 60°C, 76%. b) TBSCl, imidazole, DMF, 50°C, 99%. c) i) $Ph_3P$, THF; $H_2O$. ii) CbzCl, $NaHCO_3$, EtOAc, 99% (from *67*). d) i) 9-BBN, THF. ii) $K_3PO_4$, [1,1′-Bis(diphenylphosphino)ferrocene]$PdCl_2$, DMF, 89%. e) i) $O_3$, DCM/MeOH (1:1), $Me_2S$. ii) $NaBH_3CN$, buffer pH 4 (AcOH/NaOAc), THF. iii) $H_2$ (40 psi), Pd/C, MeOH, 53% (from *70*). f) *6N* HCl, MeOH, 83%.

After the publication of this work mannojirimycin and some derivatives were synthesized starting from halocyclohexadienes. Johnson reported a synthesis of an aza-*C*-disaccharide[35], on which the iminosugar moiety is mannojirimycin, starting from (1*S*-*cis*)-3-bromo-3,5-cyclohexadiene-1,2-diol *65* (scheme 9), obtained by fermentation of bromobenzene with *Pseudomonas putida*[33]. This diol was converted to the azide *66* by the method described above (scheme 8) by Hudlicky[32]: protection of the diol as an acetonide, epoxidation with *m*-CPBA, epoxide opening with lithium chloride and $S_N2$-type displacement of the chloride by azide. After protection of the alcohol group as a silyl ether (*67*), the azide function was reduced to an amine with triphenylphosphine and protected as a carbamate producing *68*. Hydroboration of the carbohydrate derivative *69* with 9-borabicyclo[3.3.1]nonane (9-BBN) gave an alkylboron complex that reacted with *68* in a Suzuki coupling reaction to originate *70*. Ozonolysis of *70* followed by dimethyl sulfide workup yielded aldehyde *71*, in equilibrium with its opened keto amino aldehyde form (not represented). The aldehyde was chemoselectively reduced with $NaBH_3CN$ at pH 4 using AcOH/NaOAc buffer to produce diol *72*, also in equilibrium with its open form (not represented). Diol *72* was converted into disaccharide *73* *via* an intermolecular reductive amination. Acidic deprotection with 6*N* HCl in methanol provided disaccharide *74* in its HCl salt form.

As an extension of his work, Johnson reported a synthesis[36] of β-1-*C*-aryl-1-deoxymannojirimycins from *68* (scheme 10). The vinyl bromine *68* was coupled with arylboronic acids, by means of a palladium-catalyzed cross-coupling reaction, to produce trisubstituted olefins *75*. These olefins underwent ozonolysis to produce the aldehyde intermediate *76* (in equilibrium with keto-aldehyde form), which upon selective reduction with $NaBH_3CN$ in aqueous AcOH/NaOAc pH 4 buffer gave the alcohol *77*, also in equilibrium with its open form. Reductive amination of compound *77* gave *78* with total diastereoselectivity. Deprotection of *78* with *6N* HCl in THF afforded the 1-deoxymannojirimycin derivatives *79*, in their HCl salt form.

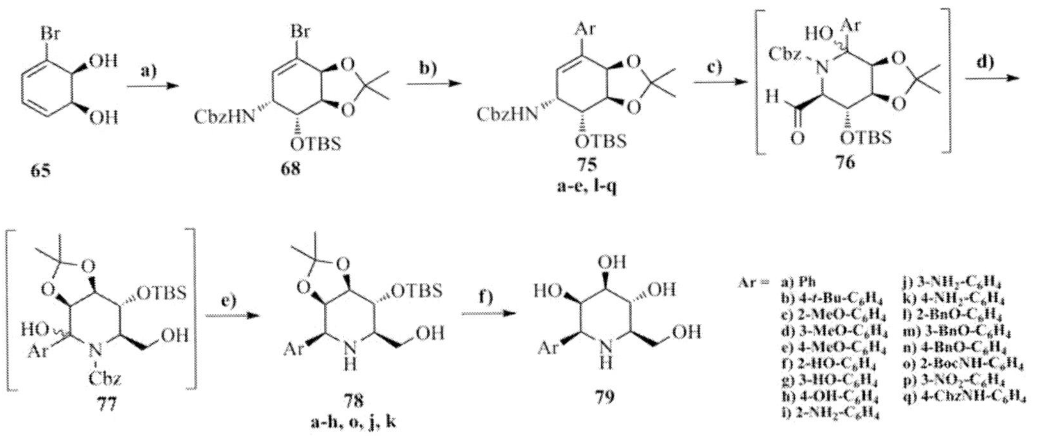


| Entry | 1 | 2 | 3 | 4 | 5 | 6 | 7 | 8 | 9 | 10 | 11 |
|---|---|---|---|---|---|---|---|---|---|---|---|
| $ArB(OH)_2$ | a | b | c | d | e | l | m | n | o | p | q |
| Product *75* | a | b | c | d | e | l | m | n | o | p | q |
| % yield | 89 | 95 | 88 | 75 | 73 | 97 | 93 | 84 | 79 | 92 | 74 |
| Product *78* | a | b | c | d | e | f | g | h | o | j | k |
| % yield | 56 | 67 | 41 | 56 | 74 | 61 | 55 | 66 | 67 | 49 | 66 |

Scheme 10. Reagents and conditions [36] a) Reported procedure[35]. b) $ArB(OH)_2$, $PdCl_2(PPH_3)_2$, 2M aq. $Na_2CO_3$, THF, Δ. c) $O_3$, DCM/MeOH 1:1; DMS. d) $NaBH_3CN$, aq. AcOH/NaOAc pH 4 buffer, THF. e) $H_2$ (40 psi), Pd/C, MeOH, 82-99%. f) 6*N* HCl, THF.

Banweel, following Hudlicky's method for the preparation of azide *62* (scheme 8) from diol *60*, reported a synthesis[37] of 1-deoxymannojirimycin *17* and a synthesis of the aza-*C*-disaccharide *87* derived from 1-deoxymannojirimycin, starting from diol *60* (scheme 11). The alcohol *62* was benzylated under standard conditions to give the benzyl ether *80*. Ozonolysis of *80* in methanol followed by reductive work-up with sodium borohydride and subsequent protection of the primary alcohol obtained, using TBSCl, afforded the azide *81*. Hydrogenolysis of this compound, using 5% Pd on carbon as catalyst, resulted in formation of lactam *82*. Conversion of *82* into the corresponding piperidine *83* was obtained upon exposure to the borane–dimethyl sulfide complex followed by cleavage of the resulted borane–amine complex with methanol in the presence of 10% palladium on carbon. Treatment of compound *83* with aqueous trifluoroacetic acid then gave the 1-deoxynojirimycin *17*, in its TFA salt form. The piperidine *83* was also used as an effective precursor to 6-*O*-α-L-rhamnopyranosyl-1-deoxymannojirimycin *87*. The former compound was *N*- and *O*-benzylated under standard conditions and the TBS group was cleaved with TBAF to give *84*. This alcohol was then coupled with the L-rhamnopyranosyl bromide, *85*, in the presence of silver triflate[38]. The coupling product *86* was deprotected in a stepwise fashion by sequential exposure to TFA for acetonide group removal, and methanolic sodium methoxide to remove the acetate groups, followed by hydrogenation in the presence of 5% Pd on carbon to cleave the benzyl groups. The aza-*C*-disaccharide *87* was obtained from diol *60* in 15 steps.

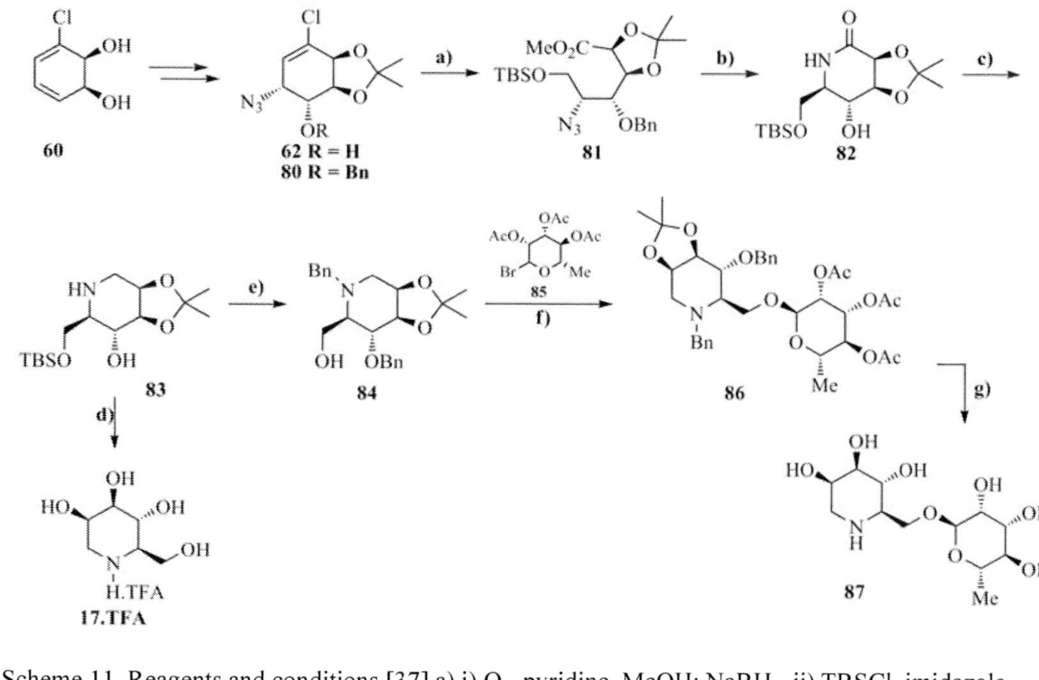

Scheme 11. Reagents and conditions [37] a) i) $O_3$, pyridine, MeOH; $NaBH_4$. ii) TBSCl, imidazole, DCM, 89% (from *80*). b) $H_2$, Pd/C, EtOAc, 86%. c) $BH_3.SMe_2$, THF; Pd/C, MeOH, 73%. d) TFA, $H_2O$. e) i) BnBr, KI, NaH, THF. ii) TBAF, THF, 89%; f) AgOTf, Molecular Sieves 4Å, DCM, 87%. g) i) TFA, DCM. ii) NaOMe, MeOH. iii) $H_2$, Pd/C, EtOH, 80%.

A different approach for the synthesis of 1-deoxymannojirimycin was reported by O'Doherty[39] (scheme 12). Starting from furfural *88* the alcohol *89* was obtained by addition of a silylated Grignard reagent that upon treatment with aqueous HCl in diethyl ether afforded vinylfuran (not represented). Asymmetric aminohydroxylation of vinylfuran by the conditions reported by Sharpless[40] gave a mixture of regioisomers *90* and *91* in 84% overall yield. The regioisomers were easily separated after selective TBS protection of the primary alcohol followed by silica gel chromatography. Compound *92* was isolated with 86% *ee* and 21% overall yield from furfural. Aza-Achmatowicz rearrangement[41] of *92* furnished *93* with 81% yield upon treatment with anhydrous *m*-CPBA. The hemiaminal *93* was subjected to chromatographic purification in silica gel and converted to the ethylaminal *94* by treatment with triethyl orthoformate. Reduction of *94* under Luche conditions[42] provided alcohol *95*. The stereochemistry of *95* was consistent with the results of Zhou[43] and others[44,45], but do not correspond to the mannose stereochemistry required. The correct stereochemistry was nevertheless obtained by conversion of alcohol *95* into its epimer *96* by Mitsunobu inversion with 94% yield. Dihydroxylation of *96* with osmium tetroxide provided triol *97* that after catalitic hydrogenation, followed by treatment with TsOH gave 1-deoxymannojirimycin *17* in its TsOH salt. The sequence of seven steps resulted in 48% overall yield from vinylfuran.

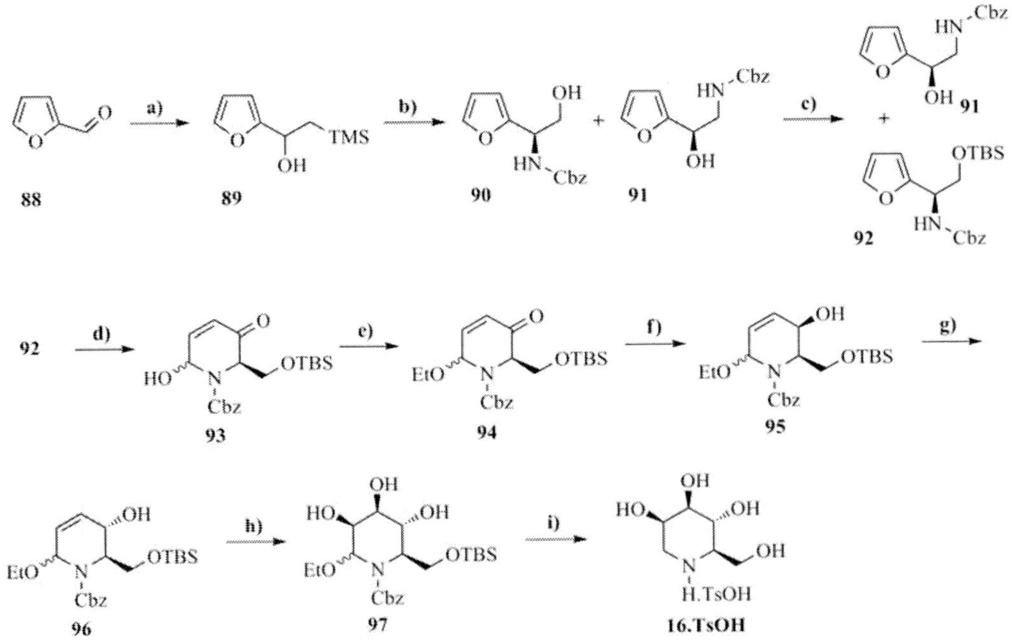

Scheme 12. Reagents and conditions [39] a) $TMSCH_2MgCl$, $Et_2O$, 0°C to rt, 90%. b) i) 1M aq. HCl, $Et_2O$; ii) CbzNHCl, NaOH,*t*-BuOCl, $(DHQ)_2PHAL$, $OsO_4$,*t*-BuOH. c) TBSCl, $Et_3N$, DMAP, DCM, 21% for *92*, 40% for *91*. d) *m*-CPBA, DCM, 0°C, 87%. e) $(EtO)_3CH$, TsOH, DCM, 87%. f) $NaBH_4$, $CeCl_3$, DCM, -78°C, 86%. g) i) $PPh_3$, DEAD, *p*-nitrobenzoic acid, THF, 0°C, 84%; ii) $Et_3N$, MeOH, 94%. h) $OsO_4$, NMO, DCM, 0°C, 92%. i) $H_2$, Pd/C, MeOH, TsOH, 94%.

More recently Fernández-Mayoralas reported another approach for the synthesis of a variety of imino sugars[46] from enantiomeric pure L- and D-diethyltartrate, catalyzed by both enantiomers of proline (scheme 13). The aldehydes *102* and *104* respectively derived from (+)-diethyl L-tartrate and (-)-diethyl D-tartrate were formed first. (+)-Diethyl L-tartrate *18* reacted with thionyl chloride[47] to form cyclic sulfide *98*. Then nucleophilic attack of the azide ion to *98* provided *99*, that after reduction of the esters groups with $NaBH_4$/LiCl in ethanol formed triol *100*. Protection of two alcohol groups was obtained by reaction of benzaldehyde dimethyl acetal with this triol providing alcohol *101*. Finally oxidation of *101* with TEMPO[48] afforded the precursor aldehyde *102*. The synthesis of the enantiomer of *102*, *104*, was obtained from (-)-diethyl D-tartrate *103*, following the same procedure.

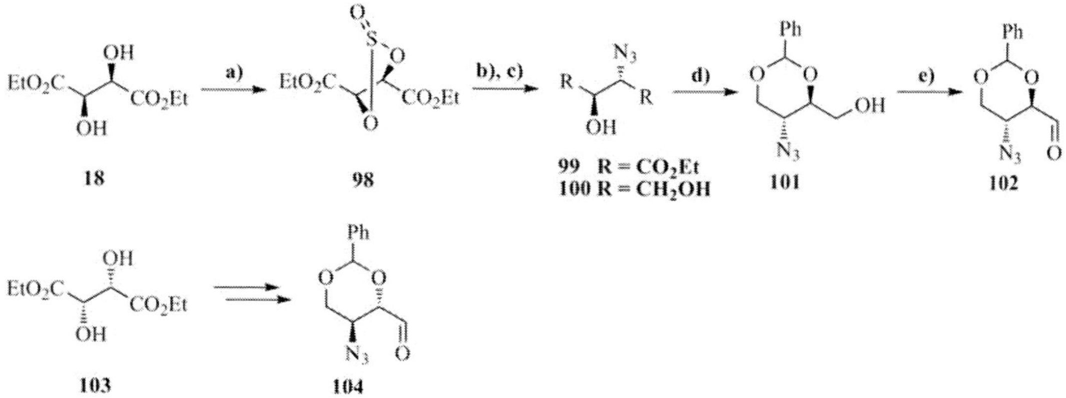

Scheme 13. Reagents and conditions [46] a) $SOCl_2$, $Et_3N$, DCM, 100%. b) $NaN_3$, DMF, 70%. c) $NaBH_4$, LiCl, EtOH, 63%. d) α,α-Dimethoxytoluene, *p*-TsOH, MeCN, 45%. e) TEMPO, trichloroisocyanuric acid, DCM, 60%.

The key step in this methodology is the proline-catalysis of the aldol reaction between the above aldehydes (*102* and *104*) with ketones *105* and *106*, as presented in scheme 14. Reaction of aldehyde *102* with ketone *105* catalized by L-proline afforded aldol *107* with 70% yield and >20:1 *de*. In the same way condensation of aldehyde *102* and ketone *106*, catalized by L-proline gave aldol *109* with 65% yield and >20:1 *de*. On the other hand, reaction of aldehyde *104*, enantiomer of *102*, with ketones *105* and *106* in the presence of D-proline formed the enantiomers of the respective aldol products obtained before, with the same *de*. The aldol products were finally hydrogenated to give imino sugar derivatives: β-homo-D-mannojirimycin *108*, β-homo-D-galactonojirimycin *113*, a mixture of α- and β-1-*C*-methyl-deoxy-D-mannojirimycin *111* and *110* (1:6 ratio), and a mixture of α- and β-1-*C*-methyl-1-deoxy-L-mannojirimycin *116* and *115* (1:6 ratio).

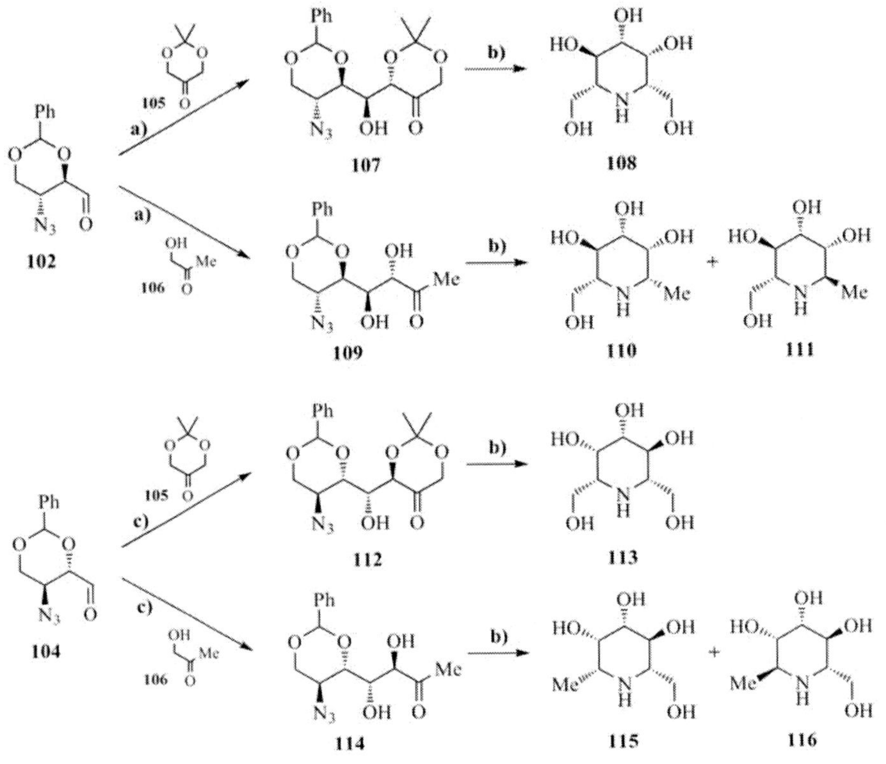

Scheme 14. Reagents and conditions [46] a) L-Proline (30% mol), DMF. b) $H_2$ (45 psi), Pd/C, MeOH/HCl. c) D-Proline (30% mol), DMF.

## Synthesys of Adenophorine and Derivatives

Lebreton synthesized 5-deoxyadenophorine[49] from the enantiomer of Garner's aldehyde *117*[50] derived from D-serine (scheme 15). The sequence started with the Horner-Wadworth-Emmons reaction forming the olefin *118* as a single isomer with 75% yield. Treatment of *118* with conc. HCl in methanol afforded the unprotected amino alcohol *119*. Reaction of this amino alcohol with propanal gave imine *120* which, after addition of allylmagnesium bromide, provided the diethylenic amino alcohol in 87:13 diastereoisomeric ratio. The styryl group introduced in the first step is important either for favoring the ring closing metathesis reaction (RCM)[51,52] and for helping to improve the diastereoselectivity in the allylation step. The absolute configuration of *121*, the major diastereoisomer of the forms reaction misture, was not determined but *trans* configuration was assumed based on similarity to products of identical reactions[53]. The minor isomer could not be separated. As the RCM reaction is known to be incompatible with free amines[54,55] the amino and the alcohol groups were protected under a oxazolidinone form. Compound *122* cyclized upon reaction with Grubbs' catalyst *129* to give the tetrahydropyridine *123*. The minor *cis* isomer formed in the allylation step was now easily separated. Epoxidation of *123* with *m*-CPBA gave *124* in 9:1 *de*, being the minor isomer removed by chromatography. The epoxide opening with acetic acid gave a mixture of monoacetate *125* and diacetate *126* in 86:14 ratio.

The regioselectivity of attack is in contrast with previous reports for the acidic opening of epoxides[56,57]. Treatment of this mixture with potassium carbonate afforded diol *127* with excellent yield. Finally hydrolysis of the oxazolidinone ring with sodium hydroxide occurs in refluxing methanol. 5-Deoxyadenophorine *128* was obtained in 10 steps and 26% overall yield from *117*.

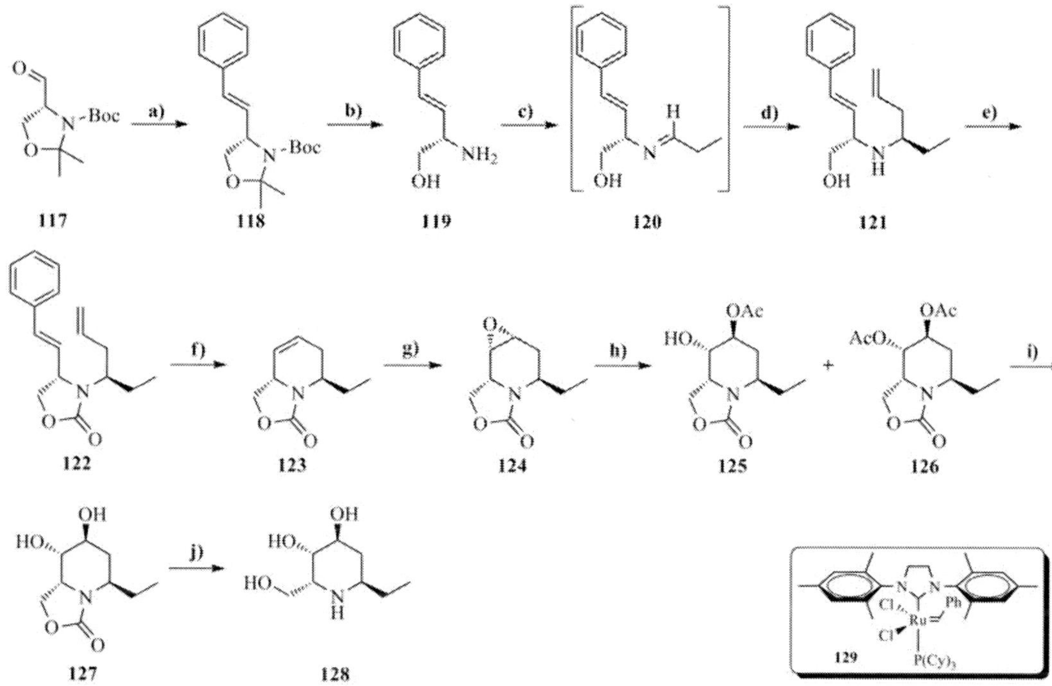

Scheme 15. Reagents and conditions [49] a) $EtO_2POCH_2Ph$, *n*-BuLi, THF, -78°C, 75%. b) Conc. HCl, MeOH, Δ, 98%. c) Propanal, $MgSO_4$, THF, rt. d) $CH_2$=$CHCH_2MgBr$, THF, $Et_2O$, -78°C, 87% (from *119*), 87/13 dr. e) CDI, $Et_3N$, DCM, 86%, 87/13 dr. f) Grubbs' catalyst *129*, DCM, 0.005M, Δ, 83% cis, 11% trans. g) *m*-CPBA, DCM, 0°C, 86%. h) AcOH, 100°C, 79% *125*, 10% *126*. i) $K_2CO_3$, MeOH, rt, 95%. j) 8*N* NaOH, MeOH, 95°C, 88%.

In the same report, Lebreton also synthesized an aza-*C*-disaccharide in which 5-deoxyadenophorine unit is included (scheme 16). Initially the diol moiety in *127* was protected by reaction with benzyl bromide. After hydrolysis of the oxazolidinone ring the amino group was selectively protected with benzylchloroformate affording *130*. Coupling of compound *130* with glucose derivative *131* in the presence of $BF_3.Et_2O$ afforded *132* as a single isomer and with 39% yield. Dissacharide *134* was formed after removal of the acetate groups with potassium in methanol and subsequent hydrogenolysis of the benzyl protecting groups. 1-*O*-β-D-Glucopyranosyl-5-deoxyadenophorine *134* was obtained in 5 steps and 26% overall yield from *127*. Coupling of *130* with acetobromo-α-D-glucose in the presence of $Hg(CN)_2$[58] gave an inseparable mixture of α- and β-dissacharides in a 3:7 ratio. Catalysis with $Ag(OTf)_2$[59] resulted in the recovery of the starting materials.

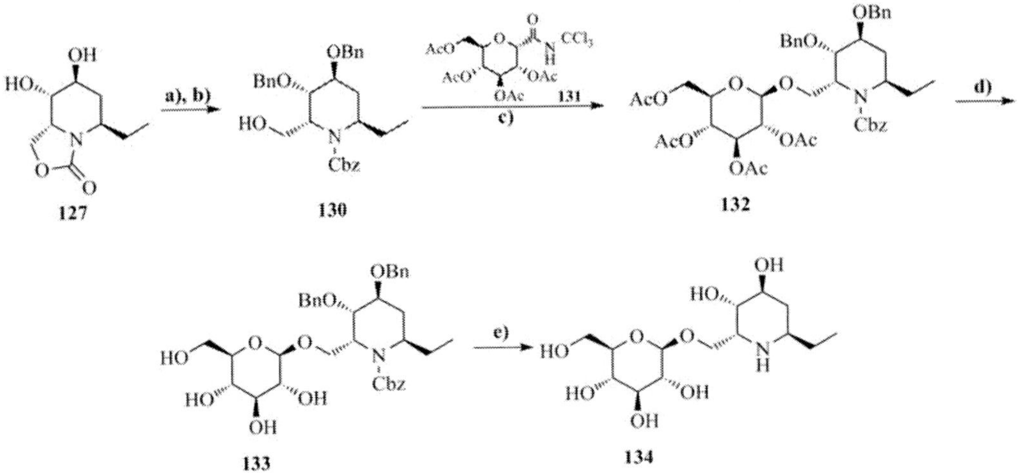

Scheme 16. Reagents and conditions [49] a) NaH, BnBr, DMF, 0°C, 93%. b) i) 8N NaOH, MeOH, Δ; ii) CbzCl, DCM. 0°C, 74%. c) $BF_3.Et_2O$, DCM, -20°C, 39%. d) $K_2CO_3$, MeOH, THF, $H_2O$, rt, 98%. e) $H_2$, $Pd(OH)_2$, MeOH, DCM, rt, 98%.

Some years later[60] the same author synthesized adenophorine *141* also from *123*, a derivative of the enantiomer of Garner's aldehyde as represented in scheme 17. Epoxidation of tetrahydropyridine *123* carried out with freshly distilled solution of dimethyldioxyrane (DMDO) in acetone, gave epoxides *124* and *135* in a 81:19 diastereoisomeric ratio. The pure epoxide *124* was isolated in 60% yield. Total regioselective opening of the epoxide ring in *124* with a selenium-boron complex, generated *in situ* from diphenyldiselenide and sodium borohydride, provided *136* as a single product with 72% yield after purification. Oxidation of *136* with hydrogen peroxide gave the allylic alcohol *137* by elimination of phenylselenic acid. The newly formed C=C was subjected to epoxidation with DMDO affording a 95:5 mixture of epoxides in which the major product was the required *trans* isomer *138*. *Cis* and *trans* isomers were separable by chromatography after benzylation of the alcohol. Epoxide *138* was obtained pure in 41% yield from *137*. These authors used also another approach in which the overall yield was improved. The alcohol function in *137* was protected first and then subjected to epoxidation with *m*-CPBA affording *138* as a 98:2 mixture of diastereoisomers. This epoxide was obtained pure after chromatography with 60% overall yield from *137*. Previous protection of the alcohol was required due to the known ability of *m*-CPBA to assist allylic alcohol epoxidation, producing the *cis* epoxide. Treatment of *138* with sulfuric acid in water:dioxane (1:1) afforded *139* as a single isomer. Cleavage of the oxazolidinone ring with sodium hydroxide in refluxing methanol, followed by hydrogenation in acidic methanol afforded adenophorine *141* as its hydrochloric salt in 8 steps and 7% overall yield from *123*.

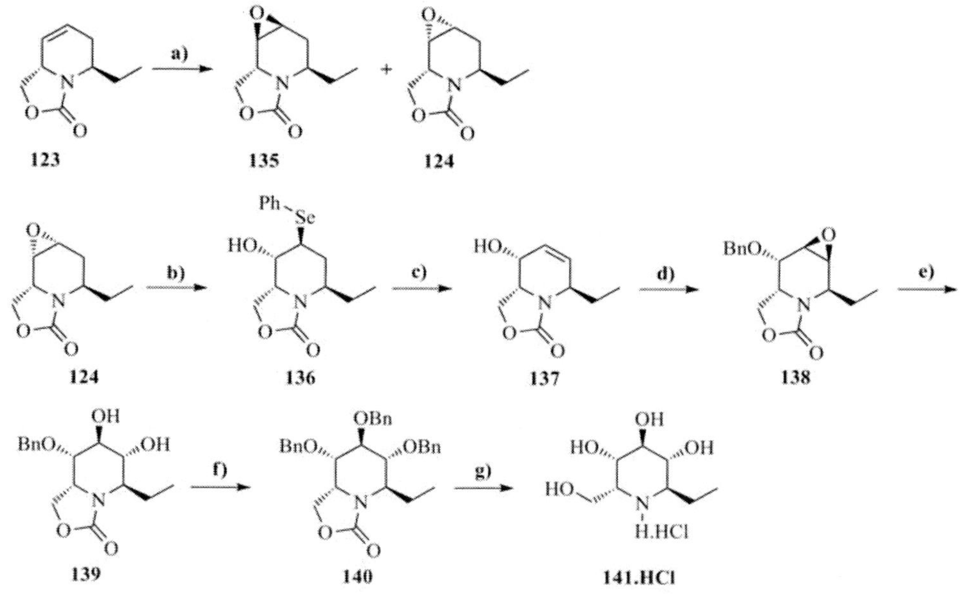

Scheme 17. Reagents and conditions [60] a) DMDO, acetone, rt, 60% for *124*. b) $Ph_2Se_2$, $NaBH_4$, EtOH, rt, 72%. c) 50% aq. $H_2O_2$, THF, rt, 63%. d) i) BnBr, NaH, THF, rt. ii) *m*-CPBA, DCM, rt, 60%. e) 1M $H_2SO_4$, $H_2O$:dioxane 1:1, 80°C. f) BnBr, NaH, DMF, rt, 60% (from *138*). g) i) 8M NaOH, MeOH, 95°C, 76%. ii) $H_2$, HCl, $Pd(OH)_2$, EtOH, rt, quant..

## Synthesis of Fagomine, Isofagomine and Derivatives

Hirai reported a synthesis[61] of fagomine *149* from 2-aminopropanol (scheme 18). First, treatment of 2-aminopropanol with di-*tert*-butyl dicarbonate and triethylamine afforded *142* with 89% yield. Swern oxidation of this alcohol, followed by Horner-Wittig reaction produced an α,β-unsaturated ester, which upon reduction with DIBAL afforded allylic alcohol *143*. The alcohol was then protected with TBSCl to give *144*. Sharpless asymmetric dihydroxylation of this protected alcohol with AD-mix-β gave a diol. Reaction of benzyl bromide with this diol followed by removal of TBS group with TsOH in methanol gave the alcohol *145* with 29% yield and 75% *ee* from *144*. The IBX oxidation[62] of the alcohol function in *145* followed by Horner-Wittig reaction and reduction of the resulting ester afforded *146*. Cyclization was promoted with $PdCl_2(MeCN)_2$[63-68] in THF at room temperature giving *147*. Ozonolysis of the double bond followed by reductive work-up with $NaBH_4$ provided *148*. Cleavage of the benzyl groups was accomplished by treatment of *148* with hydrochloric acid in methanol and hydrogenation to afford fagomine *149*.

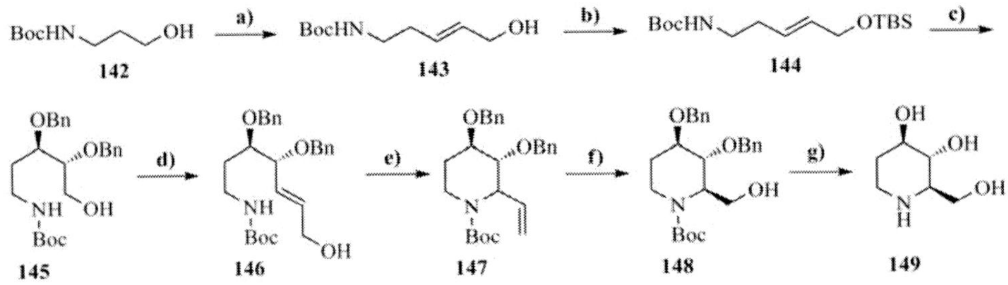

Scheme 18. Reagents and conditions [61] a) i) $(COCl)_2$, DMSO, $Et_3N$, DCM. ii) $(EtO)_2POCH_2CO_2Et$, NaH, THF, -78°C. iii) DIBAL, THF, -78°C, 83%. b) TBSCl, imidazole, DMF, 89%. c) i) AD-mix-β, $CH_3SO_2NH_2$,*t*-BuOH/$H_2O$. ii) BnBr, NaH, $Bu_4NI$, THF. iii) *p*-TsOH, MeOH, 29%. d) i) IBX, THF/DMSO. ii) $(EtO)_2POCH_2CO_2Et$, NaH, THF. iii) DIBAL, THF, 78°C, 40%. e) $PdCl_2(MeCN)_2$, THF, rt, 90%. f) $O_3$, DCM/MeOH (1:4), -78°C; $NaBH_4$, -78°C, 80%. g) i) Aq. HCl, MeOH, 70°C, 64%. ii) $H_2$, Pd/C, AcOH, 87%.

Takahata synthesized[69] isofagomine *161* starting from butadiene monoxide *150* and allylamine *151* (scheme 19). The regioselective opening of the epoxide with the amine, followed by protection of the secondary amine formed afforded the metathesis precursor *153* together with *152* in a 3.3:1 ratio. After separation *153* was treated with Grubbs' catalyst *156* to give the expected ring-closing metathesis product *154*. This alcohol, obtained as a racemic mixture, was treated with lipase PS (*Pseudomonas capacia*) immobilized on ceramic particles according to the method developd by Ogasawara[70], to afford acetate *(-)-155* in 49% yield, along with unreacted alcohol *(+)-154* in 48% yield. The enzymatic hydrolysis of *(-)-155* with the same lipase afforded alcohol *(-)-154* in 98% yield. The alcohol function in (+)-*154* was protected with TBDPSCl to form *157* before the epoxidation. The best epoxidation procedure was obtained with (triflouromethyl)methyldioxirane generated *in situ* from Oxone and 1,1,1-trifluoroacetone[71]. The *anti* epoxide *159* was obtained in 72% yield and the *syn* isomer *158* in 17% yield, after separation. The nucleophilic opening of the epoxide *159* with the organocuprate $(CH_2{=}CH)_2CuCNLi_2$[72] in the presence of $BF_3.Et_2O$ afforded *160*. Treatment of *160* with $OsO_4$ followed by oxidative cleavage of the diol with $NaIO_4$ afforded an aldehyde which was reduced with sodium borohydride to give the alcohol. Cleavage of the silyl group with 10% HCl in dioxane gave isofagomine *161* in 85% yield from *160*.

Takahata had also reported[57] a synthesis of all the isomers of fagomine from Garner's aldehyde enantiomer *117*[75] (scheme 20). Treatment of this aldehyde with methyltriphenylphosphonium iodide in the presence of sodium bis(trimethylsilyl)amide gave olefin *162*. Hydrolysis of the oxazolidine ring with TsOH in methanol, followed by silylation of the hydroxyl group gave olefin *163*. Alkylation of the amine with 4-bromo-1-butene was attempted without success[1], but the reaction was possible after removal of Boc protecting group with trifluoroacetic acid in DCM. The free amine smoothly reacted with 4-bromo-1-butene in the presence of $K_2CO_3$ in acetonitrile. Boc group was reintroduced providing *164* in 60% overall yield. RCM reaction of *164* with Grubbs' catalyst *156*[76] gave tetrahydropyridine *165* used as the starting material for the synthesis of fagomine and its three diastereomers.

[1] i) NaH, DMF, 60°C; KOH. ii) cat. N-BuNI, THF, reflux. iii) KOH, 18-crow-6-ether, THF, reflux.

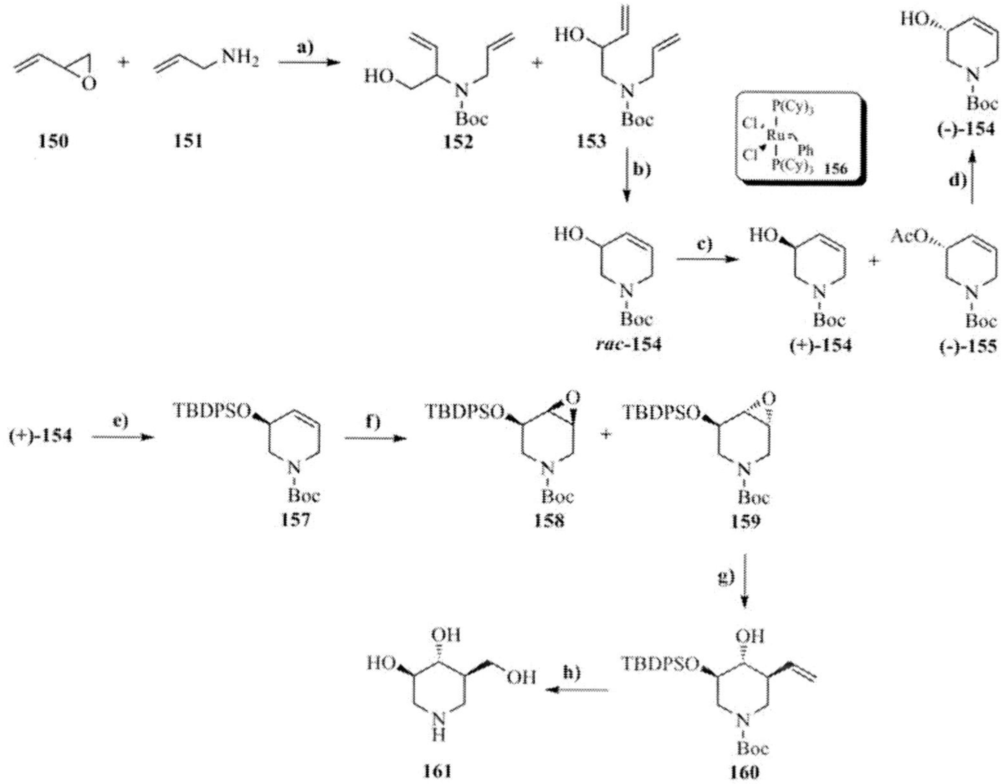

Scheme 19. Reagents and conditions [69] a) i) $H_2O$, 15°C to 100°C. ii) $BocO_2$, $NEt_3$, DCM, rt, 66% for *153*, 20% for *152*, [73,74]. b) Grubbs' catalyst *156*, DCM, rt, 99%, [74]. c) Vinyl acetate, lipase PS (immobilized on ceramic particules)/tBuOMe, 40°C, 49% for *155*, 48% for *154.* d) Lipase PS (immobilized on ceramic particules)/0,1 M phosphate buffer (pH 7), acetone, 40°C, 98%. e) TBDPSCl, imidazole, DMAP, DCM, rt, 99%. f) Oxone, aq. $Na_2EDTA$, $NaHCO_3$, $CF_3COCH_3$, MeCN, 0°C, 72% for *158*, 17% for *159*. g) $(CH_2{=}CH)_2CuCNLi_2$ (5 eq.), $BF_3.Et_2O$, (2 eq.), $Et_2O$, -78°C, 74%. h) i) $OsO_4$ (cat.), $NaIO_4$/50% EtOH, rt. ii) $NaBH_4$/50% EtOH, rt. iii) HCl 10%, dioxane, Δ, 85%.

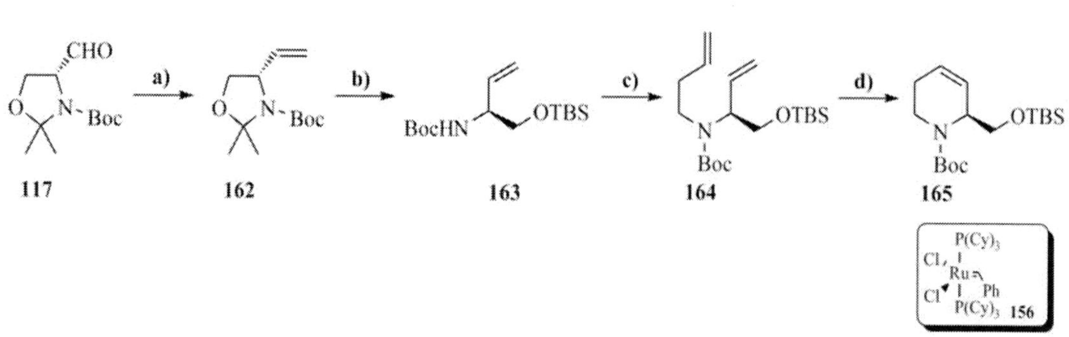

Scheme 20. *Reagents and conditions [57,77]* a) $Ph_3PCH_3I$, $NaN(TMS)_2$, THF, 63%. b) *i)* *p*-TsOH.$H_2O$, MeOH. *ii)* TBDPSCl, DMAP, imidazole, DCM, 72%. c) *i)* TFA, DCM. *ii)* 4-bromo-1-butene, $K_2CO_3$, MeCN. *iii)* $(Boc)_2O$, $Et_3N$, DCM. d) Grubbs' catalyst *156*, DCM, 97%.

Epoxidation of *165* with (trifluoromethyl)methyldioxirane, using the procedure reported by Denmark[78] gave epoxides *166* and *167* in 2:1 ratio (scheme 21), separated by chromatography. The diastereoselectivity of the epoxidation was low, but each isomer was obtained pure in 90% overall yield after purification. Acid hydrolysis of both epoxides *166* and *167* gave the corresponding amino sugars. Epoxide *166* open with complete regioselective by giving fagomine *149* in 75% yield. On the other hand, opening of the epoxide *167* gave a 1.3:1 mixture of fagomine *149* and 3,4-di-*epi*-fagomine *168* in 77% combined yield. The two isomeric compounds were separated. The nucleophilic attack of $H_2O$ to oxirane *166* occured at C-4 rather than at C-3 because the stereochemical directing effect of the hydroxymethyl group at C-2. This effect is not observed in the nucleophilic attack of $H_2O$ to compound *167* due to an *anti* relationship between the hydroxymethyl group at C-2 and the direction of the $H_2O$ attac. Nevertheless cleavage of epoxide ring in *167* in aq. KOH and 1,4-dioxane gave a 1:5 mixture of *149* and *168*, obtained in 99% combined yield.

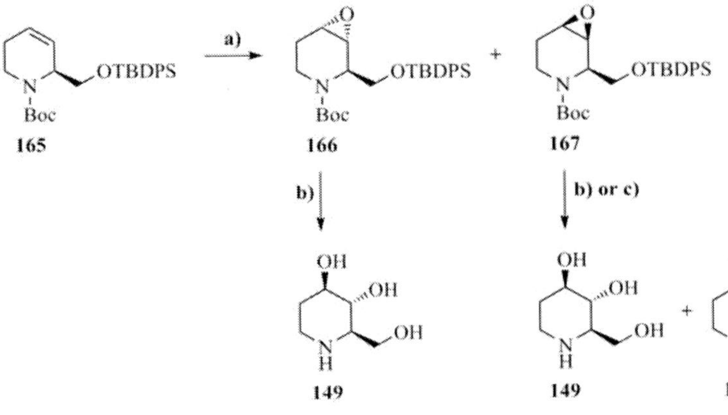


Scheme 21. *Reagents and conditions [57,77]* a) Oxone, $CF_3COCH_3$, $NaHCO_3$, aq. $Na_2EDTA$, $CH_3CN$, 60% for *166*, 30% for *167*. b) *i)* $H_2SO_4$, dioxane, $H_2O$. *ii)* Ion-exchange resin (Amberlite IRA-410), 75%. c) *i)* 0.3 M aq. KOH, dioxane, Δ. *ii)* 6*N* aq. HCl, MeOH, 60°C. *iii)* Ion-exchange resin, 99%.

3-*epi*-Fagomine *170* and 4-*epi*-fagomine *172* were obtained by *cis*- dihydroxylation of the double bond in compound *165* as shown on scheme 22. Under modified Upjonh conditions[79] dihydroxylation with catalytic amount of $K_2OsO_4.2H_2O$ and NMO of *165* gave diol *169* as a single isomer in 92% yield. Acidic deprotection, followed by treatment with Dowex resin, afforded the 3-*epi*-fagomine *170*. Surprisingly, hydroxylation of *165* with both AD-mix-α and –β reagents provided *169*, as a single isomers in both cases, in 94% and 96% yield, respectively. The bulky TBDPSO group is forcing the hydroxylation to occour by approach of reagents by the less hindered face of the tetrahydropyridine ring. In another approach silyloxy group was first removed and the so formed alcohol *171* was dihydroxilated whith $K_2OsO_4.2H_2O$ to afford also 3-*epi*-fagomine *170* after acidic removal of the Boc protecting group in 87%. The oxidation of *171* was also accomplished under Donohoe's conditions. According to Donohoe's work[80,81] a bidentade and reactive complex of osmium tetroxide and TMEDA (tetramethylethyldiamine) directs dihydroxylation of homoallylic alcohols. After deprotection of the amine group and separation by ion-exchange resin chromatography *172* and *170* were obtained as a 2:1 mixture in 86% combined yield.

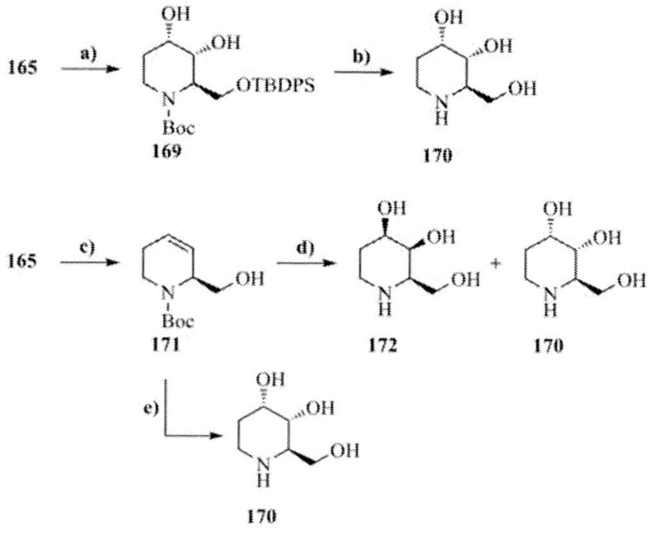

Scheme 22. Reagents and conditions [57,77] a) Cat. $K_2OsO_4.2H_2O$, NMO, acetone, $H_2O$, 92%. b) *i)* 10% HCl, dioxane. *ii)* Dowex 1X2 ($OH^-$). c) TBAF, THF, 97%. d) *i)* $OsO_4$, TMEDA, DCM. *ii)* HCl 35%, MeOH. e) *i)* Cat. $K_2OsO_4.2H_2O$, NMO, acetone, $H_2O$. *ii)* 10% HCl, dioxane, 87%.

Pandey reported a synthesis[82-84] of isofagomine *161* (scheme 23) starting from aldehyde *174* which was prepared from D-(-)-tartaric acid *173* following Kibayashi procedure[17] (Scheme 3). The aldehyde was then converted to alkyne *175* by treatment with tetrabromomethane and triphenylphosphine[85] followed by treatment with *n*-BuLi. Deprotection of the alcohol function with TBAF and its convertion into bromine *176* was accomplished by Apple's[86] methodology. Refluxing a solution of *176* in acetonitrile with *N*-[(trimethylsilyl)methyl]benzylamine in the presence of potassium carbonate, afforded *177*. Photoinduced electron transfer (PET) cyclization of this amine, using the protocol developed before by Pandey[87,88], produced *178*. Hydroboration of *178* gave the alcohol *179* in 45% yield. Hydrolisys of the acetal ring followed by catalytic hydrogenation afforded the isofagomine *161*. L-Isofagomine *180*, the enantiomer of *161*, was obtained with the same procedure starting from L-(+)-tartaric acid.

Ganem[89] had synthesized isofagomine *161* from methyl nicotinate *181* (scheme 24). Reduction of *181* with sodium borohydride and phenyl chloroformate afforded 1,2-dihydropyridine *182* in 90% yield. Reaction of the dihydropyridine *182* with *m*-CPBA afforded regioselectively *183* in 92% yield. Reduction of *183* with trimethylsilyl triflate and borane-THF complex gave alcohol *184* as a mixture of enantiomers. Oxidation of this alcohol with chromium (VI) oxide gave the achiral enone *185*, which was asymmetrically reduced with lithium borohydride using (-)-*N*-methylephedrine as the chiral auxiliary[90], affording alcohol (+)-*184* with 85% yield and 83% *ee*. Hydrolysis of the ester moiety gave hydroxyacid *186*, which after hydroboration[91] and hydrolysis of the protecting group gave isofagomine *161*.

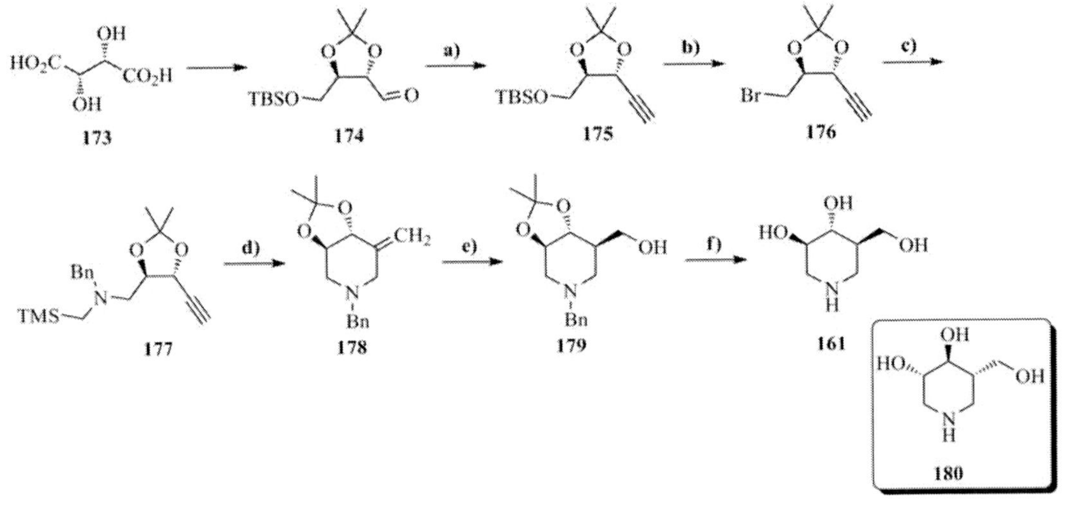

Scheme 23. Reagents and conditions [82-84] a) *i)* $CBr_4$, $Ph_3P$, DCM, 0°C, 65%. *ii)* *n*-BuLi, THF, -78°C, 90%. b) *i)* TBAF, THF, 0°C, 85%. *ii)* $CBr_4$, $Ph_3P$, DCM, 0°C, 80%. c) $PhCH_2NHCH_2TMS$, $K_2CO_3$, MeCN, Δ, 65%. d) hν, 1,4-dicyanonaphthalene, *i*-PrOH, 60%. e) 9-BBN, THF, 0°C; NaOH, $H_2O_2$, 0°C, 45%. f) *i)* HCl, MeOH; $NH_4OH$, 100%. *ii)* $Pd(OH)_2/C$, $H_2$ (75 psi), EtOH, 95%.

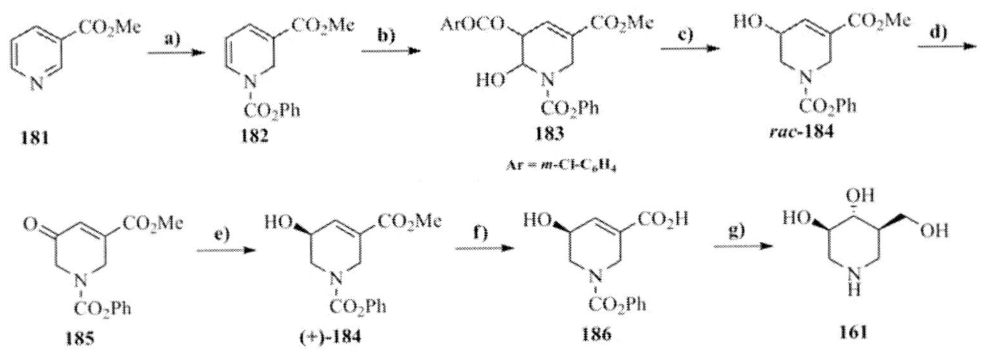

Scheme 24.Reagents and conditions [89] a) $NaBH_4$, PhOCOCl, MeOH, -78°C, 97%. b) *m*-CPBA, DCM, -70°C; 0°C, 92%. c) TMSOTf, $BH_3$.THF, -70°C to 0°C, 77%. d) $CrO_3$, acetone. e) $LiAlH_4$, (-)-N-methylephedrine, 85% (from *184*), 83% ee. f) 1N HCl, Δ, 99%. g) i) $BH_3$.THF, NaOAc, $H_2O_2$, 70%. ii) LiOH, 95%.

## Synthesis of other Imino Sugars

Hong[92] had developed a synthetic strategy for obtaining 1-deoxy imino sugars, based on compounds *193* and *194*. Schemes 25 refers to theirs preparation, obtained in 45% overall yield (1:1 ratio) from *187*[93]. Initial protection of the amine group with TsCl gave *188*. Aldol condensation of *188* with acrolein in the presence of LDA produced a 3:4 mixture of *cis* and *trans* hydroxy-esters *189*. The ester function was reduced with lithium aluminum hydride to the corresponding alcohol giving *190*. Reaction of *190* with 2,2-dimethoxypropane gave acetals *191* and *192* with 55% and 36% yields after chromatographic separation.

Treatment of each of these compounds with Grubbs' catalyst *156* gave racemic tetrahydropyridines *194* and *193* respectively. Compounds *193* and *194* were treated further to produced different 1-deoxypyranose framework types.

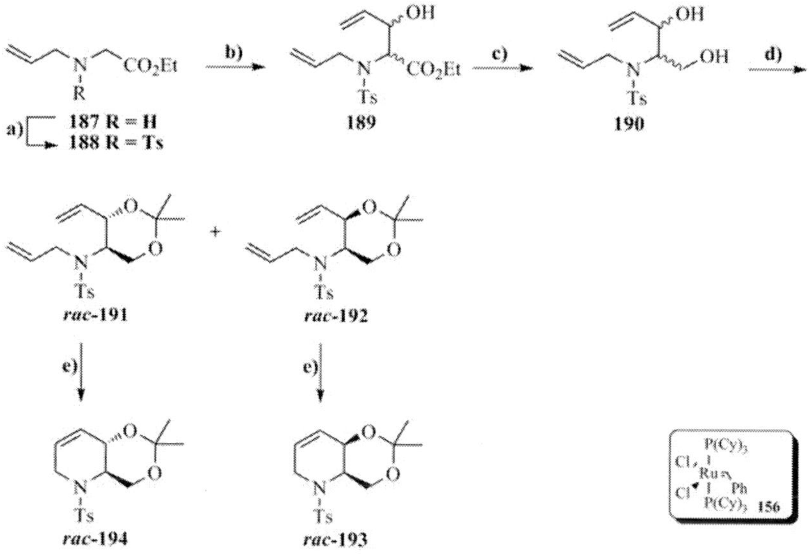

Scheme 25. Reagents and conditions [92] a) TsCl, $Et_3N$, DMAP, DCM, 89%. b) i) LDA, THF; ii) Acrolein, THF, 87%. c) $LiAlH_4$, THF, 87%. d) $Me_2C(OMe)_2$, *p*-TsOH, $C_6H_6$, 55% for *191*, 36% for *192*. e) Grubbs' catalyst *156*, DCM, 87% for *194*, 90% for *193*.

Asymmetric dihydroxylation of *194* with Hsung-Vedejs AD-mix-β[94,95] gave a mixture of diastereomers *(+)195* and *(-)196* with 42% and 35% yields after separation. Similarly, reaction of *194* with Hsung-Vedejs AD-mix-α[94,95] gave a mixture of diastereomers *(+)196* and *(-)195* with 40% and 50% yields after separation. Treatment of each of these compounds with Red-Al resulted in the cleavage of the tosyl groups. Finally the removal of the acetonides with HCl in methanol gave the imino sugars 1-deoxyallonojirimycin *199* (D-form) and *201* (L-form) and 1-deoxymannojirimicyn *16* (D-forma) and *200* (L-form), according to scheme 26.

Garner's aldehyde and its enantiomer are very useful starting materials for asymmetric synthesis, as is demonstrated along in this text. From Garner's aldehyde enantiomer *117* Takahata[96] synthesized 1-deoxygulonojirimycin *207* in 6 steps and 35% overall yield (scheme 27). Vinyl metals like vinyl zinc bromide, formed *in situ* from vinyl lithium and zinc dibromide[97,98], added diastereoselectivelly under controlled conditions to aldehyde *117* to give a 5:1 mixture of *syn* and *anti* diastereomers *202*. After recrystallization the *syn*-alcohol was obtained with 92% *de*. Treatment of *syn*-alcohol *202* with gaseous HCl in chloroform afforded acetonide *203* in 69% yield, together with the recovery of starting material (24%). *N*-Allylation of *203* by reaction with allyl iodide in the presence of NaH gave the diolefin *204*. RCM reaction of *204*, using first generation Grubbs' catalyst *156*, gave the expected tetrahydropyridine *205* that is a suitable starting material for the synthesis of various imino sugars. Under modified Upjohn conditions[79], *205* was treated with $K_2OsO_4.2H_2O$ to give diol *206* as a single isomer. Deprotection of this compound with HCl in methanol, followed by treatment with Dowex resin (50WX8 $H^+$ form) gave 1-deoxygulonojirimycin *207*.

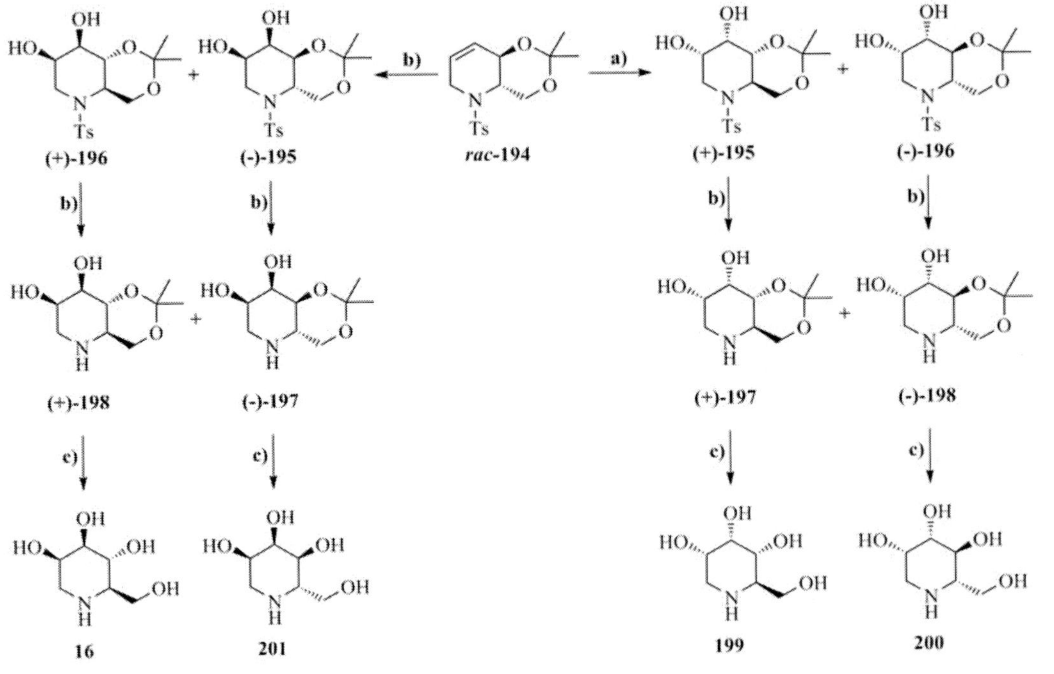

Scheme 26. Reagents and conditions [92] a) AD-mix-β, $H_2O$:*t*-BuOH (1:1.8), 42% for *(+)-195,* 35% for *(-)-196.* b) AD-mix-α, $H_2O$:t-BuOH (1:1.8), 40% for *(+)-196*, 50% for *(-)-195*. c) Red-Al ($NaAlH_2(OCH_2CH_2OCH_3)_2$), THF, 48% for *(+)-197*, 52% for *(-)-198*, 37% for *(+)-198,* 48% for (-)-*197.* d) Conc. HCl, MeOH, 83% for *199,* 90% for *200,* 89% for *(-)-17,* 78% for *201.*
*AD-mix-α* – alcene: $K_3Fe(CN)_6$: $K_2CO_3$: $MeSO_2NH_2$: $(DHQ)_2PHAL$: $OsO_4$ (1:3:3:2:0.1:0.05). *AD-mix-β* – alcene: $K_3Fe(CN)_6$: $K_2CO_3$: $MeSO_2NH_2$: $(DHQP)_2PHAL$: $OsO_4$ (1:3:3:2:0.1:0.05)[95].

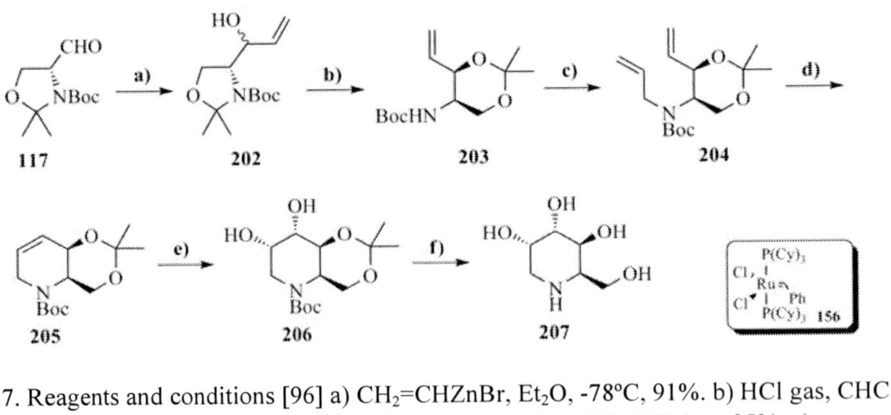

Scheme 27. Reagents and conditions [96] a) $CH_2$=CHZnBr, $Et_2O$, -78ºC, 91%. b) HCl gas, $CHCl_3$, rt, 69%. c) $CH_2$=$CHCH_2I$, NaH, THF, 0ºC, 76%. d) Grubbs' catalyst *156,* DCM, rt, 95%. e) $K_2OsO_4.2H_2O$, NMO, acetone:$H_2O$, rt, 85%. f) i) 6*N* HCl, MeOH, rt; ii) Dowex 50WX8 ($H^+$), 90%.

Under different conditions *205* was transformed into three types of iminosugars, *209*, *210* and *213* (scheme 28). By epoxidation with (trifluoromethyl)methyldioxirane *205* afforded *208* as a single isomer. Treatment of this epoxide with aqueous KOH in dioxane results in a regioselective opening of the ring by nucleophilic attack to the less hindered carbon. Subsequent treatment of the afforded diol with HCl in methanol and Amberlite resin gave 1-deoxyidonojirimycin *209*. Epoxidation of compound *205* on its other face was accomplished after hydrolysis of the acetonide with TsOH in methanol followed by treatment with *m*-CPBA. Acetonide protection was remade by treatment of the afforded epoxide with dimethoxypropane in the presence of catalytic amount of pyridine *p*-toluenesulfonate (PPTS). Treatment of *212* with sulfuric acid in dioxane resulted in the regioselective opening of the epoxide, together with acetonide and Boc group cleavage, affording 1-deoxygalactonojirimycin *213*, isolated after ion-exchange resin chromatography. Catalytic hydrogenation of *205,* followed by treatment with HCl in methanol and neutralization with sodium hydroxide gave piperidine *210*.

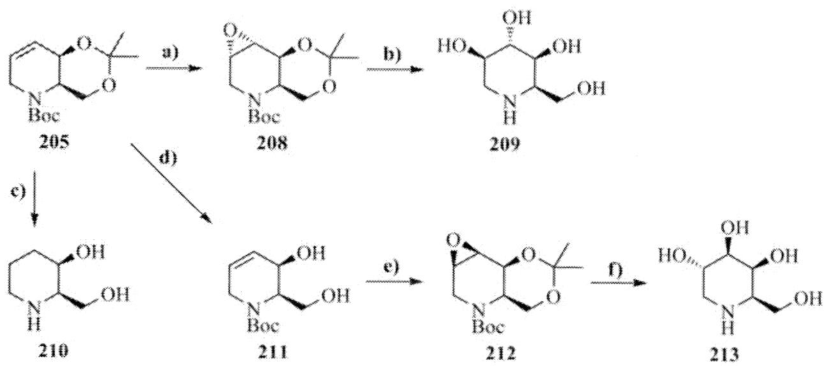


Scheme 28. Reagents and conditions [96] a) Oxone, $CF_3COCH_3$, $NaHCO_3$, aq. $Na_2EDTA$, $CH_3CN$, 0°C, 99%. b) i) 0.3 M KOH, 1,4-dioxane, $H_2O$, Δ; ii) 6*N* HCl, MeOH, rt; iii) Amberlite IRA-410 ($OH^-$), 87%. c) i) $H_2$, Pd/C (10%), MeOH, rt; ii) 5*N* HCl, MeOH, 60°C, 88%. d) *p*-TsOH.$H_2O$, MeOH, rt, 97%. e) i) *m*-CPBA, $NaH_2PO_4$, DCM, 0°C; ii) 2,2-dimethoxypropane, PPTS cat., acetone, rt, 53%. f) i) $H_2SO_4$, 1,4-dioxane, $H_2O$, Δ; ii) Amberlite IRA-410 ($OH^-$); iii) Dowex 1X2 ($OH^-$), 83%.

Han reported a synthesis of four different imino sugars[99] starting from olefin *214* (scheme 29) readily prepared from ethyl 4-bromocrotonate and *p*-methoxyphenol. Based on his own research on substrate-based methodology for the regioselective control of the Sharpless asymmetric aminohydroxylation (AA) of olefins[100], Han had prepared the olefin *214*. Osmium catalyzed AA reaction of *214* afforded the aminoalcohol *215* with an excellent regioselectivity (>20:1). After purification *215* was isolated with >99% *ee*. Protection of the hydroxyl group with *p*-methoxybenzyl chloride, followed by reduction of the ester function with $LiAlH_4$ and silylation of the afforded alcohol with TBDPSCl was obtained *216*. *N*-Allylation of *216* with allyl bromide and KH, catalysed by 18-crown-6-ether[101], furnished *217*. Desilylation of *217* with TBAF and Dess-Martin oxidation[102] of the produced alcohol gave aldehyde *218*. The transformation of the carbonyl into olefin used a modified Horner-Wadsworth-Emmones protocol[103] furnishing *219*. The RCM reaction of the diolefin *219* was accomplished by treatment of this compound with first generation Grubbs' catalyst *156*, affording tetrahydropyridine *220* in 80% yield. Tetrahydropyridine *220* is structurally similar to *205* synthesized by Takahata (Scheme 27) and so suitable for the synthesis of the same imino sugars compounds obtained by Takahata. Starting from *220* Han reported the synthesis of 1-deoxygulonojirimycin *207* and 1-deoxyidonojirimycin *209*, both synthesized by Takahata (scheme 27 and 28). Reaction of *220* with osmium tetroxide gave diol *221* as a single product. The diol was protected by reaction with dimethoxypropane for purification reasons. The acetonide was then treated with ceric ammonium nitrate (CAN) and aqueous HCl to provide 1-deoxygulonojirimycin *207* in its HCl salt form. The reaction of *221* with thionyl chloride followed by oxidation of the resulting cyclic sulfite[104] gave cyclic sulfate *222*. Treatment of this sulfate with sodium benzoate in DMF gave *223*, by regioselective nucleophilic attack. Deprotection of *223* by reaction with CAN followed by aqueous HCl afforded 1-deoxyidonojirimycin *209* in its HCl salt form.

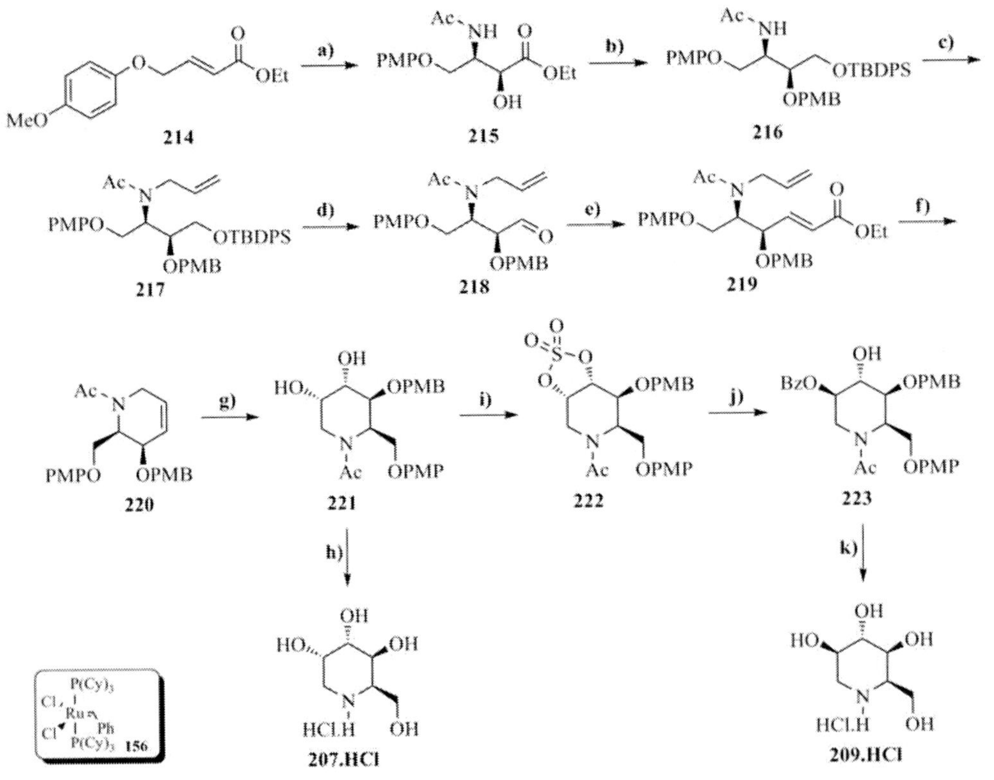

Scheme 29. Reagents and conditions [99] a) $K_2OsO_4.2H_2O$ (5% mol), $(DHQD)_2PHAL$ (6% mol), LiOH, *n*-bromoacetamide,*t*-BuOH:$H_2O$, 3:2, 4°C, 70%. b) i) NaH, PMBCl, DMF, 0°C, 80%; ii) $LiBH_4$, $Et_2O$, quantitative; iii) TBDPSCl, $NEt_3$, DMAP, DCM, 25°C, quantitative. c) KH, 18-crown-6-ether, allyl bromide, THF, 25°C, 95%. d) i) TBAF, THF, 25°C, quantitative; ii) Periodinane, DCM, 25°C, 90%. e) Triethyl phosphonoacetate, LiBr, DBU, THF, 25°C, 94%. f) Grubbs' catalyst *156*, toluene, 90°C, 80%. g) $OsO_4$, NMO,*t*-BuOH:$H_2O$ 1:1, 96%. h) i) $Me_2C(OMe)_2$, PPTS, DCM, 25°C, 94%; ii) CAN, MeCN:$H_2O$ 4:1, 0°C, 92%; iii) 6*N* HCl, 120°C, 98%. i) i) $SOCl_2$, $NEt_3$, DCM, -15°C; ii) $RuCl_3$, $NaIO_4$, MeCN:DCM:$H_2O$ 1:1:1, 25°C, 85%. j) i) NaOBz, DMF, 105°C; ii) 20% aq. $H_2SO_4$:DCM (1:1), 25°C, 85%. k) i) CAN, MeCN:$H_2O$ 4:1, 0°C; ii) 6*N* HCl, 120°C, 90%.

Selective cleavage of *p*-methoxybenzyl group in tetrahydropyridine *220* to give tetrahydropyridinol *224* was effected with 5% trifluoroacetic acid in DCM[105] (Scheme 30). Mitsunobu reaction[106] of *224* provided an inverse configuration on 3-carbon to which hidroxyl group was attached. The benzoyl was cleaved upon treatment with $K_2CO_3$ in methanol to give the free alcohol that was silylated with TBDPSCl and formed *225*. The bulky steric effect of TBDPS group was shown to improve the selectivity of the procedding dihydroxylation step. A mixture of diol *227* and *226* in a 1:10 ratio was formed upon treatment with osmium tetroxide; the diols were separated by chromatography after being acetylated. Deprotection of triacetate derived from *226* with CAN and aqueous HCl afforded 1-deoxymannojirimycin *17*, in its HCl salt form. The mixture of diols *227* and *226* was also coverted into the corresponding cyclic sulfates, which were separated by chromatography. The cyclic sulfate *228* was subjected to ring opening reaction with sodium benzoate to give *229*. Deprotection of *229* gave 1-deoxyaltronojirimycin *230*, in the HCl salt form.

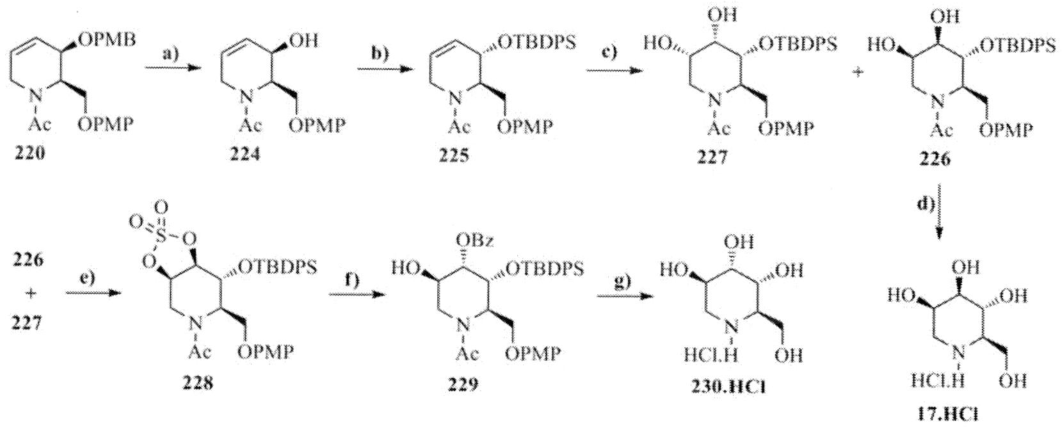

Scheme 30. Reagents and conditions [99] a) TFA, DCM, 25°C, 85%. b) i) DIPAD, $Ph_3P$, benzoic acid, THF, 0°C, 80%, ii) $K_2CO_3$, MeOH, 25°C, quantitative; iii) TBDPSCl, imidazole, $NEt_3$, DMF, 60°C, 93%. c) $K_2OsO_4.2H_2O$, NMO,*t*-BuOH:$H_2O$ 1:1, 94%. d) i) $Ac_2O$, DMAP, $NEt_3$, DCM, 25°C, 80%; ii) CAN, MeCN:$H_2O$ 4:1, 0°C; iii) 6*N* HCl, 120°C, 92% for the two steps. e) i) $SOCl_2$, $NEt_3$, DCM, -15°C; ii) $RuCl_3$, $NaIO_4$, MeCN:DCM:$H_2O$ 1:1:1, 25°C, 66%. f) i) BzONa, DMF, 105°C; ii) 20% aq. $H_2SO_4$, DCM (1:1), 25°C, 80%. g) i) CAN, MeCN:$H_2O$ 4:1, 0°C, 90%; ii) 6*N* HCl, 120°C, 98%.

Some years before Kazmaier had reported[107] the synthesis of 1-deoxy-L-altronojirimycin *239* from *N*-tosylglycine benzyl ester *231* represented in scheme 31. Deprotonation of *231* with an excess of LDA and subsequent reaction with $SnCl_2$ gave the intermediate *232*. Addition of aldehyde *21*[17,108] to *232* gave *233* with high induced stereoselectivity on C-2 (>95% *de*). Due to C-1 labile configuration under the reaction conditions, a 1:1 mixture of two isomers was obtained. The six-membered ring structure was produced under Mitsunobu cyclization conditions achieved after protection of the free hydroxyl group with 3,4-dihydro-2*H*-pyran, giving *235*. Cleavage of silyl group with TBAF gave alcohol *236*, which was treated with an excess of Mitsunobu's reagent to afford *237* as a single diastereoisomer in 92% yield. The use of an excess of Mitsunobu's reagent was crucial for the stereoselectivity of this cyclization reaction. Under those reaction conditions the equilibrium between both C-1 isomers should be faster than the cyclization itself being 1*R*-

diastereomer formation, more favorable than the formation of 2*S*-diastereomer. Diastereoisomers *236* were also separated and cyclized with stoichiometric amount of $PPh_3$/DEAD giving *237* and 1-epi-*237* respectively. Treatment of *237* with an excess of Red-Al resulted in a simultaneous cleavage of the *N*-tosyl group and reduction of the ester moiety to give the amino alcohol *238*. Further deprotection with Dowex (50 WX8) and ammonia gave 1-deoxy-L-altronojirimycin *239*.

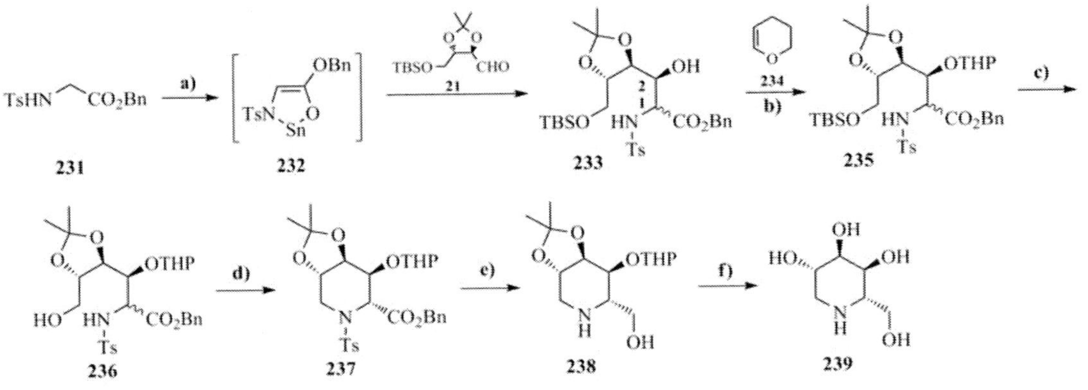

Scheme 31. Reagents and conditions [107] a) LDA, $SnCl_2$, THF, -78°C. b) 3,4-Dihydro-2H-pyran, cat. $H^+$, DCM, 83%. c) TBAF, THF, 82%. d) $PPh_3$, DEAD, THF, rt, 92%. e) Red-Al, THF, Δ. f) i) Dowex 50WX8, Δ; ii) 2*N* $NH_3$, 44%.

The synthesis of aldehydes *3*, reported by Vogel[8] was previously describe in scheme 1. Reduction of aldehyde *(+)-3* with lithium aluminum hydride[9,109,110] in ether gave the amino diol *(-)-240* (scheme 32). Protection of the amine group by a modified Oku´s procedure [111] gave carbamate *(-)-241*. Acidic hydrolysis of the acetonide furnished furanose *(+)-242*, which after hydrogenation provided 1-deoxy-L-idonojirimycin *244*. Starting from the enantiomer of *(+)-3*, *(-)-3*, following the same sequence of reactions was obtained 1-deoxy-D-idonojirimycin *209*. Compound *209* can also be obtained in a more direct way from acetonide *(+)-240* by hydrolysis with aqueous TFA and conversion of the bicyclic piperidine *243* formed into *209*, upon hydrogenation.

The aldehydes *102* and *104* were synthesized by Fernández-Mayoralas[46] and presented above in scheme 13. The condensation reactions of these aldehydes with ketones *105* and *106* were partially presented above (scheme 14): the reactions of aldehyde *102* when catalyzed by L-proline gave D-mannojirimycin derivatives; reactions of aldehyde *104* when catalyzed by D-proline afforded L-mannojirimycin derivatives. Fernández-Mayoralas also reported that D-allonojirimycin derivatives were obtained starting from aldehyde *102* by its reactions with ketones *105* and *106* in the presence of D-proline, as mixtures of α- and β- derivatives (scheme 33). When *104* was reacted with the same ketones (*105* and *106*) in the presence of L-proline, L-allonojirimycin derivatives were obtained as mixtures of α- and β- isomers.

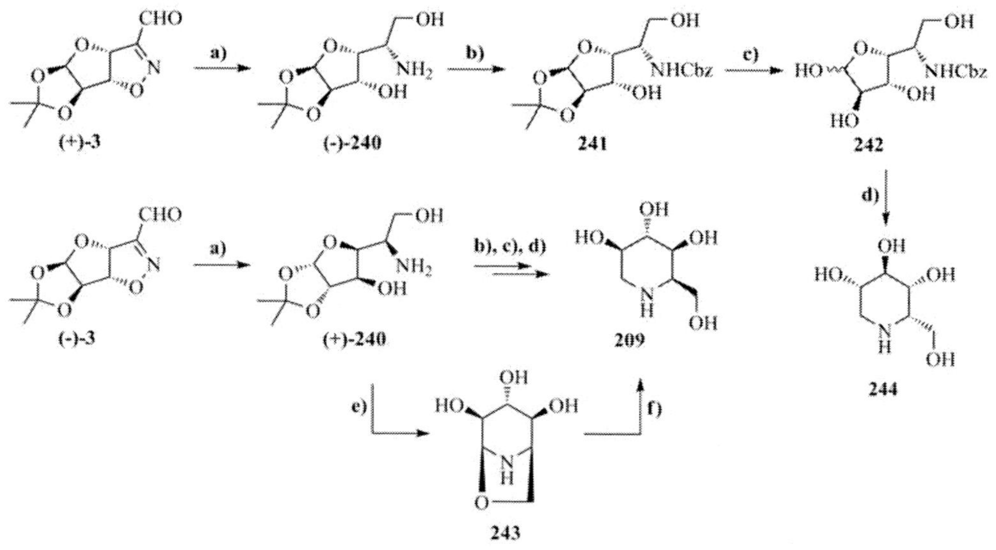

Scheme 32. Reagents and conditions [8] a) $LiAlH_4$, $Et_2O$, 25°C, 66%. b) CbzCl, $NaHCO_3$, EtOH 50%, 25°C, 81%. c) TFA:$H_2O$ 8:1, 4°C, 81%. d) $H_2$, Pd/C, MeOH, 25°C, 95%. e) TFA:$H_2O$ 9:1, 25°C; Dowex 1 ($OH^-$ form), 93%. f) $H_2$, Pd/C, $H_2O$, 25°C, 77%.

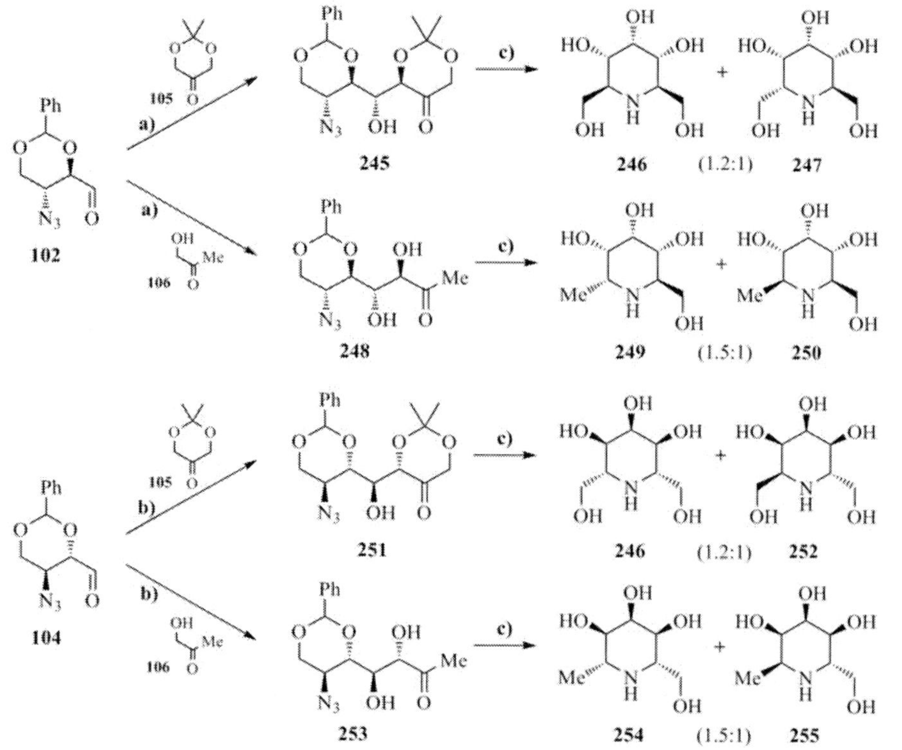

Scheme 33. Reagents and conditions [46] a) D-Proline (30% mol), DMF, 60% for *245*; 40% for *248*. b) L-Proline (30% mol), DMF, 60% for *251*; 40% for *253*. c) $H_2$ (45 psi), Pd/C, MeOH/HCl, 68% (*246* and *247*); 50% (*249* and *250*); 68% (*246* and *252*); 50% (*254* and *255*).

Hegedus[112] synthesized 1-deoxy-6-homo-D-galactonojirimycin *265* starting from diethyl L-tartrate *18* in 12 steps and 30% overall yield as shown in scheme 34. The diol moiety of *18* was protected with cyclohexanone and the ester groups reduced to the corresponding alcohols with lithium aluminum hydride to give *256*. Silylation of this diol with 1 eq. of TBSCl gave the monosilylated compound. Dess-Martin oxidation of the hydroxyl group gave aldehyde *257*. The allylstannane *258*, obtained from propagyl bromide and optically active diphenyloxazolidinone in 54% overall yield[113] was added to aldehyde *257* in the presence of $BF_3.Et_2O$. Compound *259* was formed with >95% *de*. Hydroboration of *259* followed by oxidation with hydrogen peroxide gave carboxylic acid *260*. Activation of the carboxylic acid with Mukaiyama reagent[114] gave the cyclization product *261*. Cleavage of the oxazolidinone ring by catalytic hydrogenation in the presence of Boc anhydride followed by removal of the silyl group with HF-Py afforded alcohol *262*. Cyclization of the primary alcohol *262* under Mitsunobu conditions gave bicyclic amino lactone *263*. Reduction of the lactone moiety with $LiAlH_4$ followed by acidic hydrolysis of the ketal gave 1-deoxy-6-homo-D-galactonojirimycin *265*, as its HCl salt form.

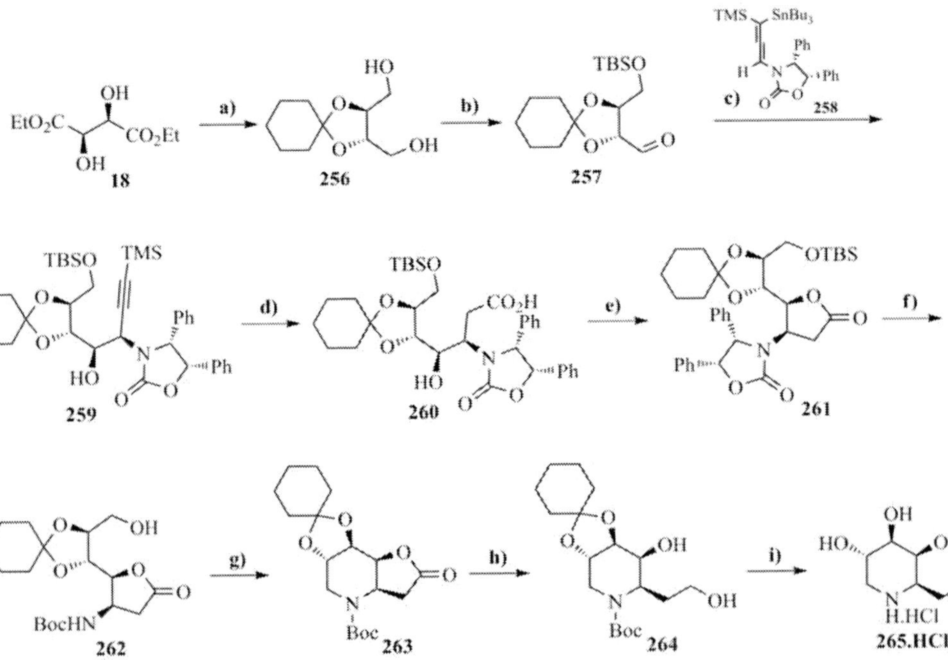

Scheme 34. Reagents and conditions [112] a) i) Cyclohexanone, $C_6H_6$, TsOH, Δ; ii) $LiAlH_4$, THF, Δ, 77%. b) i) NaH, TBSCl, THF, 25°C, 93%; ii) Dess-Martin (periodinane), DCM, 25°C, 83%. c) $BF_3.Et_2O$, DCM, -70°C, 86%. d) i) $Cy_2BH$, THF, 0°C to 25°C; ii) $H_2O_2$, aq. $NaHCO_3$, 0°C to 25°C, 88%. e) Mukaiyama's reagent, $Et_3N$, DCM, 25°C, 99%. f) i) $H_2$ (80 psi), cat. $Pd(OH)_2$, $Boc_2O$, THF, 25°C, 99%; ii) HF.Py, MeCN, 25°C, 99%. g) DEAD, $Ph_3P$, THF, -20°C to 25°C, 82%. h) $LiAlH_4$, THF, 20°C, 89%. i) HCl, MeOH, 25°C, 94%.

Altenbach[115] reported a synthesis of 1,3-dideoxy imino sugar *276* starting from L-furylglycine *266*[116] (scheme 35). Protection of the amine group by reaction with TsCl gave *267*, which was reduced with lithium aluminum hydride affording amino alcohol *268*. Oxidative ring expansion occurred by reaction of *268* with NBS[117] furnishing dihydropyridone *269* with total diastereoselectivity and 83% yield. Refluxing *269* in toluene, with a catalytic amount of TsOH, gave the intramolecular cyclization product *270*. Reaction of *270* with benzyl alcohol in the presence of $NEt_3$ led to the Michael addition product *271* also with total stereoselectivity. Reduction of ketone group under Luche conditions[118] formed a mixture of alcohols *273* and *272* in a 1:5.2 ratio. The crude mixture of alcohols *273* and *272* was treated with acetic anhydride in the presence of catalytic amount of DMAP affording the unreacted alcohol *273* and acetate *274* that was chromatographically separated. Reaction of acetate *274* with Red-Al formed *275* isolated by chromatography [Dowex-50 ($H^+$)]. Catalytic hydrogenation of *275* gave (2*S*,3*S*,5*S*)-2-(hydroxymethyl)piperidine-3,5-diol *276*.

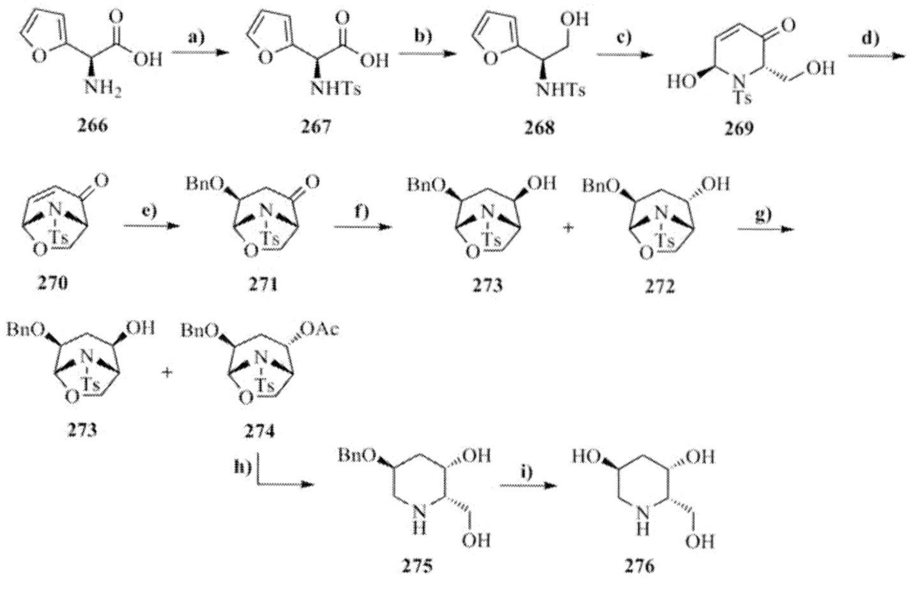

Scheme 35. Reagents and conditions [115] a) NEt(*i*-Pr)$_2$, TsCl, $CH_3CO_2Et$, rt, 61%. b) i) $LiAlH_4$, $Et_2O$, -15°C; ii) 1N aq. HCl, $CH_3CO_2Et$, 62%. c) NBS, $CH_3CO_2Na$, THF:$H_2O$ (4:1), -5°C, 83%. d) $C_6H_6$, *p*-TsOH, Δ, 87%. e) BnOH, $NEt_3$, rt, 69%. f) i) MeOH, $CeCl_3.7H_2O$, 10°C; ii) $NaBH_4$, MeOH, -5°C, 77%. g) $(CH_3CO)_2O$, DMAP, DCM, rt, 93%. h) i) Red-Al, $CH_3OCH_2CH_2OCH_3$, Δ; ii) 0.75N aq. HCl; iii) Dowex 50X8, 54%. i) $H_2$, Pd/C, $H_2O$, 92%.

A synthesis of the piperidine-3,4,5-triol *288* was recently reported by Rao[119], starting from 2-bromo-4′-methylacetophenone (scheme 36). At first, nucleophilic displacement of bromide was obtained with sodium azide in the presence of β-cyclodextrin giving *278*. Reduction of ketone group with $BH_3.Me_2S$ and (*S*)-diphenyl prolinol gave alcohol *279* with 94% yield and 100% *ee*. The same reduction was accomplished using sodium borohydride in water, affording *279* with 95% yield and 80% *ee*. The alcohol function was protected as an ether by treatment with MOMCl. The product formed, *280*, was subjected to oxidative cleavage with $RuCl_3$ and $NaIO_4$ to give methyl ester *281*. Reduction of the azide group with $PPh_3$ in the presence of β-cyclodextrin in water gave the corresponding amine, which was treated with Boc anhydride to give *282*. Reduction of the ester function using sodium borohydride in the presence of lithium chloride gave alcohol *283*. Swern oxidation[120] of *283* followed by olefination of the afforded aldehyde with $PH_3PCH_3I$ gave *284*. *N*-allylation of *284* was obtained by reaction with allyl bromide in the presence of sodium hydride with formation of the target starting material for RCM reaction, *285*. RCM reaction of di-olefin *285*, catalyzed by first generation Grubbs' catalyst *156*, gave tetrahydropyridine *286*. Reaction with catalytic amount of osmium tetroxide and *N*-methylmorfoline *N*-oxide afforded diol *287* as a single isomer in 82% yield. Treatment of *287* with HCl in methanol provided (3*R*,5*R*)-piperidine-3,4,5-triol *288*, in its HCl salt form.

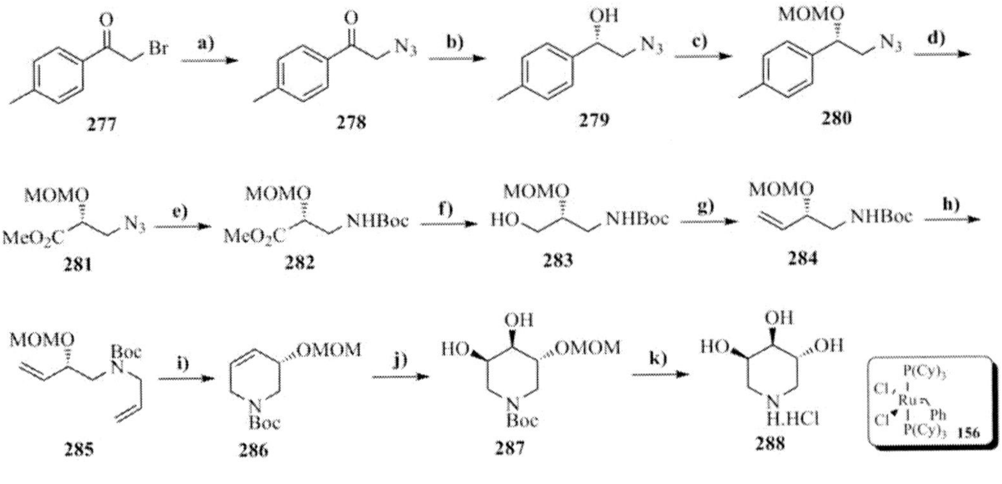

Scheme 36. Reagents and conditions [119] a) $NaN_3$, β-cyclodextrin/$H_2O$, rt, 99%. b) (*S*)-Diphenyl prolinol, $BH_3.Me_2S$, THF, 40°C, 94%. c) MOMCl, DIPEA, DMAP, DCM, 0°C to rt, 94%. d) i) $RuCl_3.H_2O$, $NaIO_4$, $MeCN/CCl_4/H_2O$ 1:1:1.5, rt; ii) $CH_2N_2$, $Et_2O$, 0°C, 85%. e) i) β-cyclodextrin/$H_2O$, $PPh_3$, rt; ii) $Boc_2O$, rt, 91%. f) $NaBH_4$, LiCl, MeOH/THF, 0°C to rt, 90%. g) i) DMSO, oxalyl chloride, $NEt_3$, DCM, -78°C; ii) $PH_3PCH_3I$,*t*-BuOK, 0°C to rt, 85%. h) NaH, allyl bromide, TBAI, THF, rt, 84%. i) Grubbs' catalyst *156*, DCM, rt, 92%. j) $OsO_4$, NMO, acetone:$H_2O$, 82%. k) *6N* HCl, MeOH, rt, 90%.

The preparation of alcohol *95* reported by O'Doherty[39] was refered before in the synthesis of 1-deoxymannojirimycin (scheme 12). Dihydroxylation of *95* by reaction with $OsO_4$/NMO afforded *289* in 96% yield (scheme 37). Hydrogenation of *289* followed by treatment with TsOH gave 1-deoxygulonojirimycin *207* in its TsOH salt form. Catalytic hydrogenation of alcohol *95* followed by treatment with TsOH gave (2*R*,3*R*)-1-(methoxymethyl)piperidin-3-ol 4-methylbenzenesulfonate *210* with 96% yield. Mitsunobu reaction of alcohol *95* afforded alcohol *96*. Treatment of *96* with TBAF followed by catalytic hydrogenation finally led to (2*R*,3*S*)-1-(methoxymethyl)piperidin-3-ol *290*.

Tetrahydropyridine *165* was prepared by Takahata[57,77] as precursor for the synthesis of the four possible isomers of fagomine (scheme 20). When *165* was reacted with (trifluoromethyl)methyl dioxirane epoxides *166* and *167* were obtained as a mixture. Epoxide *167* was obtained free of its isomer *166* by epoxidation of alcohol *171*, previously formed by cleavage of the silyl group, with *m*-CPBA (scheme 38). The reaction of each epoxide with Super-Hydride, followed by deprotection with HCl in methanol and treatment with Dowex, gave the coorresponding (hydroxymethyl)piperidinol. In the case of epoxide *166* the nucleophilic attack was region and stereoselective, affording (2*S*,3*R*)-1-(methoxymethyl)piperidin-3-ol *290*, *syn* oriented to the silyloxymethyl group, with the configuration of the newly formed alcohol governed by the incoming hydride attacking the less hindered face of the epoxide. On the contrary, the nucleophilic attack to epoxide *167* is not fully regioselective and is *anti* oriented to the silyloxymethyl group, affording a 1:5 mixture of (2*R*,3*R*)-1-(methoxymethyl)piperidin-3-ol *210* and (2*S*,4*R*)-1-(methoxymethyl)piperidin-4-ol *292*.

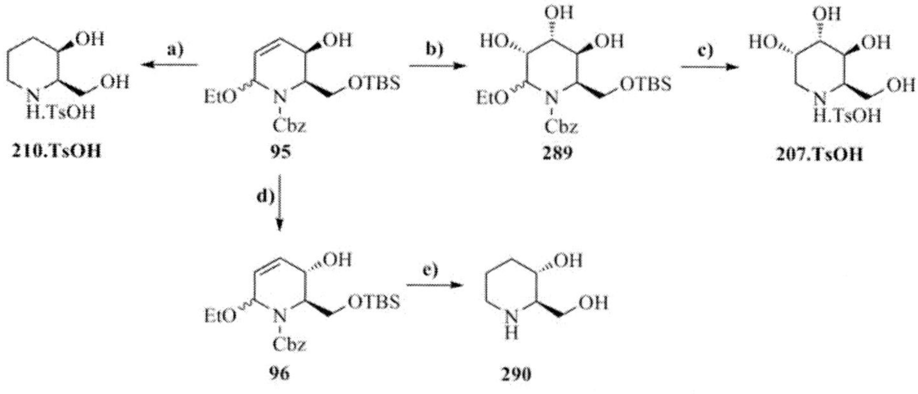


Scheme 37. Reagents and conditions [39] a) $H_2$, Pd/C, TsOH, MeOH, 96%. b) $OsO_4$, NMO, DCM/$H_2O$, 0°C, 96%. c) $H_2$, Pd/C, TsOH, MeOH, 99%. d) i) *p*-Nitrobenzoic acid, $PPh_3$, DEAD, 84%; ii) MeOH, $Et_3N$, 94%. e) i) TBAF, 86%; ii) $H_2$, Pd/C, MeOH, 61%.

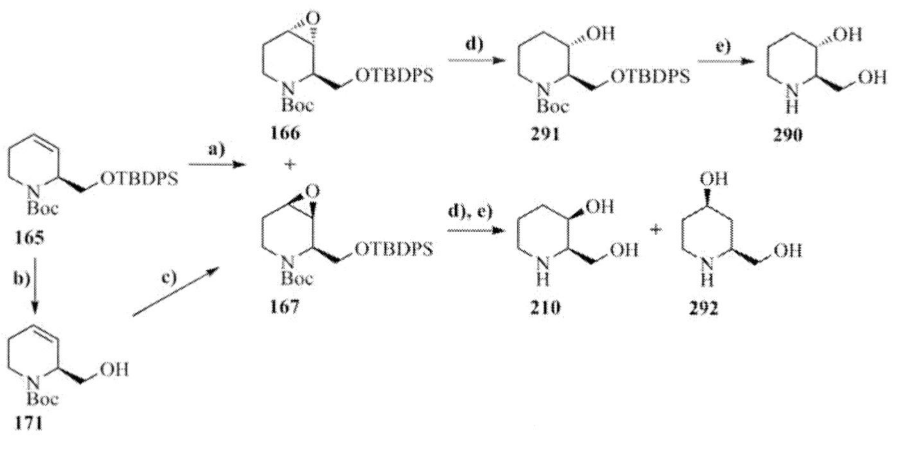


Scheme 38. Reagents and conditions [57,77] a) Oxone, $CF_3COCH_3$, $NaHCO_3$, aq. $Na_2EDTA$, MeCN, 96%. b) TBAF, THF, 97%. c) *m*-CPBA, $NaHCO_3$, DCM, 64%. d) Super-Hydride, THF, 96% for *291*. e) *i*) HCl 35%, 1,4-dioxano; *ii*) Dowex 1X2 ($OH^-$), 100% for *290*; 79% for *210+292* from *167*.

The synthesis of *178* reported by Pandey[82-84] from D-tartaric acid, was presented above (scheme 23). Catalytic hydrogenation of *178* followed by protection of the amine group with gave a 4:1 mixture of diastereomers (scheme 39). After diastereomeric separation followed by removal of Boc protecting group with HCl in methanol, the major isomer *293* afforded 5'-deoxy-4-*epi*-isofagomine *294* in its HCl salt form. 4-Hydroxy-4-*epi*-isofagomine *298* was also obtained from *178*. First treatment with osmium tetroxide resulted in the formation of *296*. After removal of benzyl group by catalytic hydrogenation, followed by cleavage of the acetal ring in HCl/methanol *298* was obtained in its HCl salt form. Functional group transformation from diol *296* led also to (3*R*,5*R*)-piperidine-3,4,5-triol *302*. Oxidative cleavage of diol *296* with $NaIO_4$ gave ketone *300*. Reduction of the carbonyl was accomplished with $NaBH_4$ furnishing alcohol *301* with 9:1 diastereoselectivity. One pot deprotection of *301* by hydrogenation in the presence of HCl followed by chromatographic purification gave (3*R*,5*R*)-piperidine-3,4,5-triol *302* in its HCl salt form. Starting from L-tartaric acid were formed compounds *295*, *299* and *303*, enantiomers of the formers *294*, *298* and *302* compounds.

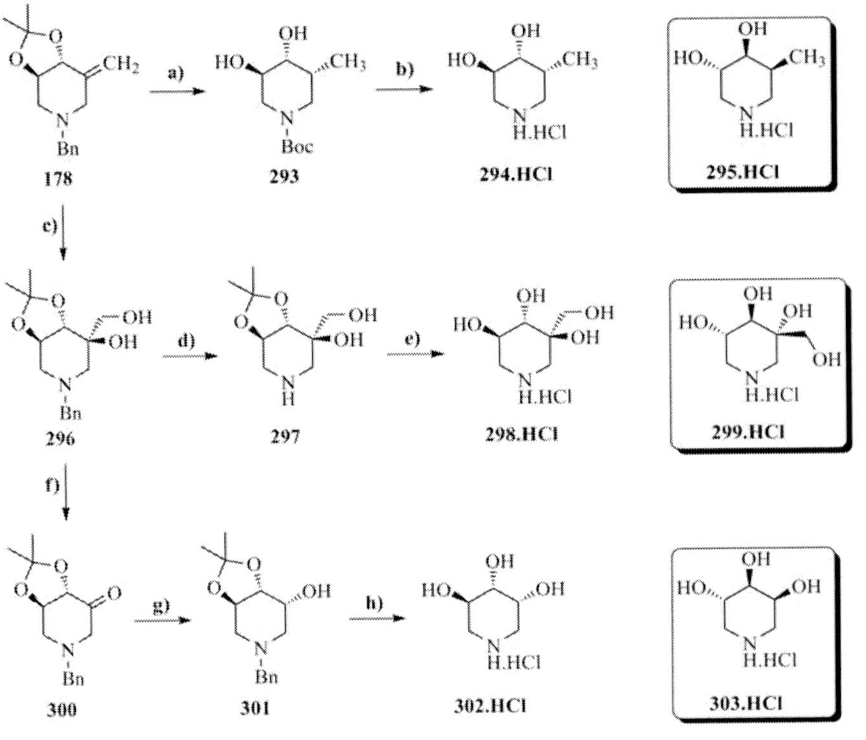

Scheme 39. Reagents and conditions [83,84] a) i) $H_2$, Pd/C, MeOH, HCl, rt, 89%; ii) $(Boc)_2O$, $Et_3N$, DCM, rt, 75%. b) HCl, MeOH, 0°C to rt, 100%. c) $OsO_4$, NMO, pyridine, acetone:$H_2O$ 9:1, 0°C to rt, 95%. d) $H_2$ (65 psi), $Pd(OH)_2$, EtOH, 90%. e) HCl, MeOH, 100%. f) $NaIO_4$, EtOH:$H_2O$ 4:1, rt, 80%. g) $NaBH_4$, MeOH, rt, 85%. h) $H_2$, $Pd(OH)_2$/C, HCl, MeOH, rt, 88%.

Takahata[69] had synthesized tetrahydropyridines *304* and *306* starting from *159*, which synthesis was already described (scheme 19). Treatment of epoxide *159* with the organocuprate $(CH_2{=}CH)_2CuCNLi_2$ gave *160* (scheme 40). Similarly *305* was obtained by treatment of *159* with $Me_2CuCNLi_2$. Hydroboration of the vinyl group in *160* with 9-borabicyclo[3.3.1]nonane (9-BBN) followed by treatment with hydrogen peroxide gave the corresponding primary alcohol; cleavage of Boc protecting group in the amine with HCl in 1,4-dioxane afforded 5'-homoisofagomine *304*. 5'-Deoxyisofagomine *306* was obtained by deprotection of *305* with HCl in 1,4-dioxane.

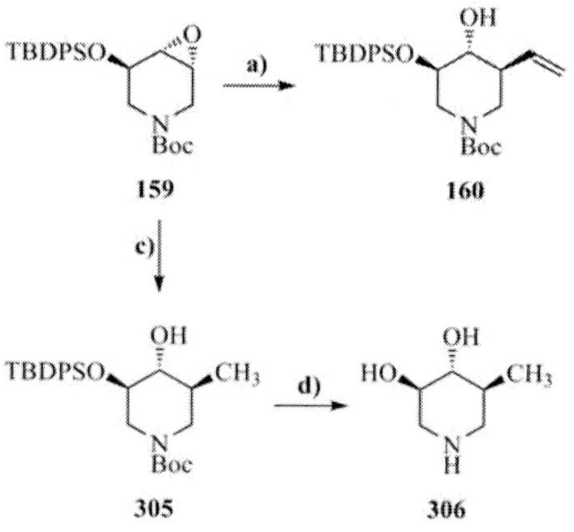

Scheme 40. Reagents and conditions [69] a) $(CH_2=CH)_2CuCNLi_2$, $BF_3.OEt_2$, -78°C, 74%. b) i) 9-BBN/THF, rt; ii) $H_2O_2$, NaOH (3M/THF), rt; iii) HCl/dioxane, Δ, 88%. c) $Me_2CuCNLi_2$, $BF_3.OEt_2$, -78°C, 71%. d) HCl/dioxane, Δ, 92%.

Comins synthesized[121] (+)-deoxoprosopinine *313* from *12* (scheme 41). The starting material *12* was obtained from 4-methoxypyridine[11] (scheme 2) and acylated with phenyl chloroformate and *n*-BuLi to give carbamate *307*. Diastereoselective acetylation of *307* with $Pb(OAc)_4$[14] afforded *308* in 57% yield and >98% *de*. Treatment of the resulting acetate *308* in refluxing formic acid followed by treatment with ammonia in methanol afforded the bicyclic compound *309*. Reduction of the ketone group with $NaBH_4$ in the presence of $CeCl_3$[122,123] followed by reaction of the afforded alcohol with $Ac_2O$ gave diacetate *310*. The allylic acetate *310* suffered then addition of allyl silane *311*, catalyzed by $BF_3.Et_2O$. After catalytic hydrogenation *312* was formed. The oxazolidinone ring opening occured by heating *312* in an ethanolic solution of KOH to finally produce (+)-deoxoprosopinine *313*.

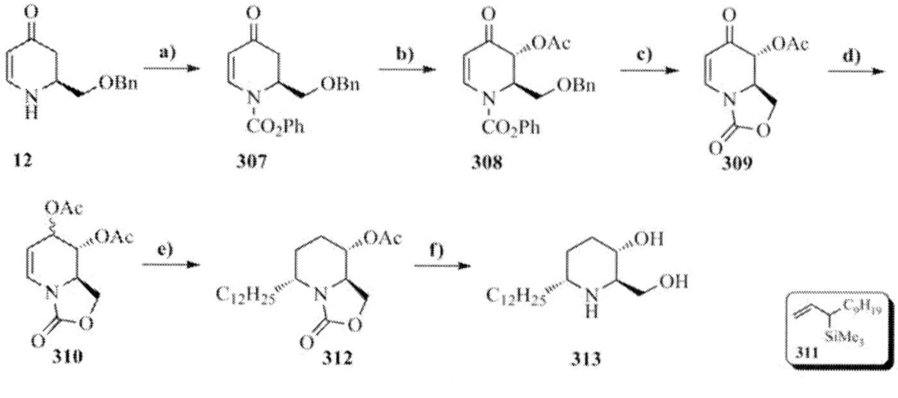

Scheme 41. Reagents and conditions [121] a) *n*-BuLi, PhOCOCl, THF, -78°C, 94%. b) $Pb(OAc)_4$, toluene, Δ, 57%. c) i) $HCO_2H$, Δ; ii) $NH_3$, MeOH, 0°C, 73%. d) i) $NaBH_4$, $CeCl_3.7H_2O$, MeOH, -40°C; ii) $Ac_2O$, $Et_3N$, DMAP, DCM, 98%. e) i) $BF_3.Et_2O$, *311*, DCM; ii) $H_2$, Pt/C, EtOH, 71%. f) KOH, EtOH, 140°C (sealed tube), 85%.

Lebreton synthesized[124] (-)-3-*epi*-deoxoprosopinine *323* from Garner's aldehyde *314* as described in scheme 42. Olefination of aldehyde *314* with diethyl benzylphosphonate gave *315*, which afforded amino-alcohol *316* upon treatment with HCl in refluxing MeOH.

Addition of tridecanal to the amine function of *316* gave imine *317* which was reacted with allyl magnesium bromide to give *318* as a mixture of *trans* and *cis* isomers in 87:13 ratio. After separation the *trans* isomer was protected by reaction with carbonyldiimidazole (CDI) in the presence on $NEt_3$ to afford *319*. RCM of *319* occurred in the presence of second-generation Grubbs' catalyst *129* to give the expected tetrahydropyridine *320*. Epoxidation of the double bond with *m*-CPBA gave *endo* epoxide *321* as the major isomer (90/10 ratio), isolated in 81% yield. Reaction of *321* with lithium triethylborohydride (Super-Hydride) results in the regioselective opening of the epoxide affording *322* in 98% yield. Removal of the oxazolidinone protection occurred with KOH in refluxing methanol with formation of (-)-3-*epi*-deoxoprosopinine *323*.

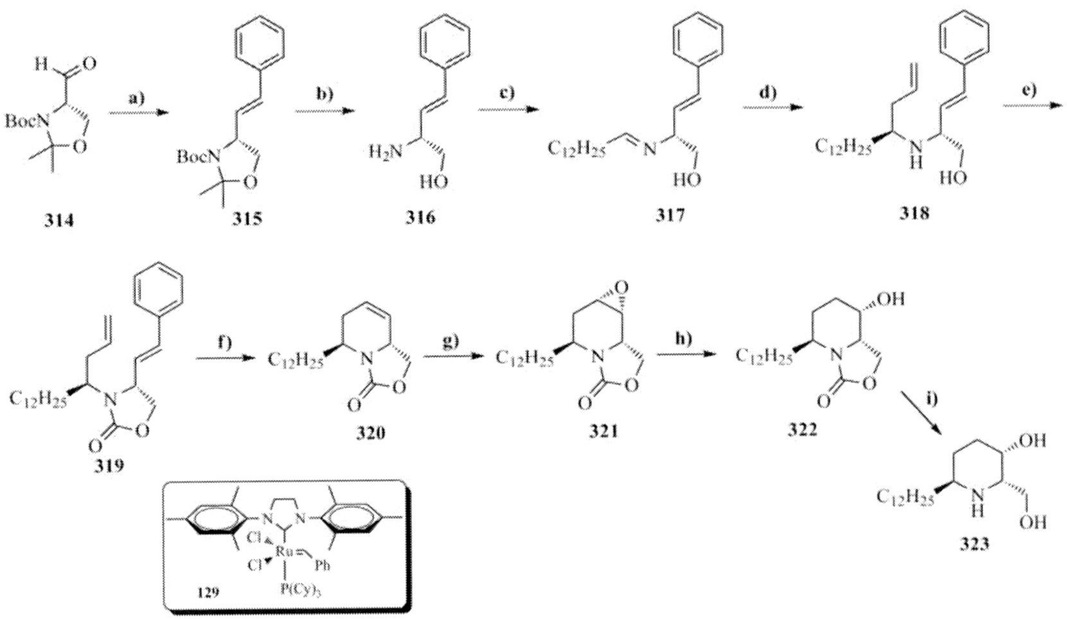

Scheme 42. Reagents and conditions [124] a) $(OEt)_2POCH_2Ph$, BuLi, THF, -78°C to rt, 75%. b) Conc. HCl, MeOH, Δ, 98%. c) $H_{25}C_{12}CHO$, THF, $MgSO_4$. d) $H_2C{=}CHCH_2MgBr$, THF, $Et_2O$, -78°C to -10°C, 56%. e) CDI, $Et_3N$, DCM, 91%. f) Grubbs' catalyst *129*, DCM, Δ, 99%. g) *m*-CPBA, DCM, rt, 81%. h) Super-Hydride, THF, 0°C to rt, 98%. i) 8N KOH, MeOH, Δ, 90%.

Lemaire reported[125] the synthesis of piperidines *331* and *335* from isoxazolines *326* and *327* (scheme 43) obtained by a 1,3-dipolar cycloaddition protocol reported by Mukaiyama[126,127]. The nitro compound *325* first generate *in situ* the reactive nitrile oxide[128] that add to the alkene *324*, prepared from D-mannitol[129], giving the mixture *326* and *327* in a 3:1 ratio. Both compounds were separated. Hydrolysis of the ketal group in *326* was obtained with AcOH. The afforded diol reacted with $SOCl_2$ to give cyclic sulfite *328*, that was further oxidized in the presence of $RuO_2$[130] with formation of the cyclic sulfate *329* in 81% yield. The isoxazoline ring opening was performed by catalytic hydrogenation; the amine initially formed then attacks the sulfate ring giving *330*. Treatment with aq. $H_2SO_4$ in dioxane led to (3*R*,4*S*,6*S*)-6-(benzyloxymethyl)piperidine-3,4-diol *331*. In a similar way (3*R*,4*R*,6*R*)-6-(benzyloxymethyl)piperidine-3,4-diol *335* was obtained from isoxazoline *327* following the same sequence of reactions. The major difference in reactivity was observed for oxidation of cyclic sulfites *328* and *332*[131]. Oxidation of *332* was accomplished by $RuCl_3$ catalysis.

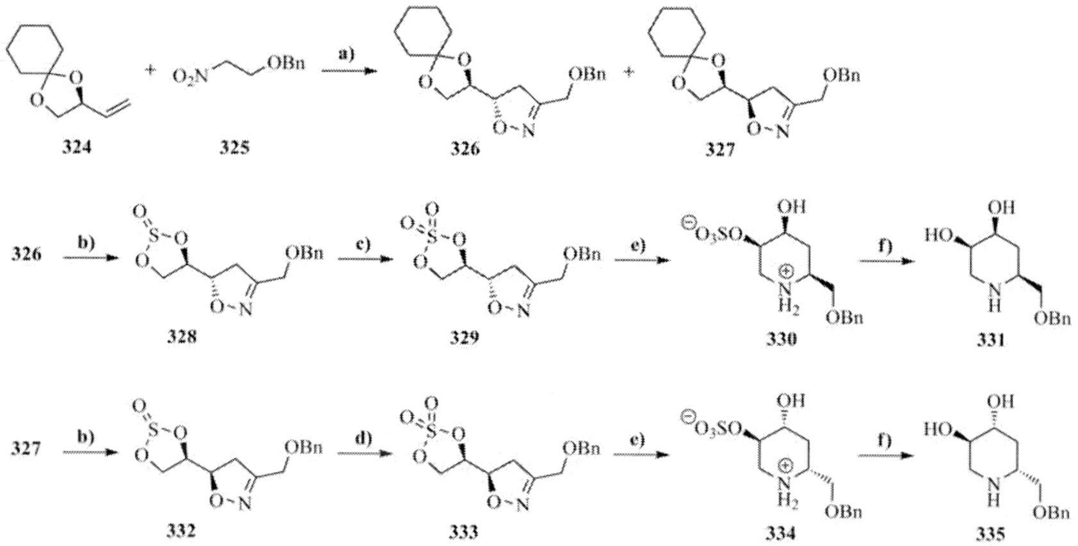

Scheme 43. Reagents and conditions [125] a) 1,4-$C_6H_4(NCO)_2$, $NEt_3$, toluene, Δ, 70%. b) i) AcOH, $H_2O$, 80°C; ii) $SOCl_2$, pyridine, DCM, 0°C to rt, 75% (*328*). c) $NaIO_4$, $RuO_2$, MeCN/$H_2O$, rt, 82%. d) $NaIO_4$, $RuCl_3$, MeCN/$H_2O$, 54% from 327. e) $H_2$, Pd/C, $Na_2CO_3$, MeOH, rt, 77% (*330*), 34% (334). f) $H_2O$, $H_2SO_4$, 1,4-dioxane, 40°C, 93% (*331*), 86% (*335*).

The synthesis of the phosphonic acid derivative *345* presented in scheme 44 was reported by Wong[132]. The sequence of reactions started with the reaction of dihydrofuran *336* with acetyl bromide. The olefin *337* was obtained in 94% yield. Substitution of the bromide ion by the phosphate group was accomplished with triethyl phosphate and subsequent treatment with acid gave hydroxyl group deprotection with formation of allylic alcohol *338*. Sharpless asymmetric epoxidation[19] of *338* afforded *339*. Swern oxidation the alcohol function in *339* followed by reaction of the resulting aldehyde with triethyl orthoformate gave *340*. The regioselective opening of the epoxide was achieved by reaction with diethylaluminum azide[133] generated *in situ*, affording hydroxyl azide *341*. Deprotection of the diethyl acetal with TFA regenerates the aldehyde function. Aldol condensation of aldehyde *342* to dihydroxyacetone[134] (DHAP) catalyzed by fucose-1-phosphate aldolase[135] gave *343* after dephosphorylation with acid phosphatase. Hydrogenation of *343* provides phosphonate *344* which was treated with TMSBr and water to give phosphonic acid *345*.

The synthesis of mannojirimycin derivative *352* described in scheme 45 was reported by Wong[132] using a similar procedure. Starting from butane-1,4-diol *346*, one of the hydroxyl groups was protected by reaction with 1 eq. of TBSCl and the other was substituted by iodide. Treatment of *347* with triethyl phosphate afforded phosphonate *348*. From *348* after desilylation with TBAF in acetic acid, was formed epoxide *349* by treatment with *m*-CPBA. It was then converted into the azido alcohol *350* as described for the transformation of *339* into *341* (scheme 44). Conversion of *350* into *351* was achieved by addition of DHAP, catalyzed by FDP aldolase, and dephosphorylation catalyzed by acid phosphatase after cleavage of the diacetal group to afford *351*. Catalytic hydrogenation of *351* gave phosphonate *352*.

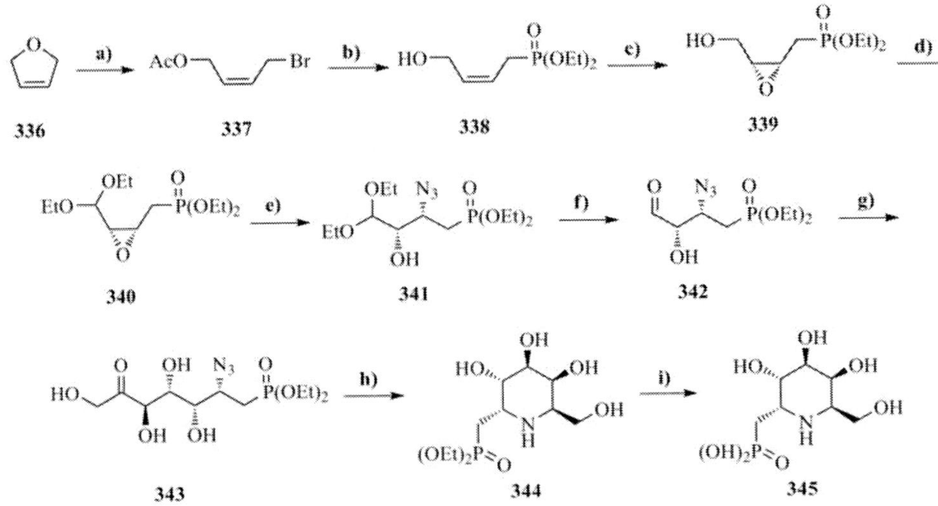

Scheme 44. Reagents and conditions [132] a) AcBr, 94%. b) i) $P(EtO)_3$, NaI, 88%; ii) EtOH, *p*-TsOH, 91%. c) (-)-diisopropil tartrate, $Ti(i\text{-}PrO)_4$, cumene hydroperoxide, 84%. d) i) $(COCl)_2$, DMSO, $Et_3N$; ii) $(EtO)_3CH$, *p*-TsOH, 65%. e) $Et_2AlCl$, $LiN_3$, 70%. f) i) TFA, DCM. g) Dihydroxyacetone phosphate (DHAP), FucA, pH 6.7; iii) Acid phosphatase, pH 4.8, 30%. h) $H_2$ (50 psi), Pd/C, 35%. i) i) TMSBr, $Et_3N$, DCM; ii) $H_2O$, THF, 79%.

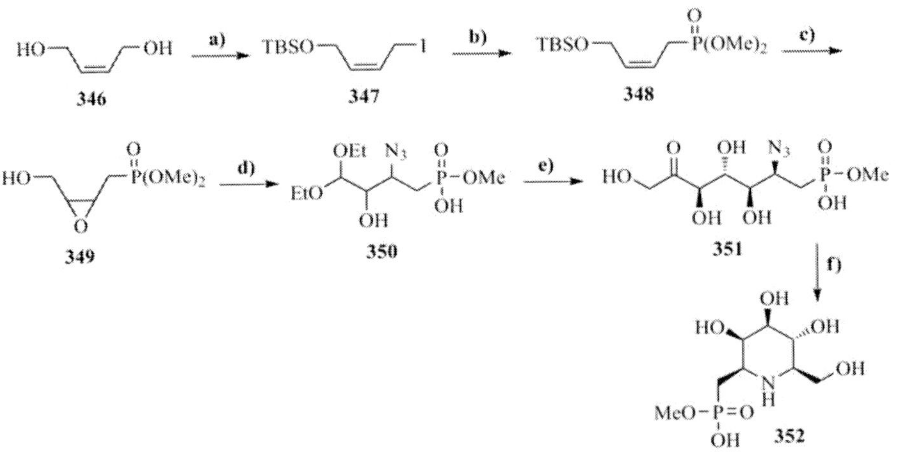

Scheme 45. Reagents and conditions [132] a) i) TBSCl, $NEt_3$, 90%; ii) $PPh_3$, $I_2$, imidazole, 94%. b) $P(OMe)_3$, NaI, 92%. c) i) AcOH, 100%; ii) *m*-CPBA, 63%. d) i) Dess-Martin; $(EtO)_3CH$, *p*-TosOH, 74%; ii) $NaN_3$, $NH_4Cl$, 98%. e) i) 0.1N HCl, 53°C; ii) Dihydroxyacetone phosphate (DHAP), FDP aldolase, pH 6.7; iii) Acid Phosphatase, pH 4.8, 85%. f) $H_2$ (50 psi), Pd/C.

Knapp synthesized[136] *362* starting from 4-methoxypyridine *353* (scheme 46). Addition of benzyl chloroformate to *353* and reduction of the afforded pyridinium salt with sodium borohydride gave *354* after hydrolysis. Treatment of *354* with NBS gave bromide *355*.

Enatioselective reduction[137,138] of *355* with the Corey-Baksi-Shibata (CBS) reagent, followed by protection of the afforded alcohol gave *357* with 89% *ee*. Reductive removal of the bromo substituent with *n*-$Bu_3SnH$ and 2,2'-azobis-(2methylpropionitrile) (AIBN) led to *358*. Dihydroxilation of *358* with osmium tetroxide formed diol *359* that was treated with thionyl chloride in the presence of imidazole. The afforded cyclic sulfite was regioselectively opened with lithium azide to give *360*. Hydrogenation of the azide function with Raney nickel in ethanol, in the presence of acetic anhydride afforded *361* in 90% yield. Deprotection of *361* with catalytic hydrogenation (Pd/C) followed by treatment with hydrochloric acid gave (2*R*,3*R*,4*R*)-2-acetamido-3,4-dihydroxypiperidine *362* in its HCl salt form.

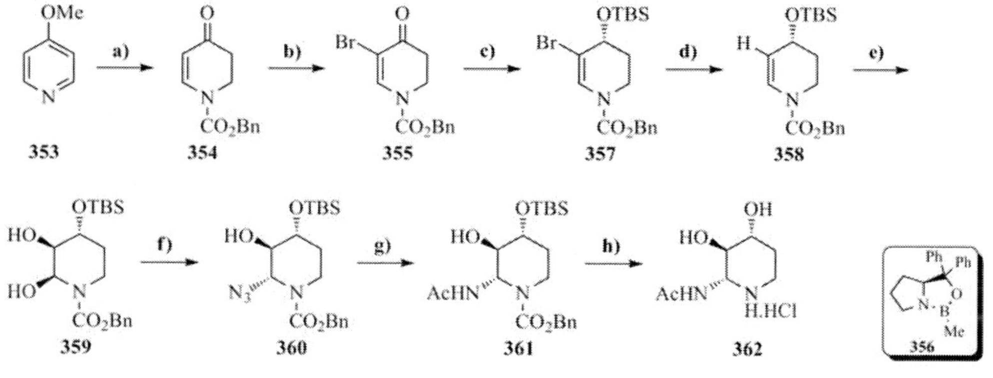


Scheme 46. Reagents and conditions [136] a) i) $NaBH_4$, $ClCO_2Bn$, MeOH, -78°C; ii) $H_2O$, -60° to 23°C, 80%. b) NBS, DCM, 85%. c) i) (*S*)-2-methyl-CBS-oxazaborolidine *356*, catecholborane, DCM, $CCl_4$, toluene; ii) TBSCl, imidazole, DMF, 65%. d) *n*-$Bu_3SnH$, AIBN, toluene, Δ, 71%. e) $OsO_4$, NMO, MeCN, acetone/$H_2O$, 93%. f) i) $SOCl_2$, imidazole, THF, -30°C; ii) $LiN_3$, DMF, 0°C, 96%. g) Ra-Ni, $H_2$, $Ac_2O$, THF, 90%. h) i) $H_2$, Pd/C, EtOH; ii) 3N HCl, 71%.

Recent literature reports the synthesis of some imino sugars bearing atypical substituent groups. Qing reported a synthesis[139,140] of difluorinated imino sugars *375* and *378* (scheme 48). Starting from trifluoroetanol *363*, first the alcohol function was protected in the form of ether (scheme 47). The product formed, *364*, was treated with LDA and monomeric formaldehyde to produce vinyl alcohol *365*. The alcohol was converted into vinyl ether *366* by addition of vinyl bromide. A sigmatropic rearrangement was produced by treatment of *366* with 2.2 eq. of LDA, affording *367*. The free hydroxyl group was mesylated and subjected to regioselective allylic substitution with $NaN_3$, catalyzed by Pd(0), affording *369*. Reduction of the azide and protection of the corresponding amine in its Cbz form gave *370*. This compound was used as starting material in imino sugar synthesis.

The asymmetric amino hydroxylation of *370* was carried out using $(DHQ)_2PHAL$ (AD-mix-α) and $(DHQD)_2PHAL$ (AD-mix-β) affording diols *371* and *372* respectively (scheme 48). Each of these diols was converted into the corresponding difluoropiperidine in two steps. Treatment of diol *371* with thionyl chloride afforded furanose *373* that on catalytical hydrogenation gave (3*S*,4*R*,6*R*)-5,5-difluoro-6-methylpiperidine-3,4-diol *375* with total selectivity. The selectivity observed can be racionalized by the stability of intermediate *374* and the steric hindrance of fluorine atoms[141]. The same reaction sequence applied to *372* gave the enantiomer of *375*, (3*R*,4*S*,6*S*)-5,5-difluoro-6-methylpiperidine-3,4-diol *378*.

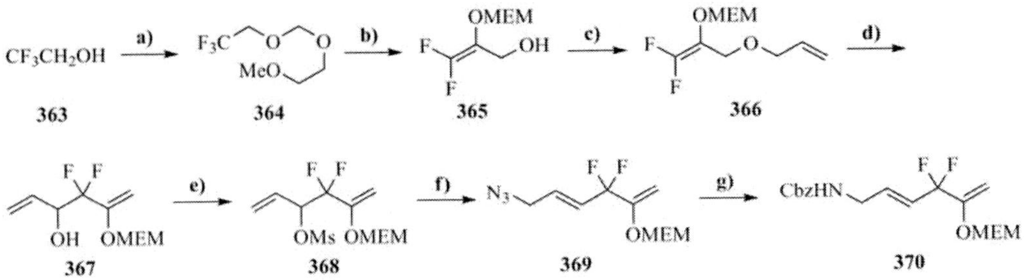

Scheme 47. Reagents and conditions [139,140] a) NaH, MEMCl, 0°C. b) LDA, $CH_2O$, -78°C. c) Allyl bromide, 50% aq. NaOH, $Bu_4NHSO_4$. d) LDA, -78°C, 32% from *364*. e) MsCl, $Et_3N$, DMAP, DCM, 96%. f) $NaN_3$, THF, $Pd(PPh_3)_4$, 96%. g) *i)* $PPh_3$, THF; *ii)* $H_2O$, Δ; *iii)* CbzCl, $K_2CO_3$, 90%.

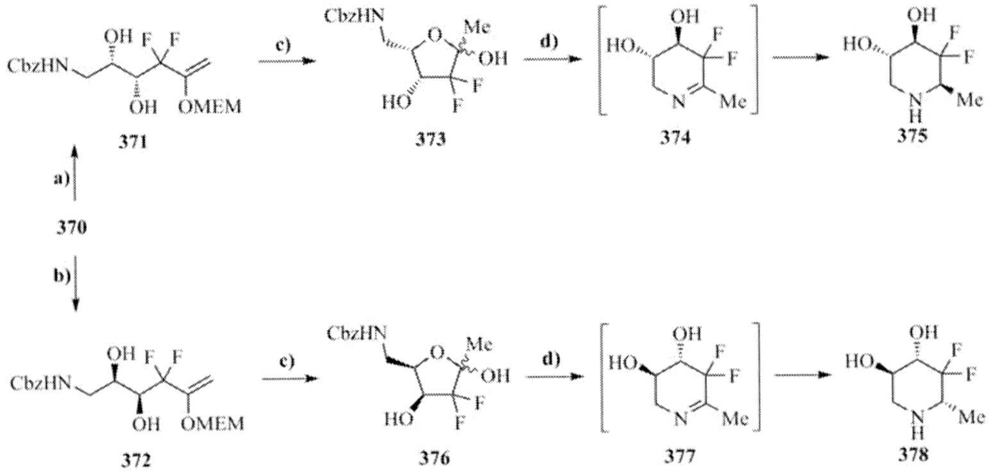

Scheme 48. Reagents and conditions [139,140] a) AD-mix-α, $MeSO_2NH_2$, rt, 55%, 82% *ee*. b) AD-mix-β, $MeSO_2NH_2$, rt, 57%, 84% *ee*. c) *i)* $SOCl_2$, MeOH, rt; *ii)* Ethyl acetate, sat. aq. $K_2CO_3$, rt. d) $H_2$, Pd/C, MeOH, 43% (*375* from *371*); 46% *(378* from *372)*.

Qing reported the synthesis of different difluorinated imino sugars[140,142] from *gem*-difluoroallylindium[143] obtained by treatment of bromodifluoropropene *380* with indium in DMF (scheme 49). Addition of aldehyde *379* to this generated intermediate gave *381* as a mixture of 7.7:1 ratio of *anti* and *syn* isomers. Protection of the alcohol as an ether followed by reaction with $OsO_4$ gave diols *383* and *384* in 1:1 ratio that were separated. Selective protection of the primary alcohol function in *383* with BzCl gave *385*, and reaction of the secondary alcohol function with $Tf_2O$ followed by nucleophilic displacement of the triflate group by $N_3$ gave azide *386*. Isopropylidene ketal protection was cleaved by aqueous acetic acid affording *387*. The resulting primary alcohol was selectively mesylated giving *388*. Reduction of azide group in *388* with $PPh_3$ results in the formation of the corresponding amine, which cyclizes under the reaction conditions to a piperidine structure. Treatment of this compound with CbzCl gave *389*. Deprotection of *389* by catalytic hydrogenation followed by treatment with $NH_3$ in methanol afforded 1,4-di-deoxy-4-difluoro-D-mannojirimycin *390*. 1,4-Di-deoxy-4-difluoro-L-allonojirimycin *391* was obtained applying the same reaction sequence to diol *384*.

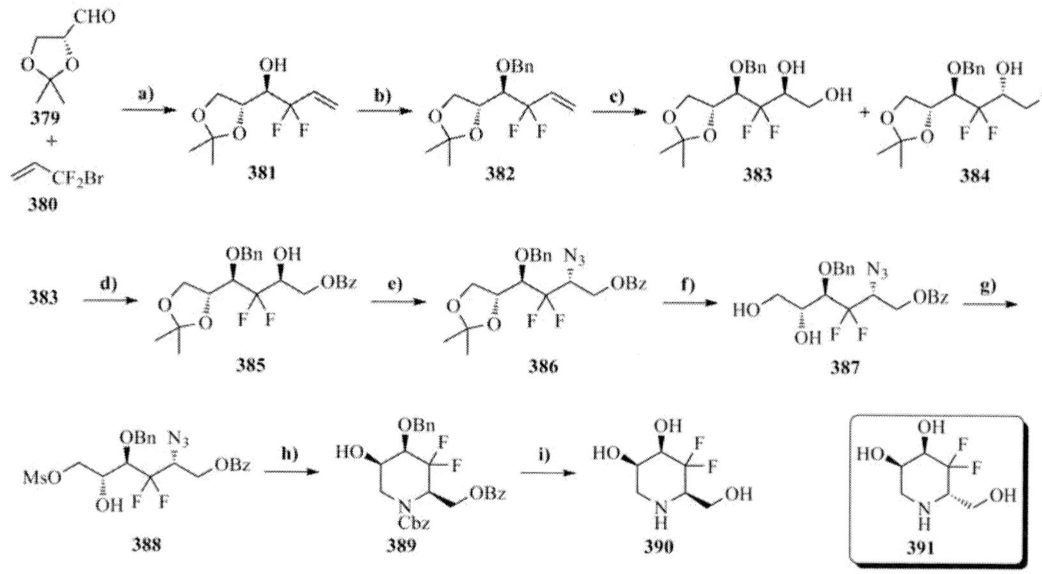

*Scheme 49. Reagents and conditions [140,142]* a) In, DMF, rt, 90%. b) NaH (1.6 eq.), BnBr, TBAI, THF, 0°C, 93%. c) $OsO_4$, NMO, acetone/$H_2O$, rt, 95%. d) BzCl, pyridine, DCM, -78°C, 90%. e) *i)* $Tf_2O$, pyridine, DCM, -25°C; *ii)* $NaN_3$, DMF, rt, 81%. f) 75% aq. AcOH, 50°C, 95%. g) MsCl, collidine, DCM, 0°C, 81%. h) *i)* $PPh_3$, THF, rt; *ii)* Sat. aq. $NaHCO_3$, 65°C; *iii)* CbzCl, rt, 82%. i) *i)* $H_2$, $Pd(OH)_2/C$, MeOH, rt, *ii)* $NH_3$/MeOH, rt, 89%.

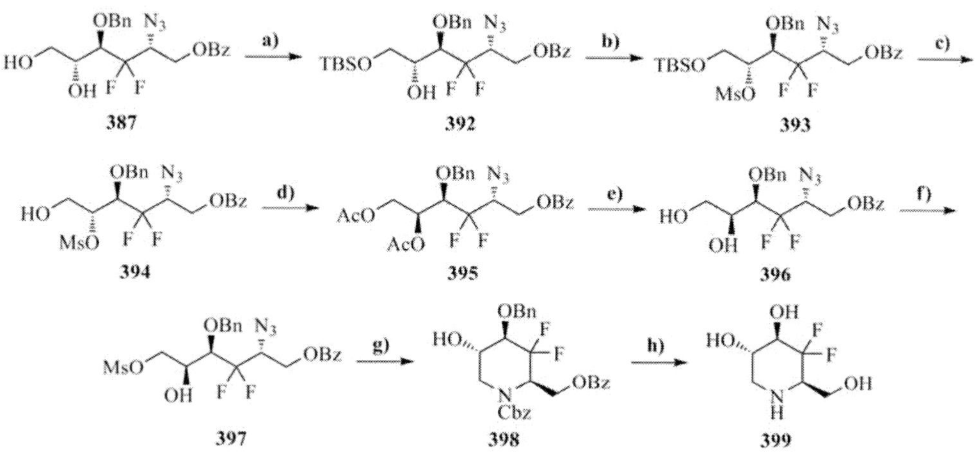

Scheme 50. Reagents and conditions [140] a) TBSCl, imidazole, DMF, rt, 86%. b) MsCl, $Et_3N$, DMAP, DCM, rt, 93%. c) AcOH/$H_2O$/THF 3:1:1, rt, 89%. d) AcOK, $Ac_2O$, 140°C, 75%. e) HCl, MeOH, DCM, rt, 97%. f) MsCl, collidine, DCM, 0°C, 81%. g) i) $PPh_3$, THF, rt, ii) Sat. aq. $NaHCO_3$, 65°C; iii) CbzCl, rt, 82%. h) i) $H_2$, $Pd(OH)_2/C$, MeOH, rt; ii) $NH_3$ sat. MeOH, rt, 89%.

Many compounds structurally similar to imino sugars have been synthesized. Córdova synthesized lactams *404* and *407* starting from *p*-anisidine *401* and benzyloxyacetaldehyde *400* as represented in scheme 51. The olefin *402* was obtained in one-pot reaction by combining aldehyde *400* and amine *401* in 3:1 proportion in the presence of (*S*)-proline. The respective Mannich adduct formed was treated with methyl diethylphosphonoacetate and DBU forming *402* as a mixture of *syn* and *anti* isomers in 4:1 ratio. The *syn* isomer, isolated in 64% yield and 95% *ee*, was treated with osmium tetroxide with the formation of diol *403* in 64% yield and 95% *ee*. Compound *403* was cyclized by treatment with acetic acid in refluxing methanol giving piperidinone *404* in 3 steps from *401*. Olefin *405* was obtained in two steps: combination of *400* and *401* (3:1) in the presence of (*S*)-proline followed by reaction of the aldehyde with $Ph_3P{=}CH_2COOMe$ to give alkene *405* in 56% and 92% *ee*. Cyclization of *405* was accomplished by its treatment with acetic acid in refluxing methanol. Dihydroxylation of *406* afforded piperidinone *407*.

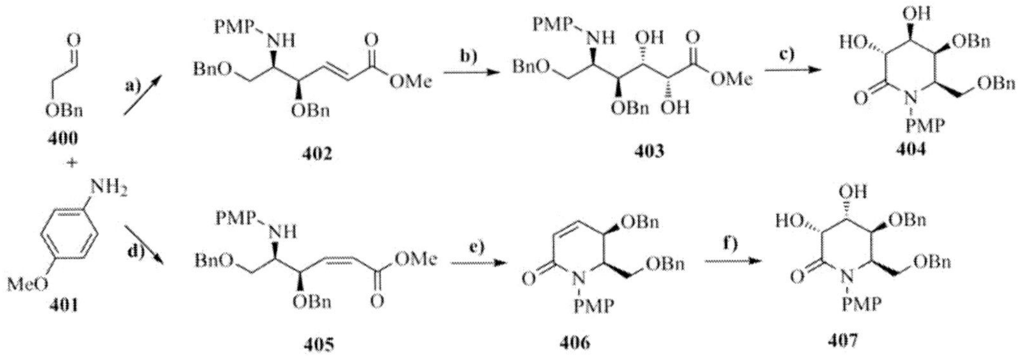

Scheme 51. Reagents and conditions [144] a) *400:401* 3:1, (*S*)-proline, DBU, LiCl, $(EtO)_2POCH_2COOMe$, DMF, rt, 64%. b) $OsO_4$, NMO, acetone/$H_2O$ 9:1, rt, 67%. c) AcOH, MeOH, Δ, 74%. d) *400:401* 3:1, (*S*)-proline, $Ph_3P{=}CH_2COOMe$, DMF, rt, 56%. e) AcOH, MeOH, Δ, 82%.f) $OsO_4$, NMO, acetone/$H_2O$ 9:1, rt, 80%.

Pandey presented the synthesis[145] of lactams *423* and *428* from L-tartaric acid (scheme 52 and 53) and lactam *424* from D-tartaric acid (scheme 52). Typically the alcohol *409* was obtained by reported methods[146] and was converted into aldehyde *410* by oxidation with 2-iodoxybenzoic acid (IBX). Coupling of *410* with *413* via reductive amination, using $NaBH(OAc)_3$ as the reducing agent[147] afforded *414*. The amino alcohol *413* was previously prepared from 3-aminopropanol *411* in two steps: treatment of *411* with Boc anhydride followed by reaction with acetaldehyde diethyl acetal catalysed by pyridinium *p*-toluenesulfonate (PPTS) afforded cyclic aminoacetal *412*; silylation of *412* followed by cleavage of the aminoacetal ring gave *413*. Reaction of *414* with formaldehyde in refluxing benzene gave product *415*. Photochemical induced cyclization of *415* catalyzed by 1,4-dicyanonaphthalene[87,88] afforded *416* in 60% yield. The olefin function was dihydroxylated giving diol *417* that was converted into alcohol *418* upon oxidative cleavage of the diol moiety with $NaIO_4$ and concomitant reduction of the afforded ketone with $NaBH_4$. The alcohol was protected in the form of benzyl ether affording *419*. Cleavage of the acetonide protection in *419* followed by treatment of the product with benzyl bromide furnished *420*. The oxazine ring was cleaved with 6N HCl in refluxing methanol to give a secondary amine that was protected by reaction with $(Boc)_2O$. The free alcohol function was oxidized to carboxylic acid with pyridinium dichromate to give *421*. After TFA/DCM treatment the Boc group was cleaved. Addition of Mukaiyama's reagent 2-chloro-1-methylpyridinium iodide afforded β-lactam[148] *422* in 52% yield. Deprotection of the hydroxyl groups occurs by catalytic hydrogenation affording *423*. β-Lactam *424*, enantiomer of *423*, was obtained by the same reaction sequence starting from D-tartaric acid.

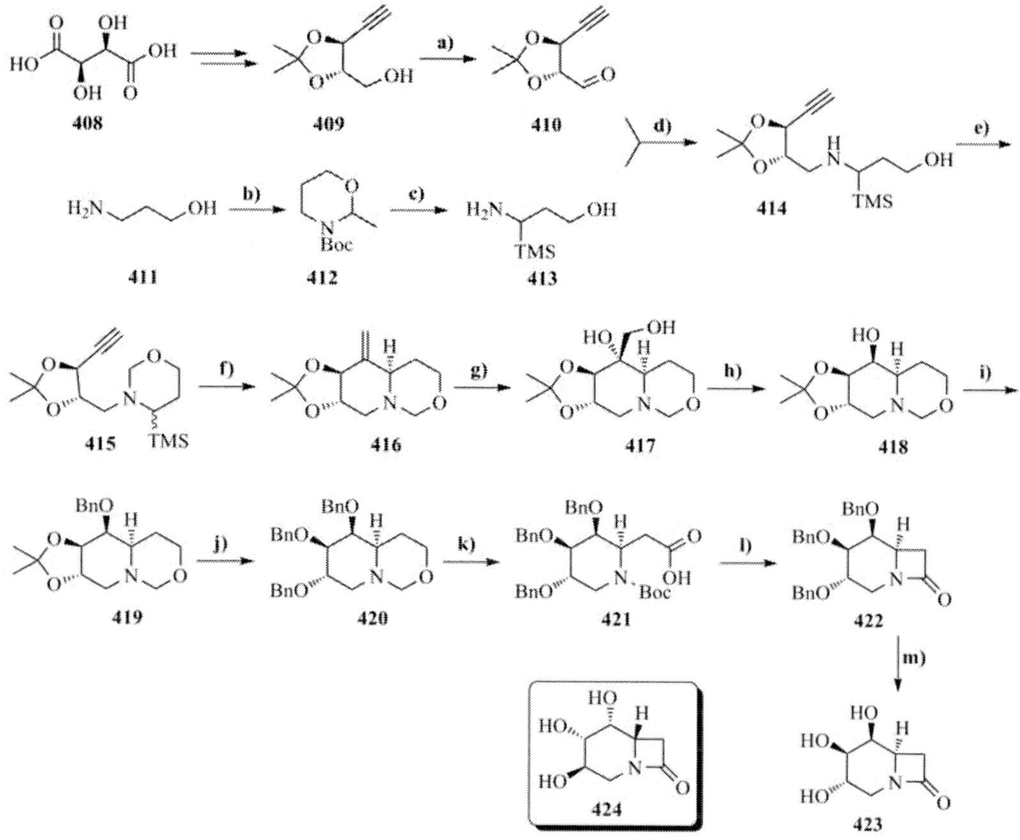

Scheme 52.Reagents and conditions [145] a) IBX, EtOAc, Δ, 85%. b) i) $(Boc)_2O$, $Et_3N$; ii) $CH_3CH(OEt)_2$, PPTS, $C_6H_6$, Δ, 77%. c) i) *s*-BuLi, TMEDA, TMSCl, -78°C to 0°C; ii) 2N HCl, dioxane, 80°C, 88%. d) $NaBH(OAc)_3$, 1,2-dichloroethane; 2N NaOH, 71%. e) $(CH_2O)n$, $C_6H_6$, Dean-Stark, 95%. f) hν, MeCN/i-PrOH 3:1, 1,4-dicyanonaphthalene, 60%. g) $OsO_4$, $K_3Fe(CN)_6$, $K_2CO_3$, pyridine, *t*-BuOH/$H_2O$ 1:1, rt, 90%. h) i) $NaIO_4$, silica gel; ii) $NaBH_4$, MeOH, rt, 82%. i) BnBr, NaH, THF, Δ, 86%. j) i) 1N HCl, MeOH, rt; ii) BnBr, NaH, TBAI, THF, Δ, 78%. k) i) 6*N* HCl, dioxane/MeOH, Δ; ii) $(Boc)_2O$, $Et_3N$, rt, DCM; iii) $Py_2Cr_2O_7$, DMF, 57%. l) i) TFA, DCM, 0°C; ii) 2-chloro-1-methylpyridinium iodide, $Et_3N$, MeCN, 60°C, 53%. m) $H_2$ (60 psi), Pd/C, MeOH, 95%.

The alcohol *425* as its congener *409* can be used in the synthesis of fused β-lactam of the same type (scheme 53). It was obtained from L-tartaric acid by reported methods[84], and was subjected to the reaction conditions applied to alcohol *409*. Compound *426* was afforded in 75% overall yield. Photochemical induced cyclization of *426* gave *427* in 60% yield. It was converted into β-lactam *428* after analogous reaction sequence to that presented in scheme 52 for the synthesis of *423*.

Chiral aziridine *429* was chosen by Lee for the synthesis of tetrahydropyridinols *436* and *439*[149] (scheme 54). The aziridine *429* was easely prepared from (*R*)-(+)-α-methylbenzylamine and ethyl 2,3-dibromopropionate[150]. Reaction of aziridine *429* with methyl dimethylphosphonate in the presence of *n*-BuLi gave compound *430*. Horner-Wadsworth-Emmons olefination of phosphonate *430* with acetaldehyde produced enone *431*. Reduction of the ketone group in *431* with $NaBH_4$ in the presence of $ZnCl_2$ afforded alcohol *432* as a single isomer[151]. On the other hand, reduction of ketone group with L-Selectride afforded alcohol *433*. Regioselective ring-opening of aziridine *432* was performed with thiophenol giving the open chain compound *434*. Reaction of *434* with allyl bromide and NaH gave di-olefin *435*, which was cyclized by treatment with first generation Grubbs' catalyst *156*, affording tetrahydropyridine *436*. Tetrahydropyridine *439* was obtained starting from alcohol *433*, using the same procedures as for the synthesis of *436*.

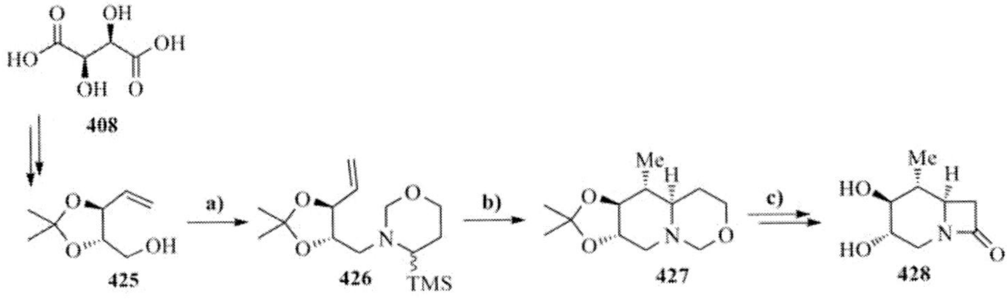

Scheme 53. Reagents and conditions [145] a) i) IBX, EtOAc, Δ; ii) *413*, $NaBH(OAc)_3$, 1,2-dichloroethane, 2N NaOH; iii) $(CH_2O)_n$, $C_6H_6$, Dean-Stark, 75%. b) hν, MeCN/i-PrOH 3:1, 1,4-dicyanonaphthalene, 60%. c) See j) to m) in scheme 52.

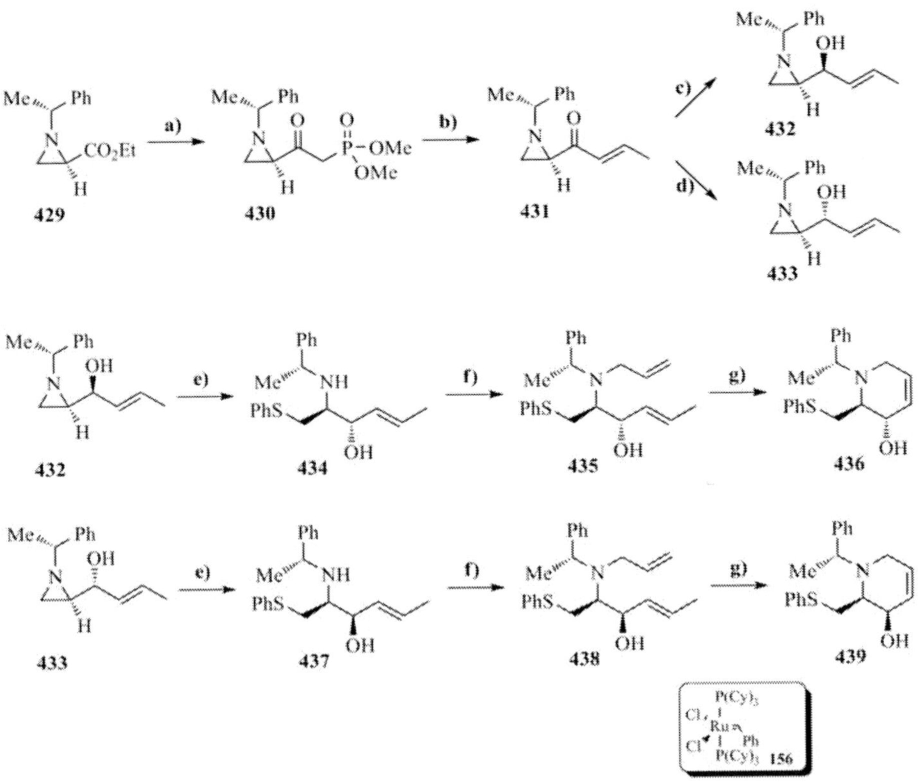

Scheme 54. Reagents and conditions [149] a) $(MeO)_2POCH_3$, *n*-BuLi, THF, -78°C, 98%. b) $CH_3CHO$, $K_2CO_3$, MeCN, rt, 75%. c) $ZnCl_2$, $NaBH_4$, MeOH, -78°C, 96%. d) L-Selectride, THF, -78°C, 40%. e) PhSH, DCM, rt, 95% (*434*), 83% (*437*). f) Allyl bromide, NaH, THF, TBAI, Δ, 94% (*435*), 73% (*438*). g) Grubbs' catalyst *156*, toluene, Δ, 91% (*436*), 74% (*439*).

## REFERENCES

[1] Asano, N.; Nash, R. J.; Molyneux, R. J.; Fleet, G. W. J. *Tetrahedron: Asymmetry* 2000, *11*, 1645-1680.

[2] Afarinkia, K.; Bahar, A. *Tetrahedron: Asymmetry* 2005, *16*, 1239-1287.

[3] Pearson, M. S. M.; Mathé-Allainmat, M.; Fargeas, V.; Lebreton, J. *Eur. J. Org. Chem.* 2005, *2005*, 2159-2191.

[4] La Ferla, B.; Nicotra, F. In *Iminosugars as Glycosidase Inhibitors*; Univ.-Prof. Dipl.-Ing. Dr. Arnold, E. S., Ed. 1999, p 68-92.

[5] Cipolla, L.; La Ferla, B.; Nicotra, F. *Curr. Topics Med. Chem.* 2003, *3*, 485-511.

[6] Paulsen, H.; Sanster, I.; Heyns, K. *Chem. Ber.* 1967, *100*, 802-815.

[7] Paulsen, H.; Todt, K. *Chem. Ber.* 1967, *100*, 512-529.

[8] Schaller, C.; Vogel, P.; Jager, V. *Carbohydr. Res.* 1998, *314*, 25-35.

[9] Muller, I.; Jager, V. *Tetrahedron Lett.* 1982, *23*, 4777-4780.

[10] Mangeney, P.; Alexakis, A.; Normant, J. F. *Tetrahedron Lett.* 1988, *29*, 2677-2680.

[11] Comins, D. L.; Fulp, A. B. *Tetrahedron Lett.* 2001, *42*, 6839-6841.

[12] Comins, D. L.; Joseph, S. P.; Goehring, R. R. *J. Am. Chem. Soc.* 1994, *116*, 4719-4728.

[13] Comins, D. L.; Salvador, J. M. *J. Org. Chem.* 1993, *58*, 4656-4661.

[14] Comins, D. L.; Stolze, D. A.; Thakker, F.; McArdle, C. L. *Tetrahedron Lett.* 1998, *39*, 5693-5696.

[15] Comins, D. L.; Huang, S.; McArdle, C. L.; Ingalls, C. L. *Org. Lett.* 2001, *3*, 469-471.

[16] Goldstein, S. W.; Overman, L. E.; Rabinowitz, M. H. *J. Org. Chem.* 1992, *57*, 1179-1190.

[17] Iida, H.; Yamazaki, N.; Kibayashi, C. *J. Org. Chem.* 1987, *52*, 3337-3342.

[18] Feit, P. W. *J. Med. Chem.* 1964, *7*, 14-17.

[19] Katsuki, T.; Sharpless, K. B. *J. Am. Chem. Soc.* 1980, *102*, 5974-5976.

[20] Behrens, C. H.; Sharpless, K. B. *Aldrichimica Acta* 1983, *16*, 67-79.

[21] Nagasawa, T.; Kuroiwa, K.; Narita, K.; Isowa, Y. *Bull. Chem. Soc. Jpn.* 1973, *46*, 1269-1272.

[22] Aoyagi, S.; Fujimaki, S.; Kibayashi, C. *J. Chem. Soc., Chem. Commun.* 1990, *1990*, 1457-1459.

[23] Overman, L. E.; Flippin, L. A. *Tetrahedron Lett.* 1981, *22*, 195-198.

[24] Lindstrom, U. M.; Somfai, P. *Tetrahedron Lett.* 1998, *39*, 7173-7176.

[25] Nakajima, N.; Horita, K.; Abe, R.; Yonemitsu, O. *Tetrahedron Lett.* 1988, *29*, 4139-4142.

[26] Bérubé, G.; Deslongchamps, P. *Can. J. Chem.* 1990, *68*, 404-411.

[27] Sharpless, K. B.; Amberg, W.; Bennani, Y. L.; Crispino, G. A.; Hartung, J.; Jeong, K. S.; Kwong, H. L.; Morikawa, K.; Wang, Z. M.; et al. *J. Org. Chem.* 1992, *57*, 2768-2771.

[28] Corey, E. J.; Guzman-Perez, A.; Noe, M. C. *J. Am. Chem. Soc.* 1995, *117*, 10805-10816.

[29] Ko, S. Y.; Lee, A. W. M.; Masamune, S.; A. Reed, L.; Barry Sharpless, K.; Walker, F. J. *Tetrahedron* 1990, *46*, 245-264.

[30] Somfai, P.; Marchand, P.; Torsell, S.; Lindstrom, U. M. *Tetrahedron* 2003, *59*, 1293-1299.

[31] Staudinger, H.; Meyer, J. *Helv. Chim. Acta* 1919, *2*, 635-646.
[32] Hudlicky, T.; Rouden, J.; Luna, H.; Allen, S. *J. Am. Chem. Soc.* 1994, *116*, 5099-5107.
[33] Hudlicky, T.; Gonzalez, D.; T. Gibson, D. *Aldrichimica Acta* 1999, *32*, 35-62.
[34] Bajwa, J. S.; Anderson, R. C. *Tetrahedron Lett.* 1991, *32*, 3021-3024.
[35] Johns, B. A.; Pan, Y. T.; Elbein, A. D.; Johnson, C. R. *J. Am. Chem. Soc.* 1997, *119*, 4856-4865.
[36] Johnson, C. R.; Johns, B. A. *J. Org. Chem.* 1997, *62*, 6046-6050.
[37] Banwell, M. G.; Ma, X.; Asano, N.; Ikeda, K.; Lambert, J. N. *Org. Biomol. Chem.* 2003, *1*, 2035-2037.
[38] Yamada, H.; Takimoto, H.; Ikeda, T.; Tsukamoto, H.; Harada, T.; Takahashi, T. *Synlett* 2001, 1751-1754.
[39] Haukaas, M. H.; O'Doherty, G. A. *Org. Lett.* 2001, *3*, 401-404.
[40] Li, G.; Angert, H. H.; Sharpless, K. B. *Angew. Chem. Int. Ed. Eng.* 1996, *35*, 2813-2817.
[41] Achmatowicz, O.; Bielski, R. *Carbohydr. Res.* 1977, *55*, 165-176.
[42] Luche, J. L. *J. Am. Chem. Soc.* 1978, *100*, 2226-2227.
[43] Yang, C.-F.; Xu, Y.-M.; Liao, L.-X.; Zhou, W.-S. *Tetrahedron Lett.* 1998, *39*, 9227-9228.
[44] Park, K. H.; Yoon, Y. J.; Lee, S. G. *J. Chem. Soc., Perkin Trans. I* 1994, 2621-2623.
[45] Wu, X. D.; Khim, S. K.; Zhang, X.; Cederstrom, E. M.; Mariano, P. S. *J. Org. Chem.* 1998, *63*, 841-859.
[46] Calderon, F.; Doyaguez, E. G.; Fernandez-Mayoralas, A. *J. Org. Chem.* 2006, *71*, 6258-6261.
[47] Byun, H.-S.; He, L.; Bittman, R. *Tetrahedron* 2000, *56*, 7051-7091.
[48] De Luca, L.; Giacomelli, G.; Porcheddu, A. *Org. Lett.* 2001, *3*, 3041-3043.
[49] Felpin, F. X.; Boubekeur, K.; Lebreton, J. *J. Org. Chem.* 2004, *69*, 1497-1503.
[50] McKillop, A.; Taylor, R. K.; Watson, R.; Lewis, N. *Synthesis* 1994, 31-33.
[51] Kirkland, T. A.; Lynn, D. M.; Grubbs, R. H. *J. Org. Chem.* 1998, *63*, 9904-9909.
[52] Garcia-Fortanet, J.; Murga, J.; Carda, M.; Marco, J. A. *Org. Lett.* 2003, *5*, 1447-1449.
[53] Felpin, F.-X.; Lebreton, J. *Tetrahedron Lett.* 2003, *44*, 527-530.
[54] Phillips, A. J.; Abell, A. D. *Aldrichimica Acta* 1999, *32*, 75-89.
[55] François-Xavier Felpin, J. L. *Eur. J. Org. Chem.* 2003, *2003*, 3693-3712.
[56] Guanti, G.; Riva, R. *Tetrahedron Lett.* 2003, *44*, 357-360.
[57] Takahata, H.; Banba, Y.; Ouchi, H.; Nemoto, H.; Kato, A.; Adachi, I. *J. Org. Chem.* 2003, *68*, 3603-3607.
[58] Liu, P. S. *J. Org. Chem.* 1987, *52*, 4717-4721.
[59] Anzeveno, P. B.; Creemer, L. J.; Daniel, J. K.; King, C. H. R.; Liu, P. S. *J. Org. Chem.* 1989, *54*, 2539-2542.
[60] Pearson, M. S. M.; Evain, M.; Mathé-Allainmat, M.; Lebreton, J. *Eur. J. Org. Chem.* 2007, *2007*, 4888-4894.
[61] Yokoyama, H.; Ejiri, H.; Miyazawa, M.; Yamaguchi, S.; Hirai, Y. *Tetrahedron: Asymmetry* 2007, *18*, 852-856.
[62] Wirth, T. *Angew. Chem. Int. Ed.* 2001, *40*, 2812-2814.
[63] Hirai, Y.; Terada, T.; Amemiya, Y.; Momose, T. *Tetrahedron Lett.* 1992, *33*, 7893-7894.
[64] Hirai, Y.; Nagatsu, M. *Chem. Lett.* 1994, *23*, 21-22.

[65] Hirai, Y.; Shibuya, K.; Fukuda, Y.; Yokoyama, H.; Yamaguchi, S. *Chem. Lett.* 1997, *26*, 221-222.
[66] Hirai, Y.; Watanabe, J.; Nozaki, T.; Yokoyama, H.; Seiji, Y. *J. Org. Chem.* 1997, *62*, 776-777.
[67] Yokoyama, H.; Otaya, K.; Yamaguchi, S.; Hirai, Y. *Tetrahedron Lett.* 1998, *39*, 5971-5974.
[68] Yokoyama, H.; Otaya, K.; Kobayashi, H.; Miyazawa, M.; Yamaguchi, S.; Hirai, Y. *Org. Lett.* 2000, *2*, 2427-2429.
[69] Ouchi, H.; Mihara, Y.; Takahata, H. *J. Org. Chem.* 2005, *70*, 5207-5214.
[70] Sakagami, H.; Ogasawara, K. *Synthesis* 2000, 521-524.
[71] Yang, D.; Wong, M.-K.; Yip, Y.-C. *J. Org. Chem.* 1995, *60*, 3887-3889.
[72] Lipshutz, B. H.; Wilhelm, R. S.; Kozlowski, J. A. *Tetrahedron* 1984, *40*, 5005-5038.
[73] Bottini, A. T.; Dev, V. *J. Org. Chem.* 1962, *27*, 968-973.
[74] Baker, S. R.; Cases, M.; Keenan, M.; Lewis, R. A.; Tan, P. *Tetrahedron Lett.* 2003, *44*, 2995-2999.
[75] Dondoni, A.; Perrone, D. *Org. Synth.* 1999, *77*, 64-77.
[76] Grubbs, R. H.; Chang, S. *Tetrahedron* 1998, *54*, 4413-4450.
[77] Banba, Y.; Abe, C.; Nemoto, H.; Kato, A.; Adachi, I.; Takahata, H. *Tetrahedron: Asymmetry* 2001, *12*, 817-819.
[78] Denmark, S. E.; Forbes, D. C.; Hays, D. S.; DePue, J. S.; Wilde, R. G. *J. Org. Chem.* 1995, *60*, 1391-1407.
[79] VanRheenen, V.; Kelly, R. C.; Cha, D. Y. *Tetrahedron Lett.* 1976, *17*, 1973-1976.
[80] Donohoe, T. J.; Mitchell, L.; Waring, M. J.; Helliwell, M.; Bell, A.; Newcombe, N. J. *Tetrahedron Lett.* 2001, *42*, 8951-8954.
[81] Donohoe, T. J. *Synlett* 2002, 1223-1232.
[82] Pandey, G.; Kapur, M. *Tetrahedron Lett.* 2000, *41*, 8821-8824.
[83] Pandey, G.; Kapur, M. *Org. Lett.* 2002, *4*, 3883-3886.
[84] Pandey, G.; Kapur, M.; Khan, M. I.; Gaikwad, S. M. *Org. Biomol. Chem.* 2003, *1*, 3321-3326.
[85] Corey, E. J.; Fuchs, P. L. *Tetrahedron Lett.* 1972, *13*, 3769-3772.
[86] Appel, R. *Angew. Chem. Int. Ed. Eng.* 1975, *14*, 801-811.
[87] Pandey, G.; Kumaraswamy, G.; Bhalerao, U. T. *Tetrahedron Lett.* 1989, *30*, 6059-6062.
[88] Pandey, G.; Devi Reddy, G.; Kumaraswamy, G. *Tetrahedron* 1994, *50*, 8185-8194.
[89] Zhao, G.; Deo, U. C.; Ganem, B. *Org. Lett.* 2001, *3*, 201-203.
[90] Kawasaki, M.; Suzuki, Y.; Terashima, S. *Chem. Lett.* 1984, *13*, 239-242.
[91] Klein, J.; Dunkelblum, E. *Tetrahedron* 1968, *24*, 5701-5710.
[92] Hong, B.-C.; Chen, Z.-Y.; Nagarajan, A.; Kottani, R.; Chavan, V.; Chen, W.-H.; Jiang, Y.-F.; Zhang, S.-C.; Liao, J.-H.; Sarshar, S. *Carbohydr. Res.* 2005, *340*, 2457-2468.
[93] Hoffmann, T.; Waibel, R.; Gmeiner, P. *J. Org. Chem.* 2003, *68*, 62-69.
[94] Vedejs, E.; Kruger, A. W. *J. Org. Chem.* 1999, *64*, 4790-4797.
[95] Zehnder, L. R.; Wei, L. L.; Hsung, R. P.; Cole, K. P.; McLaughlin, M. J.; Shen, H. C.; Sklenicka, H. M.; Wang, J.; Zificsak, C. A. *Org. Lett.* 2001, *3*, 2141-2144.
[96] Takahata, H.; Banba, Y.; Ouchi, H.; Nemoto, H. *Org. Lett.* 2003, *5*, 2527-2529.
[97] Coleman, R. S.; Carpenter, A. J. *Tetrahedron Lett.* 1992, *33*, 1697-1700.

[98] Williams, L.; Zhang, Z.; Feng, S.; Carroll, P. J.; Joullie, M. M. *Tetrahedron* 1996, *52*, 11673-11694.
[99] Singh, O. V.; Han, H. *Tetrahedron Lett.* 2003, *44*, 2387-2391.
[100] Han, H.; Cho, C.-W.; Janda, K. D. *Chem. Eur. J.* 1999, *5*, 1565-1569.
[101] Oppolzer, W.; Mirza, S. *Helv. Chim. Acta* 1984, *67*, 730-738.
[102] Dess, D. B.; Martin, J. C. *J. Org. Chem.* 1983, *48*, 4155-4156.
[103] Blanchette, M. A.; Choy, W.; Davis, J. T.; Essenfeld, A. P.; Masamune, S.; Roush, W. R.; Sakai, T. *Tetrahedron Lett.* 1984, *25*, 2183-2186.
[104] Gao, Y.; Sharpless, K. B. *J. Am. Chem. Soc.* 1988, *110*, 7538-7539.
[105] Kempson, J.; Pattenden, G. *Synlett* 1999, 533-536.
[106] Mitsunobu, O. *Synthesis* 1981, 1-28.
[107] Grandel, R.; Kazmaier, U. *Tetrahedron Lett.* 1997, *38*, 8009-8012.
[108] Marshall, J. A.; Beaudoin, S. *J. Org. Chem.* 1994, *59*, 6614-6619.
[109] Jager, V.; Muller, I. *Tetrahedron* 1985, *41*, 3519-3528.
[110] Jager, V.; Muller, I.; Paulus, E. F. *Tetrahedron Lett.* 1985, *26*, 2997-3000.
[111] Kayakiri, H.; Oku, T.; Hashimoto, M. *Chem. Pharm. Bull.* 1991, *39*, 1397-1401.
[112] Achmatowicz, M.; Hegedus, L. S. *J. Org. Chem.* 2004, *69*, 2229-2234.
[113] Ranslow, P. B. D.; Hegedus, L. S.; delosRios, C. *J. Org. Chem.* 2004, *69*, 105-111.
[114] Mukaiyama, T. *Angew. Chem. Int. Ed. Eng.* 1979, *18*, 707-721.
[115] Ostrowski, J.; Altenbach, H.-J.; Wischnat, R.; Brauer, David J. *Eur. J. Org. Chem.* 2003, *2003*, 1104-1110.
[116] Williams, R. M. In *Synthesis of Optically Active a-Amino Acids*; Pergamon Press: Oxford, 1989.
[117] Ciufolini, M. A.; Hermann, C. Y. W.; Dong, Q.; Shimizu, T.; Swaminathan, S.; Xi, N. *Synlett* 1998, 105-114.
[118] Luche, J.-L.; Rodriguez-Hahn, L.; Crabbé, P. *J. Chem. Soc., Chem. Commun.* 1978, 601-602.
[119] Reddy, M. S.; Narender, M.; Rao, K. R. *Tetrahedron* 2007, *63*, 331-336.
[120] Mancuso, A. J.; Swern, D. *Synthesis* 1981, 165-184.
[121] Comins, D. L.; Sandelier, M. J.; Abad Grillo, T. *J. Org. Chem.* 2001, *66*, 6829-6832.
[122] Luche, J. L.; Gemal, A. L. *J. Am. Chem. Soc.* 1979, *101*, 5848-5849.
[123] Comins, D. L.; Hong, H.; Salvador, J. M. *J. Org. Chem.* 1991, *56*, 7197-7199.
[124] Felpin, F.-X.; Boubekeur, K.; Lebreton, J. *Eur. J. Org. Chem.* 2003, *2003*, 4518-4527.
[125] Gallienne, E.; Gefflaut, T.; Bolte, J.; Lemaire, M. *J. Org. Chem.* 2006, *71*, 894-902.
[126] Mukaiyama, T.; Hoshino, T. *J. Am. Chem. Soc.* 1960, *82*, 5339-5342.
[127] Kantorowski, E. J.; Brown, S. P.; Kurth, M. J. *J. Org. Chem.* 1998, *63*, 5272-5274.
[128] Gefflaut, T.; Martin, C.; Delor, S.; Besse, P.; Veschambre, H.; Bolte, J. *J. Org. Chem.* 2001, *66*, 2296-2301.
[129] Bergmeier, S. C.; Stanchina, D. M. *J. Org. Chem.* 1999, *64*, 2852-2859.
[130] Kim, B. M.; Sharpless, K. B. *Tetrahedron Lett.* 1989, *30*, 655-658.
[131] Lemaire, M.; Veny, N.; Gefflaut, T.; Gallienne, E.; Chênevert, R.; Bolte, J. *Synlett* 2002, 1359-1361.
[132] Mitchell, M.; Qaio, L.; Wong, C.-H. *Adv. Synth. Catal.* 2001, *343*, 596-599.
[133] Benedetti, F.; Berti, F.; Norbedo, S. *Tetrahedron Lett.* 1998, *39*, 7971-7974.
[134] Jung, S.-H.; Jeong, J.-H.; Miller, P.; Wong, C.-H. *J. Org. Chem.* 1994, *59*, 7182-7184.

[135] Garcia-Junceda, E.; Shen, G.-J.; Sugai, T.; Wong, C.-H. *Bioorg. Med. Chem.* 1995, *3*, 945-953.

[136] Knapp, S.; Yang, C.; Pabbaraja, S.; Rempel, B.; Reid, S.; Withers, S. G. *J. Org. Chem.* 2005, *70*, 7715-7720.

[137] Corey, E. J.; Link, J. O. *Tetrahedron Lett.* 1989, *30*, 6275-6278.

[138] Corey, E. J.; Bakshi, R. K. *Tetrahedron Lett.* 1990, *31*, 611-614.

[139] Wang, R. W.; Qing, F. L. *Org. Lett.* 2005, *7*, 2189-2192.

[140] Wang, R. W.; Qiu, X. L.; Bols, M.; Ortega-Caballero, F.; Qing, F. L. *J. Med. Chem.* 2006, *49*, 2989-2997.

[141] Kajimoto, T.; Chen, L.; Liu, K. K. C.; Wong, C. H. *J. Am. Chem. Soc.* 1991, *113*, 6678-6680.

[142] Zhang, X.; Xia, H.; Dong, X.; Jin, J.; Meng, W. D.; Qing, F. L. *J. Org. Chem.* 2003, *68*, 9026-9033.

[143] Kirihara, M.; Takuwa, T.; Takizawa, S.; Momose, T.; Nemoto, H. *Tetrahedron* 2000, *56*, 8275-8280.

[144] Liao, W.-W.; Ibrahem, I.; Cordova, A. *Chem. Commun.* 2006, 674-676.

[145] Pandey, G.; Dumbre, S. G.; Khan, M. I.; Shabab, M.; Puranik, V. G. *Tetrahedron Lett.* 2006, *47*, 7923-7926.

[146] Pandey, G.; Kapur, M. *Synthesis* 2001, 1263-1267.

[147] Abdel-Magid, A. F.; Carson, K. G.; Harris, B. D.; Maryanoff, C. A.; Shah, R. D. *J. Org. Chem.* 1996, *61*, 3849-3862.

[148] Maison, W.; Kosten, M.; Charpy, A.; Kintscher-Langenhagen, J.; Schlemminger, I.; Lützen, A.; Westerhoff, O.; Martens, J. *Eur. J. Org. Chem.* 1999, *1999*, 2433-2441.

[149] Lee, H. K.; Im, J. H.; Jung, S. H. *Tetrahedron* 2007, *63*, 3321-3327.

[150] Häner, R.; Olano, B.; Seebach, D. *Helv. Chim. Acta* 1987, *70*, 1676-1693.

[151] Yun, J. M.; Sim, T. B.; Hahm, H. S.; Lee, W. K.; Ha, H. J. *J. Org. Chem.* 2003, *68*, 7675-7680.

In: Stereochemistry Research Trends
Editor: M. A. Horvat and J. H. Golob

ISBN 978-1-60456-825-7


*Chapter 2*

# STEREOCHEMICAL STUDIES IN INSECT CHEMISTRY: SYNTHESIS AND PROPERTIES OF CHIRAL RESOLVING AGENTS

***Akio Ichikawa*[1] *and Hiroshi Ono*[2]**
[1]National Institute of Agrobiological Sciences, 1-2 Owashi, Tsukuba, Ibaraki 305-8634, Japan
[2]National Food Research Institute, 2-1-12 Kannondai, Tsukuba, Ibaraki 305-8642, Japan

## ABSTRACT

Both enantiomers of a pheromone from the tea tussock moth were synthesized from (*S*)- and (*R*)-citronellol to examine the stereochemistry of the natural product. The absolute configurations of hydroxy-fatty acids isolated from the resinous secretion of lac insects were determined by NMR. The novel chiral resolving agents, MαNP acid, MβNP acid, and M9PP acid, were synthesized by asymmetric induction, using L-(–)-menthol. The absolute configurations of these agents and the conformations of their ester derivatives were determined by X-ray crystallography and NMR. The utilities of the single-enantiomer MαNP and M9PP acids were demonstrated by enantioresolution of insect pheromones. Both MαNP and M9PP acids exhibited superior resolution ability in HPLC and powerful shielding in $^1$H NMR.

## 1. INTRODUCTION

We studied the stereochemistry of chiral insect metabolites. Stereochemistry is important for the biological activity of many semiochemicals or compounds carrying information between organisms [1,2]. Additionally, the enantiopurity of semiochemicals affects their biological activity; thus, stereochemistry must be determined with enantiomeric excess (ee). Minor enantiomer contaminants in agrochemicals and pharmaceuticals may cause undesired effects. Thus, the determination of the ee and preparation of enantiopure compounds are important in both industry and academia.

We studied the novel chiral resolving agents 2-methoxy-2-(1-naphtyl)propanoic acid (M□NP acid; Figure 1-1), 2-methoxy-2-(2-naphtyl)propanoic acid (M□NP acid), and 2-methoxy-2-(9-phenanthryl)propanoic acid (M9PP acid). These agents possess non-racemizable quaternary chiral centers and powerful resolving abilities, and are promising candidates for stereochemical studies. In this chapter, we report the development process and use of each agent.

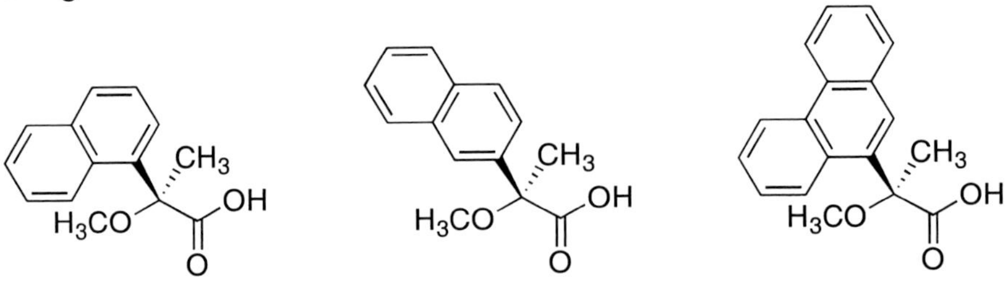


**(*S*)-MαNP acid** **(*S*)-MβNP acid** **(*S*)-M9PP acid**

Figure 1-1. Structures of novel chiral resolving agents.

# 2. Pheromone from Tea Tussock Moth

## 2.1. Structure of Tea Tussock Moth Pheromone [3]

The tea tussock moth, *Euproctis pseudoconspersa* (Strand), is a pest of the Theaceae family (e.g., tea, Japanese camellia, sasanqua camellia). Because the larvae have venomous spicules on their backs, *E. pseudoconspersa* is also a public health nuisance [4]. Wakamura et al. isolated the sex attractant pheromone of *E. pseudoconspersa*, a mixture of one major and two minor components [3]. Gas chromatography-mass spectrometry (GC-MS) analysis revealed that dimethylpentadecyl isobutyrate is the major component of this mixture. However, the analysis did not reveal the positions of the two methyl groups.

For this reason, we established the following reaction conditions to prepare chloride derivatives from the natural pheromone: (1) solvolysis with KOH in aq. EtOH, followed by (2) chlorination with methanesulfonyl chloride (MsCl) in *N,N*′-dimethylformamide (DMF). A synthetic analog, (*RS*)-10-methylhexadecyl isobutyrate, was used as a model compound, because of the limited amount of the natural pheromone.

GC-MS analysis of the chloride derived from the major natural pheromone component show diagnostic fragments at *m/z* 188 (11%) and 189 (9%) with isotopic ion peaks at *m/z* 190 (5%) and 191 (3%; Figure 2-1). This suggested that one methyl group branched at the 10-position of the pentadecyl chain [5,6]. The position of the other methyl group was estimated to be at the 12-, 13-, or 14-position. The other regioisomer, 10,11-dimethylpentadecyl isobutyrate, was excluded as a candidate, because we expected this compound to show the fragment peaks at *m/z* 188–191, with strong intensity from the cleavage between the carbon atoms at the 10- and 11-positions.

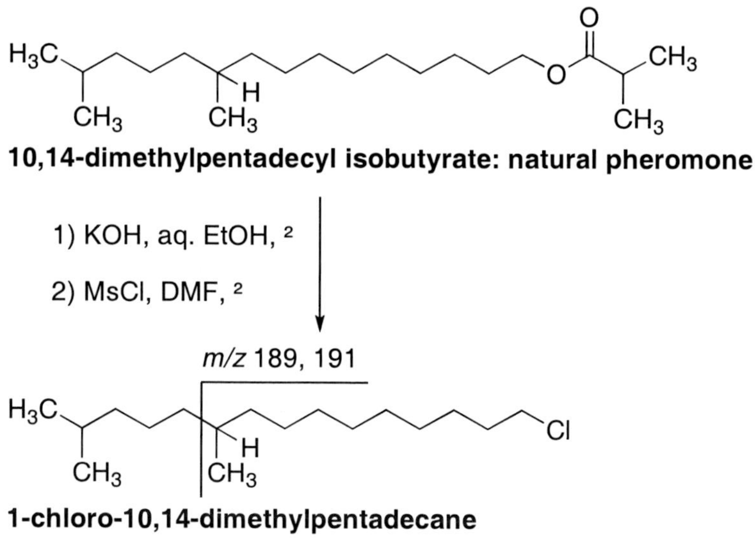


Figure 2-1. Preparation of a chloride derivative from the major component of a sex attractant pheromone from the tea tussock moth.
Natural chlorine isotope variation, $^{35}Cl$:$^{37}Cl$ = 76:24.

Comparing the retention time and the mass spectrum with three synthetic regioisomers in GC-MS, Wakamura et al. identified the major pheromone component as 10,14-dimethylpentadecyl isobutyrate [3]. Using the chlorides prepared above, they also identified the minor pheromone components [3].

## 2.2. Asymmetric Synthesis of Tea Tussock Moth Pheromone [7, 8]

The major pheromone component from the tea tussock moth *E. pseudoconspersa* has a chiral center at the 10-position; both enantiomers were synthesized from monoterpene citronellol, a fragrance found in rose, geranium, and Java citronella oils [9] (Figure 2-2). Previously, members of our research group reported on marine terpenoids [10–13]. Based on this research, (*S*)- and (*R*)-citronellol were selected as starting materials.

The double bond in (*S*)-(–)-citronellol was reduced with $LiAlH_4$ and $CoCl_2$ in tetrahydrofuran (THF), yielding (*S*)-(–)-3,7-dimethyl-1-octanol [14]. Subsequent oxidation with pyridinium chlorochromate (PCC) in $CH_2Cl_2$ yielded (*S*)-(–)-3,7-dimethyloctanal.

The Wittig reaction [15] of (*S*)-(–)-3,7-dimethyloctanal and an ylide prepared from 7-hydroxyheptyltriphenylphosphonium bromide yielded (*S*)-10,14-dimethyl-7-pentadecen-1-ol, but the geometry of the double bond was not determined. Diimide reduction [16] yielded (*R*)-10,14-dimethyl-1-pentadecanol, in which the stereochemistry was changed from *S* to *R*. Acylation with isobutyric anhydride in pyridine gave (*R*)-10,14-dimethylpentadecyl isobutyrate. (*S*)-10,14-Dimethylpentadecyl isobutyrate was synthesized from (*R*)-(+)-citronellol in a similar manner.

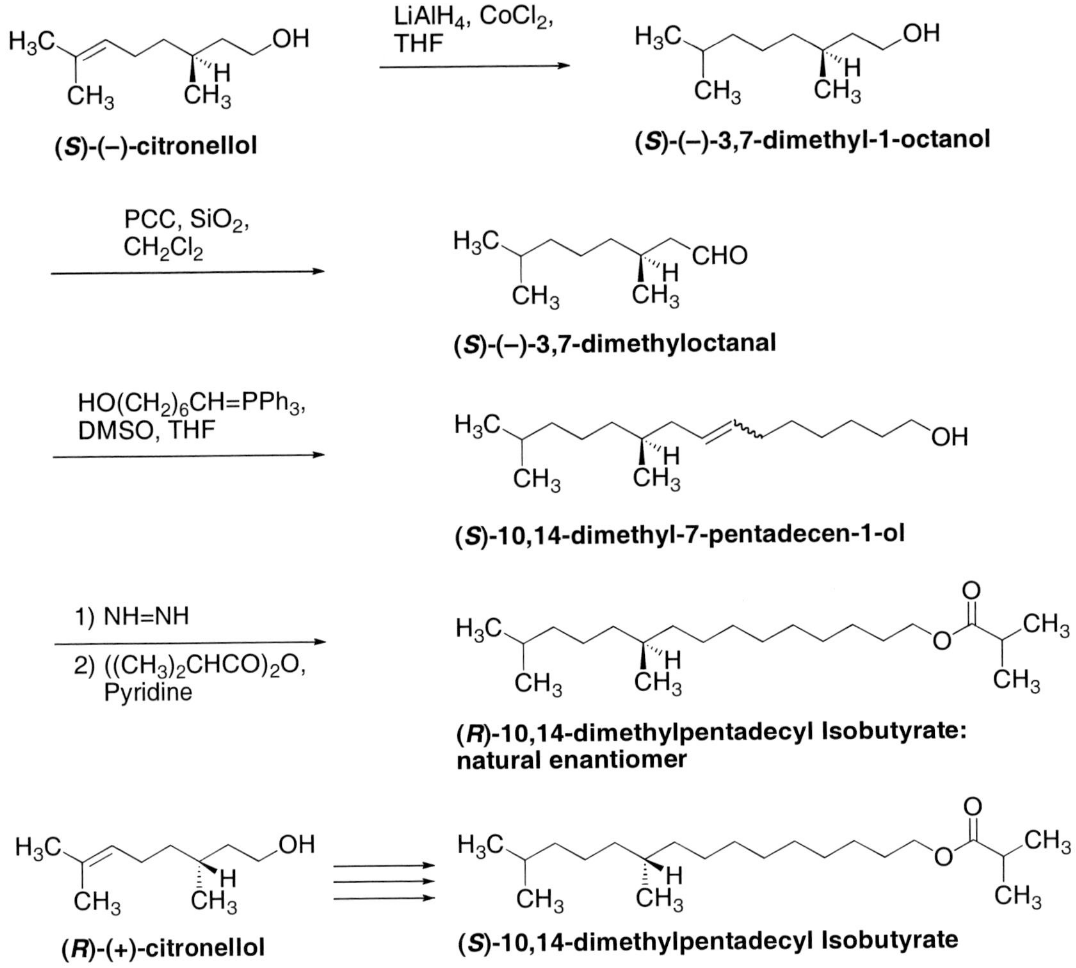


Figure 2-2. Asymmetric synthesis of a sex attractant pheromone from the tea tussock moth.

Wakamura et al. clarified the electroantennographic (EAG) activity of the natural and synthetic pheromones [7,8]. The dose-response curve of the (*R*)-enantiomer was higher than that of the (*S*)-enantiomer, and almost identical to that of natural pheromone. These results suggest that the structure of the major sex attractant pheromone for *E. pseudoconspersa* is (*R*)-10,14-dimethylpentadecyl isobutyrate, which was later confirmed by Zhao et al. [17].

## 2.3. Enantiopurity of Citronellol [7]

We were interested in the enantiopurity of the pheromone synthesized above. The reaction conditions did not affect the enantiopurity of the starting materials; thus, the enantiopurity of the products was identical to those of (*S*)- and (*R*)-citronellol. However, the attached information on the ee of citronellol was unclear. Indeed, it stated, "(*S*)-(−)-□-citronellol, 99%, $[□]_D^{20}$ −5.3 (neat)." The absolute value of the specific rotation was too small to determine the exact ee.

For this reason, (*S*)-(–)- and (*R*)-(+)-3,7-dimethyl-1-octanol were acylated with (+)-MTPA chloride, yielding (*R*)-MTPA esters **A** and **B**, respectively (Figure 2-3). However, the diastereomeric esters **A** and **B** were inseparable by HPLC. Additionally, it was not possible to determine the ee from $^1H$ NMR spectra (270 MHz, $CDCl_3$). The (*R*)-MTPA esters of (*S*)- and (*R*)-citronellol (esters **C** and **D**, respectively) were also inseparable by HPLC and indistinguishable by $^1H$ NMR.

Bergot et al. reported a method for determining the ee of citronellol [18]. This method required two steps to prepare the diastereomeric amides **E** and **F**: first, the Jones oxidation of citronellol, followed by condensation with (*R*)-(+)-1-(1-naphthyl)ethylamine. Then, they analyzed the diastereomeric amides by HPLC. A disadvantage of this method is the chromium waste produced by the Jones oxidation.

Using BINAP–Rh(I) catalysts, Noyori et al. reported the asymmetric synthesis of (*S*)- and (*R*)-citronellal as intermediates in menthol synthesis [19,20], thereafter referred to as the Takasago process. They also reported the preparation of (*S*)- and (*R*)-citronellol (both 98% ee) by enantioselective hydrogenation of geraniol and nerol using BINAP–Ru(II) catalysts [21].

It is important to consider the influence of the minor enantiomer in biological activity. We required a powerful chiral resolving agent to clarify the stereochemistry of the acyclic insect metabolites.

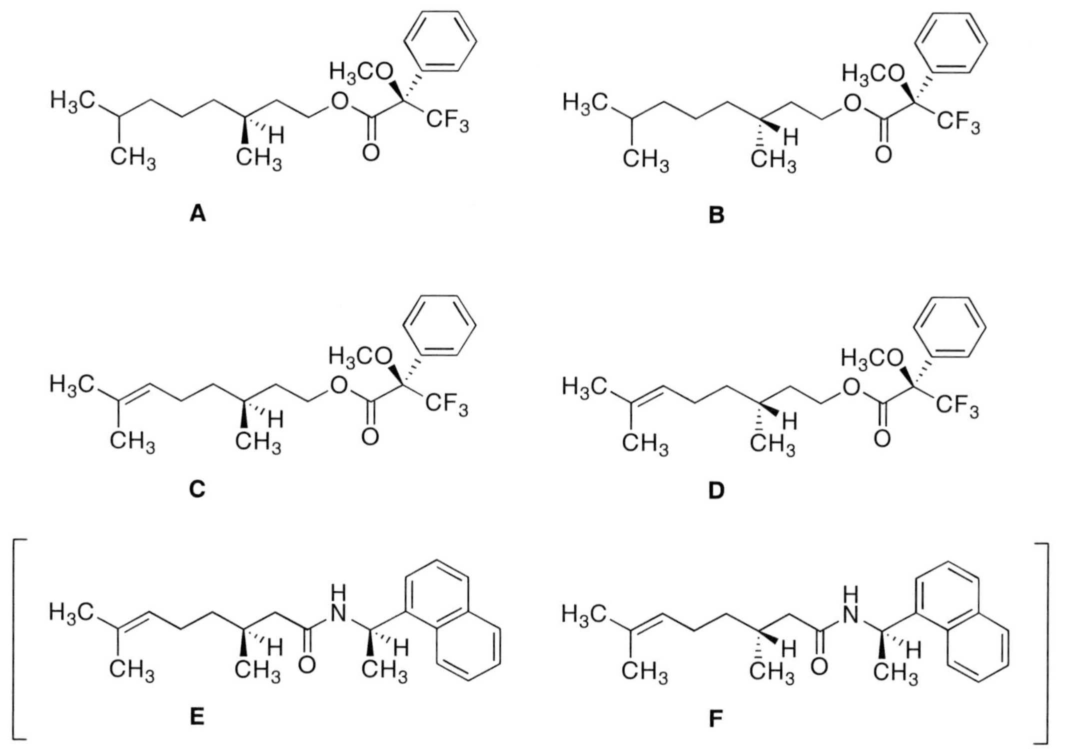


Figure 2-3. Structures of diastereomeric esters and amides.

## 3. Lac Resins: Absolute Configurations of Hydroxy-Fatty Acids [22]

The lac insect is a species of the scale insect genus *Laccifer* (= *Kerria*). They secrete lac lesins onto the host trees. These lac resins are then refined to obtain shellac, which is produced in India, Thailand, Myanmar, and China. Shellac is widely used as a coating or adhesive for many products, including fruits, tablets, and electronics. The major components of shellac are 9,10,16-trihydroxyhexadecanoic acid (aleuritic acid; Figure 3-1), 6-hydroxytetradecanoic acid (butolic acid), terpenoids (jalaric acid, laccijalaric acid), and their oligomers [2].

6-Hydroxytetradecanoic acid was isolated from the lac resin secreted by *L. sindica*. 8-Hydroxytetradecanoic acid and 8-hydroxyhexadecanoic acid were isolated from the lac resin secreted by *L. chinensis*.

Kusumi and Ohtani reported the modified Mosher's method to determine the absolute configuration of secondary alcohols and primary amines [23,24]. High-field shifts in $^1$H NMR were observed at the $R_2$ and $R_1$ moieties of (*R*)- and (*S*)-MTPA (3,3,3-trifluoro-2-methoxy-2-phenylpropanoic acid) esters **X** and **Y**, respectively (Figure 3-2); therefore, negative and positive $\Delta\delta$ (= $\delta$[(*S*)-MTPA] – $\delta$[(*R*)-MTPA]) values were observed at the $R_1$ and $R_2$ moieties, respectively.

The hydroxy-fatty acids were methylated with trimethylsilyldiazomethane, and condensed with (*R*)- and (*S*)-MTPA acid using 1,3-dicyclohexylcarbodiimide (DCC) and 4-dimethylaminopyridine (DMAP) in $CH_2Cl_2$. Their $\Delta\delta$ values in $^1$H NMR were determined by COSY spectra (Figure 3-1, **A**, **B**, and **C**). However, a full assignment was not determinable for the overlapping methylene signals.

Kusumi and Riguera independently developed the novel chiral anisotropic reagents 1- and 2-naphthylmethoxyacetic acid (1NMA and 2NMA) [25,26]. Large high-field shifts were observed at the $R_1$ substituent of the (*R*)-2NMA ester **Z** (Figure 3-3), and a single 2NMA ester could be used to assign the absolute configuration of the secondary alcohol.

Kusumi et al. kindly prepared the (*R*)-2NMA esters of the hydroxy-fatty acids isolated from the resin secreted by *L. chinensis* (Figure 3-1, **D** and **E**). Because of strong shielding by the 2-naphthyl group, full assignments of the methylene signals were established in $^1$H NMR (400 MHz, $CDCl_3$), and the absolute configurations of both hydroxy-fatty acids were determined to be *R*. They also identified minor hydroxy-fatty acids from the resin as the (*R*)-2NMA esters [27].

The (*R*)-2NMA used in this study contained a minor (*S*)-enantiomer. Most complex natural products are single-enantiomers; as a result, the ee of chiral resolving agents is not always an important problem. However, simple insect metabolites are not always single-enantiomers; thus, we required a powerful chiral resolving agent with a non-racemizable chiral center. In 2003, Arita et al. reported the synthesis of enantiopure 1NMA and 2NMA using L-phenylalaninol [28].

Considering the structure of aleuritic acid (Figure 3-1), we also examined the application of the modified Mosher's method to the *syn*- and *anti*-glycols [22,29–31].

Presently, polymers made from lactic acid (2-hydroxypropanoic acid) are of interest as biodegradable materials (Figure 3-3). The structures of hydroxy-fatty acids isolated from lac resins may be useful in the design of novel biodegradable and sustainable materials.

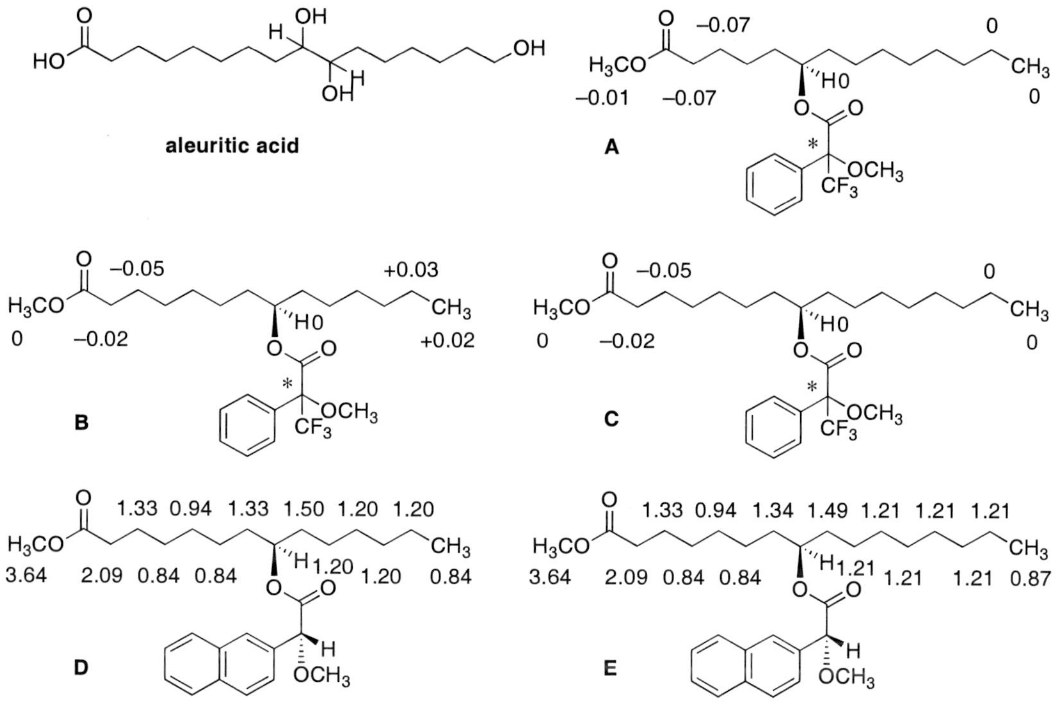


Figure 3-1. Structure of aleuritic acid, and $^{1}H$ NMR data for MTPA or (*R*)-2NMA esters of hydroxy-fatty acids, isolated from lac resins.
The Δδ (= δ[(*S*)-MTPA] – δ[(*R*)-MTPA]) values were assigned for esters **A**, **B** and **C**. $^{1}H$ NMR chemical shifts were assigned for esters **D** and **E**.

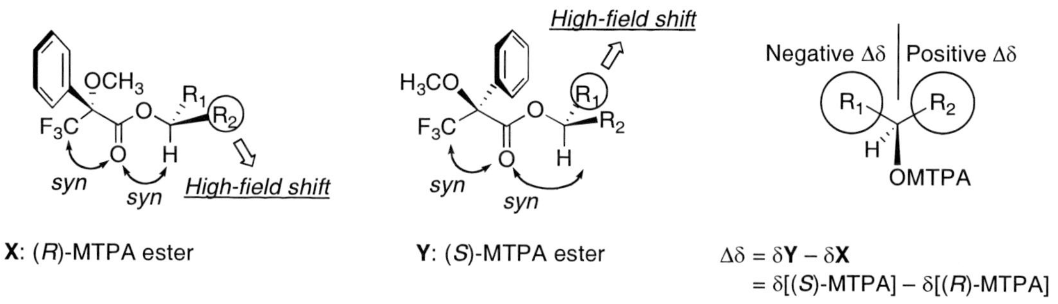


Figure 3-2. Conformations of the (*R*)- and (*S*)-MTPA esters, and distribution of the negative and positive Δδ values in $^{1}H$ NMR.

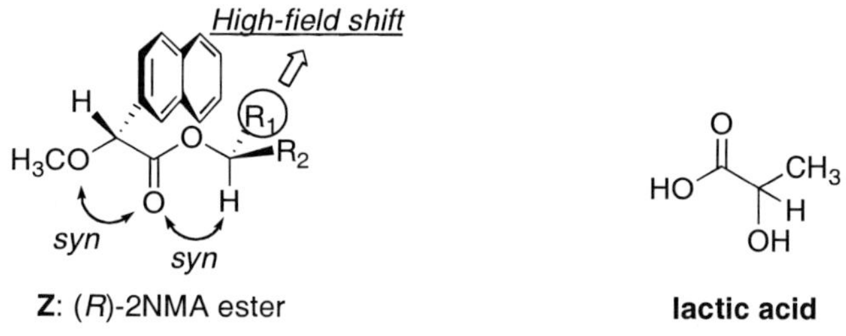


Figure 3-3. Conformation of (*R*)-2NMA ester and structure of lactic acid.

# 4. RACEMIZATION

Partial racemization has been reported for 2-methoxy-2-phenylacetic acid (MPA) in esterification [32]. It is probable that the racemization of 2NMA and MPA originates in their carbon acid structures, –CH–C=O (Figure 4-1) [33].

Similar isomerization has been reported for amino acids [34]. The aspartic acid (Asp) residue is the most racemizable amino acid found in proteins [34,35]. Fujii reported the presence of D-Asp in tissues from elderly patients (e.g., alpha A-crystallin from elderly human lens). An inversion and racemization mechanism for the Asp residue via a cyclic intermediate has been reported [35].

Thalidomide is another important example of a carbon acid. Hashimoto et al. investigated the rate of racemization of thalidomide under physiological conditions [36,37]. They also investigated the effects of the non-racemizable thalidomide analog (*R*)- and (*S*)-methylthalidomide on TNF-α production regulation. Takeuchi et al. synthesized (*R*)- and (*S*)-3-fluorothalidomide as isosteric analogs of thalidomide [38].

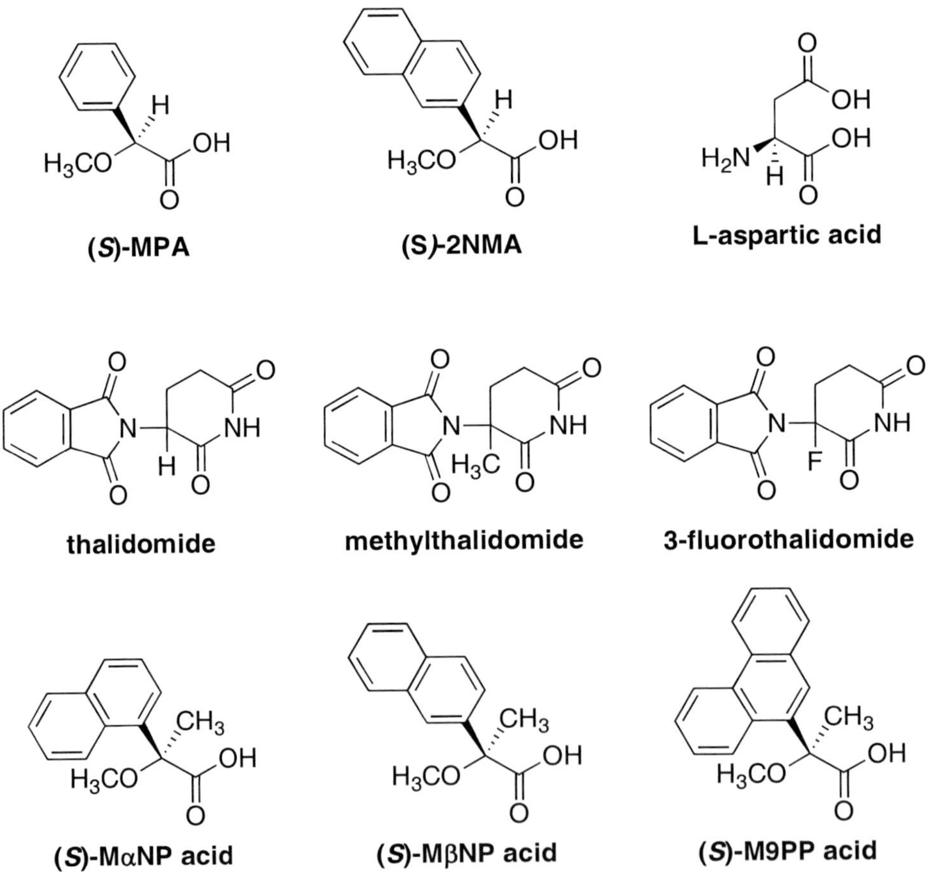


Figure 4-1. Structures of chiral resolving agents, L-aspartic acid, thalidomide, methylthalidomide, and 3-fluorothalidomide.

Such strategies may also be applicable to chiral resolving agents. As mentioned, Kusumi and Riguera independently reported 1NMA and 2NMA as chiral anisotropic reagents for NMR studies [25,26,28,39].

This prompted us to develop 2-aryl-2-methoxypropanoic acid (MαNP acid, MβNP acid, and M9PP acid) as non-racemizable chiral resolving agents. Goto et al. synthesized (–)-MαNP acid and (–)-MβNP acid by fractional crystallization with (+)-α-methylbenzylamine, and used them for the resolution of amino acid enantiomers as diastereomeric amides [40,41].

Because of the prices of 1-bromonaphthalene and 2-bromonaphthalene, which are necessary for preparing the Grignard reagents, we started the study of MαNP acid. Dr. D. Ibarra, a post-doctoral fellow from Germany, kindly taught us how to prepare the Grignard reagents. Currently, THF solutions of 1- and 2-naphthylmagnesium bromide are available from Aldrich.

## 5. Prelog's Method: Absolute Configuration of MαNP Acid [42–45]

Chikamatsu and Naemura referred to Prelog's method for determining the absolute configurations of α-hydroxy acids [46,47]. Specifically, this involves Grignard additions to L-menthol pyruvate, selecting the *Si*-face of the ketone carbonyl group.

The addition of 1-naphthylmagnesium bromide to L-menthol pyruvate yielded α-hydroxy esters **A** and **B** at 39 and 26%, respectively (Figure 5-1) [44]. Their NMR signals were assigned using two-dimensional (2D) NMR spectra (COSY, CH COSY, HOHAHA, HMBC, and NOESY; Figure 5-2). Hydroxy protons were always observed in $^1$H NMR spectra. IR spectra suggested that esters **A** and **B** formed intramolecular hydrogen bonds. Large high-field shifts were observed at the 2-position of ester **A** (0.51 ppm), and at the isopropyl group of ester **B** (0.46, 0.26, and 0.13 ppm). Negative Δδ (= δ**B** – δ**A**) values were observed at the isopropyl group, and the 4- and 5-positions of the menthol moiety (Figure 5-3). In contrast, positive Δδ values were observed at the 1- and 2-positions. Considering the shielding provided by the 1-naphthyl groups, the stereochemistry of the acyl moieties of esters **A** and **B** were assigned as *S* and *R*, respectively (Figure 5-2). This assignment is consistent with the stereoselectivity of the Grignard reaction reported by Prelog [47].

(+)- and (–)-MαNP acids were prepared by fractional crystallization [41]. Ester **A** was linked with (+)-MαNP acid via the methyl ester; thus, the absolute configuration was determined as (*S*)-(+)- and (*R*)-(–)-MαNP acid. Based on these results, the previously assigned stereochemistry [48] must now be revised.

(*S*)-(+)- and (*R*)-(–)-MαNP acid were condensed with L-(–)-menthol, yielding esters **C** and **D**, respectively (Figure 5-1, 5-2). Esters **C** and **D** were also prepared from (*RS*)-MαNP acid by esterification, or from the mixture of Grignard adducts, **A** and **B**, by methylation.

The Δδ (= δ*R* – δ*S* = δ**D** – δ**C**) values of the MαNP esters **C** and **D** were almost equal to those of esters **A** and **B** (= δ*R* – δ*S* = δ**B** – δ**A**) in the menthol moieties (Figure 5-3); thus, the carbonyl group and the methoxy group of the MαNP moiety were *syn*. A similar conformation was reported for 1NMA ester [49]. Solvolysis of esters **C** and **D** yielded (*S*)-(+)- and (*R*)-(–)-MαNP acid, respectively.

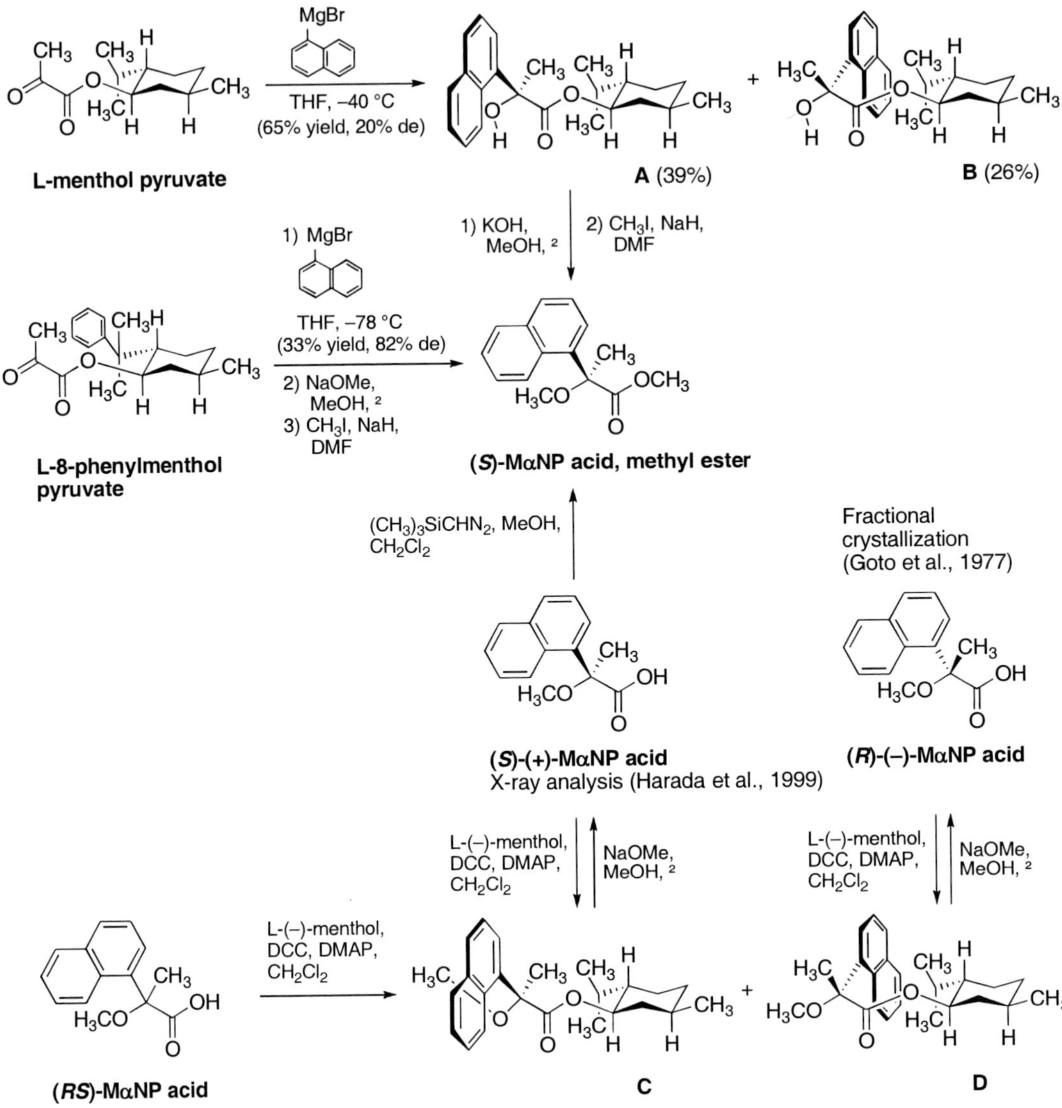


Figure 5-1. Synthesis of (*S*)-(+)- and (*R*)-(–)-MαNP acid and their derivatives.
Because of the urea byproduct, 1-ethyl-3-(3-dimethylaminopropyl)carbodiimide hydrochloride (EDC) was used for esterification in subsequent studies.

Harada et al. unambiguously determined the absolute configuration of (*S*)-(+)-MαNP acid by X-ray crystallography [50]. They also concluded that the *syn* conformations of MαNP esters were stabilized by the intramolecular bifurcated hydrogen bond CH···O [51].

The separation factor ($\alpha$) for esters **C** and **D** was 1.73 in normal-phase HPLC (Figure 5-4). These results indicated that the single-enantiomer MαNP acid could efficiently resolve racemic menthol via esterification.

L-(–)-menthol was acylated with coumarin-3-carboxylic acid, and analyzed by enantioselective HPLC, using a Chiralcel OD-RH column (MeCN:$H_2O$ = 13:7). The opposite enantiomer was not detected; that is, the L-(–)-menthol used in this study was enantiopure (> 99% ee) [52].

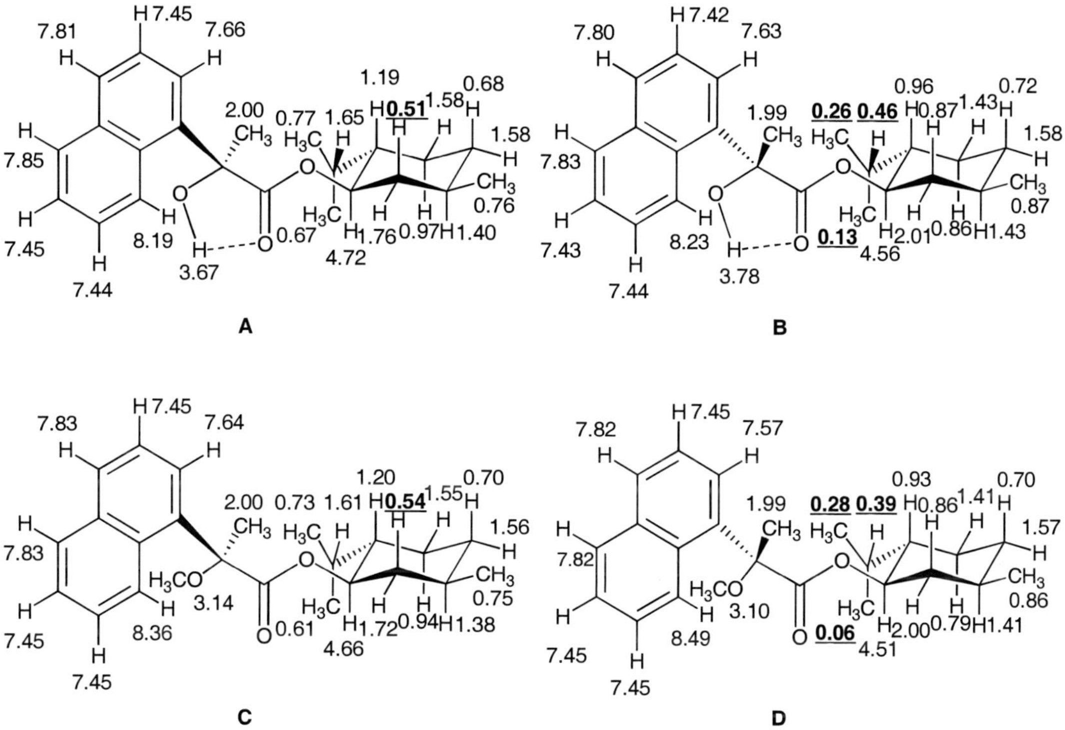

Figure 5-2. $^1$H NMR chemical shifts for esters **A**, **B**, **C**, and **D**.
Esters **A** and **B**, 600 MHz, $CDCl_3$; esters **C** and **D**, 400 MHz, $CDCl_3$. Typical high-field shifts were underlined. The prochiral methyl groups of esters **A** and **B** were assigned based on NOESY spectra. The NMR spectra of esters **A** and **B** were determined by Dr. S. Hiradate (NIAES).

Whitesell et al. reported that the Grignard reaction with the pyruvate ester of L-(–)-8-phenylmenthol [53] afforded high chemical yield derivative α-hydroxyesters with excellent levels of asymmetric induction [54]. Using Whitesell's method, the methyl ester of (*S*)-(+)-MαNP acid was also prepared (Figure 5-1) [44]. The *cisoid* conformation of the pyruvate moiety could explain this selectivity [54]: the Grignard reagent selected the opposite sides of isopropyl and phenylisopropyl groups (i.e., *Si*-face; Figure 5-5).

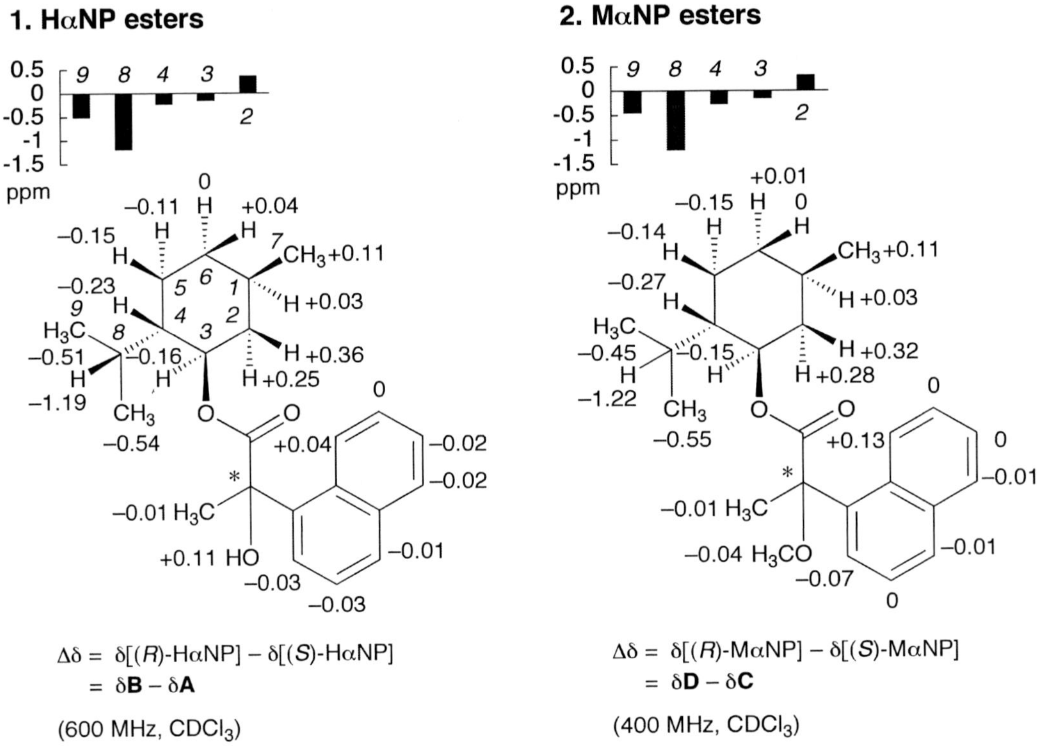


Figure 5-3. Δδ (= δ$R$ – δ$S$) values for L-menthol esters in $^{1}$H HMR.
Black bars represent the Δδ values at the 2-, 3-, 4-, 8-, and 9-positions in the menthol moieties.
Abbreviation: HαNP acid, 2-hydroxy-2-(1-naphthyl)propanoic acid.

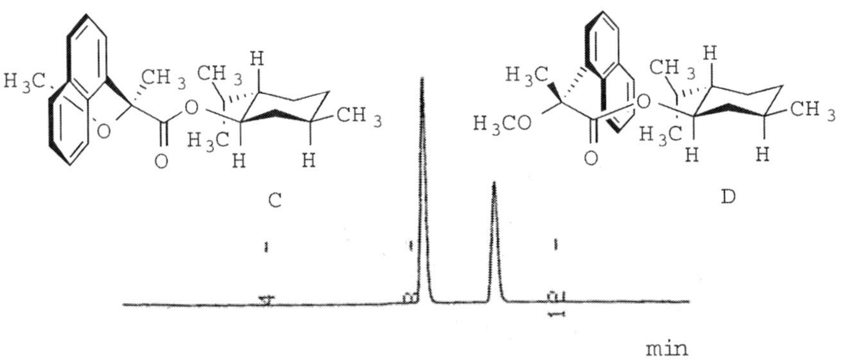


Figure 5-4. HPLC profile of L-menthol MαNP esters.
Column, Silica SG80 (250 mm × 4.6 mm I.D.); mobile phase, hexane:EtOH = 199:1; flow rate, 0.5 mL/min; detection, UV 260 nm; $\alpha = (t_2 - t_0)/(t_1 - t_0) = 1.73$; $t_0$, octadecylbenzene. Esters **C** and **D** of this lot were prepared by the methylation of a mixture of Grignard adducts **A** and **B**. The mobile phase was switched to hexane/EtOAc in subsequent studies.

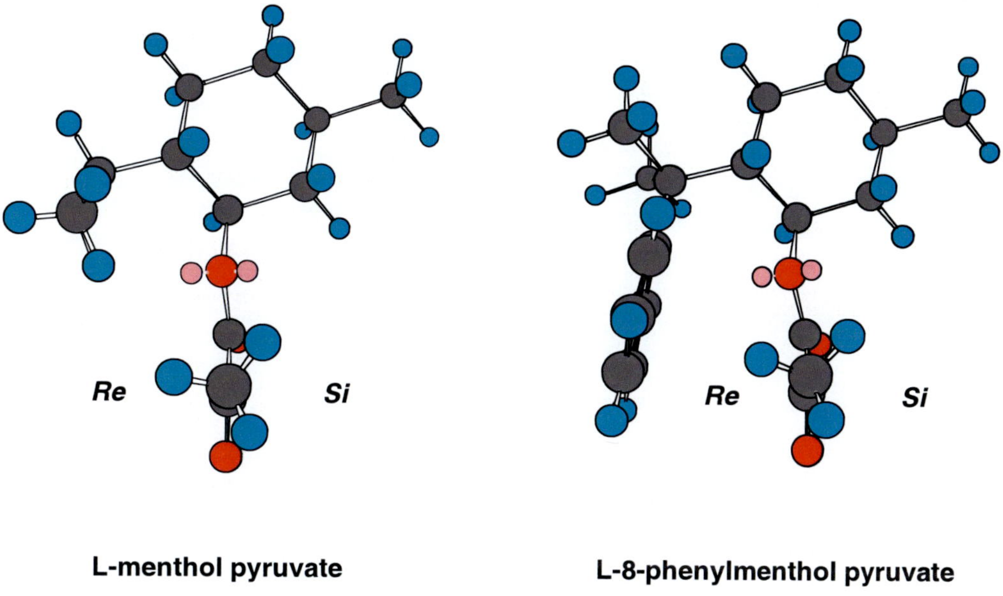


Figure 5-5. Conformations of L-menthol pyruvate and L-8-phenylmenthol pyruvate.

## 6. MαNP Esters of Citronellol [55]

In our synthetic study of the sex attractant pheromone of *E. pseudoconspersa*, we were unable to determine the ee of citronellol using MTPA; thus, we used a single-enantiomer MαNP acid. (*R*)-(–)-MαNP acid was methylated by trimethylsilyldiazomethane, and analyzed by enantioselective HPLC (Chiralcel OD-RH, MeCN/$H_2O$); the ee was up to 99%.

(*S*)-(–)- and (*R*)-(+)-citronellol were condensed with the enantiopure (*R*)-(–)-MαNP acid, using DCC and DMAP in $CH_2Cl_2$, yielding esters **X** and **Y**, respectively (Figure 6-1). Methyl groups in the 3-positions of the alcohol moieties were observed as doublets at 0.66 and 0.55 ppm in $^1$H NMR spectra, respectively; thus, the diastereomeric excess (de) of esters **X** and **Y** was readily determined from the $^1$H NMR spectra. Figure 6-2 shows the $^1$H NMR signals of the crude ester **Y**. The minor diastereomer **X**, derived from (*S*)-(–)-citronellol, was also observed. The $^{13}$C satellite bands [56] were used as the internal standards for determining de, because the area of each satellite band represented 0.5% of the major enantiomer **Y**.

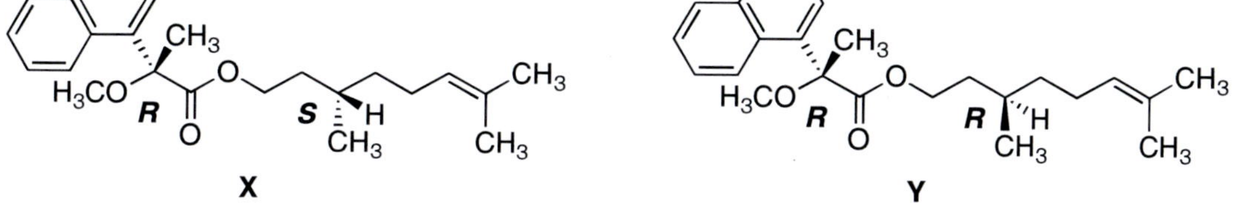


Figure 6-1. Structures of the (*R*)-MαNP esters of citronellol.

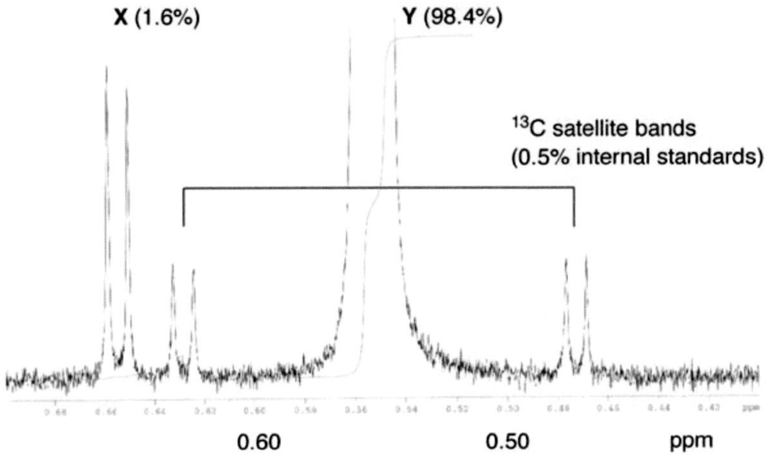


Figure 6-2. The $^1$H NMR spectrum of (*R*)-citronellol (*R*)-MαNP ester (800 MHz, $CDCl_3$).

Esters **X** and **Y** were only marginally separated by normal-phase HPLC (Figure 6-3). The minor diastereomer **X** (Figure 6-2) was separated from the major diastereomer **Y** by preparative HPLC, using a pre-packed glass column (Kusano C.I.G. HS, silica gel). This study prompted us to use single-enantiomer MαNP acid for the preparation of enantiopure alcohols. Ohrui et al. reported highly potent chiral labeling reagents for the chiral discrimination of anteiso fatty alcohols [57,58].

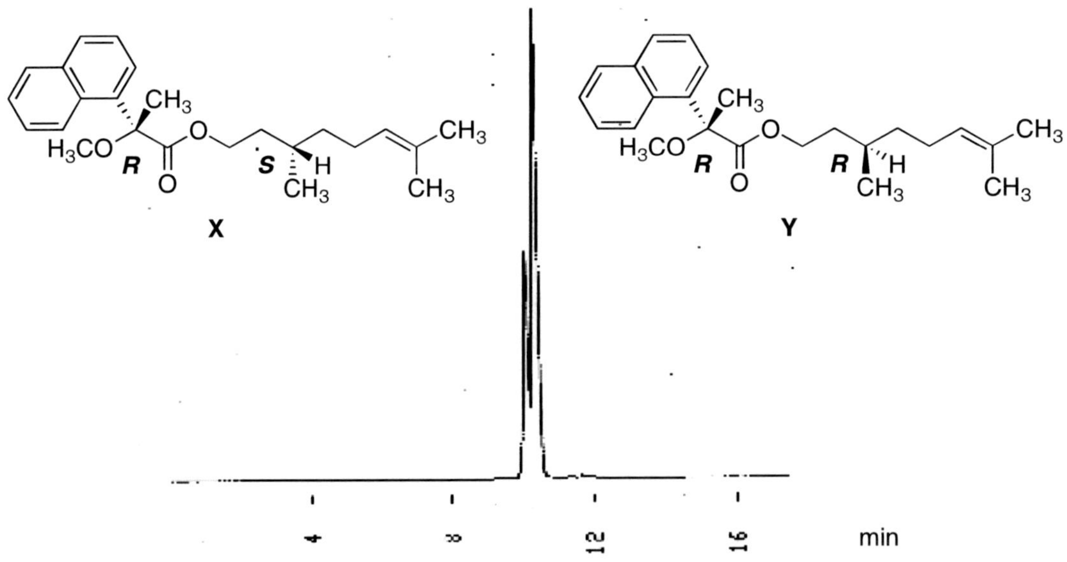


Figure 6-3. HPLC profile of the citronellol (*R*)-MαNP esters.
Esters **X** (2.5 μg) and **Y** (5 μg) were co-injected as ethanol solutions. Column, Silica SG80 (250 mm × 4.6 mm I.D.); mobile phase, hexane:EtOH = 99.4:0.6; flow rate, 0.5 mL/min; detection, UV 300 nm.

# 7. The MαNP Method [45,51,59–64,80,81]

Considering these results and the crystal structure of the M9PP ester [62], we propose the following predominant conformation for (*S*)-MαNP ester **A** (Figure 7-1).

1. The hydrogen atom at the α-position of alcohol moiety and the carbonyl group are *syn*.
2. The carbonyl group and the methoxy group are *syn*.
3. The methyl group of the MαNP moiety and the aromatic hydrogen atom at the 2-position are *syn*.

The same was true for (*R*)-MαNP ester **B**. The $R_2$ substituent of (*S*)-MαNP ester **A** and the $R_1$ substituent of (*R*)-MαNP ester **B** are highly shielded by 1-naphthyl groups. The stereochemistry of the secondary alcohol could be readily elucidated from the Δδ (= δ[(*R*)-MαNP ester] – δ[(*S*)-MαNP ester]) values for the alcohol moiety. The Δδ values of the MαNP esters were clearly larger than those of the MTPA esters [45,63].

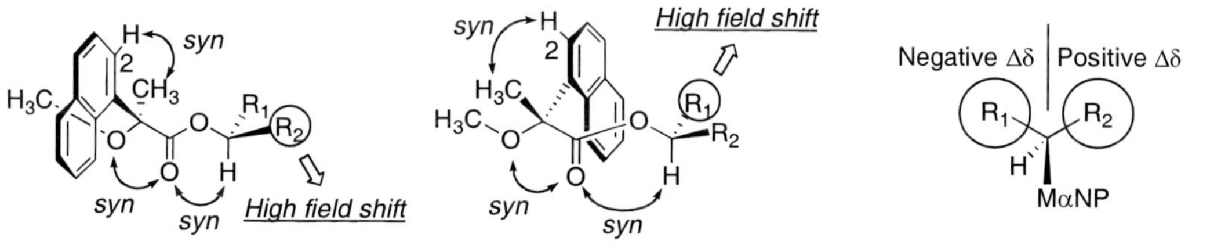


**A**: (*S*)-MαNP ester

**B**: (*R*)-MαNP ester

Δδ = δ**B** – δ**A**
= δ[(*R*)-MαNP] – δ[(*S*)-MαNP]

Figure 7-1. Preferred conformations of the (*S*)- and (*R*)-MαNP esters, and the distribution of negative and positive Δδ values in $^1$H NMR.

MαNP acid is also useful for the preparation of single-enantiomer secondary alcohols (Figure 7-2). The two diastereomeric esters prepared from single-enantiomer MαNP acid and racemic alcohol (esters **A** and **C**) were readily separable by silica gel HPLC. The enantiomeric esters **B** and **C** (Figures 7-1 and 7-2, respectively) showed identical chemical shifts in $^1$H NMR; thus, the stereochemistry of the diastereomeric esters **A** and **C** could be determined by comparison of their $^1$H NMR signals, assigned using 2D NMR spectra. Solvolysis of the diastereomeric esters yielded single-enantiomer secondary alcohols and MαNP acid, demonstrating that MαNP acid is reusable [63,64].

In the case of simple aliphatic alcohols [i.e., 6-methyl-5-hepten-2-ol (sulcatol), 3-octanol, 1-octen-3-ol, 2-methyl-4-heptanol, 2-methyl-4-octanol, and β-ionol], the substituent $R_1$ of the first-eluted ester **A** in normal-phase HPLC was larger and more hydrophobic than $R_2$ [63,81].

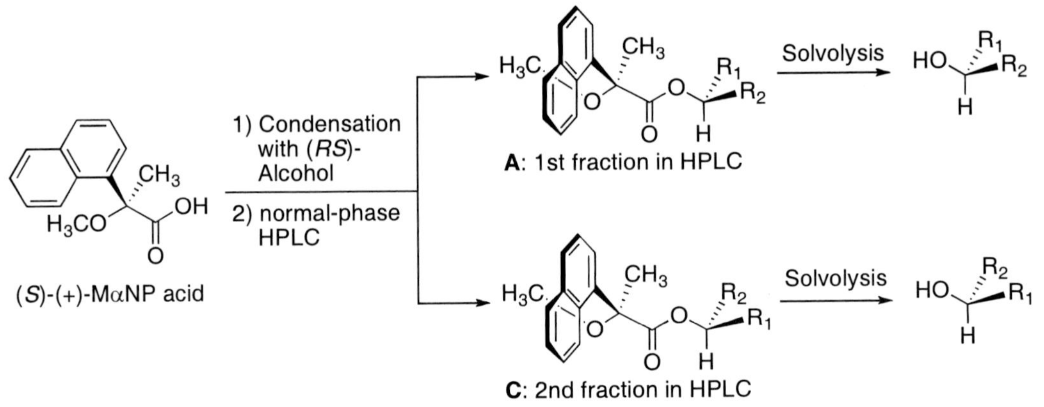


Figure 7-2. Enantioresolution of racemic secondary alcohol using (*S*)-(+)-MαNP acid.

# 8. APPLICATION OF THE MαNP METHOD

## 8.1. Standard Alcohol: Methyl (*R*)-3-Hydroxypentanoate [63]

The conformational hypothesis regarding the MαNP esters was verified using a standard compound, methyl (*R*)-3-hydroxypentanoate. Methyl (*R*)-3-hydroxypentanoate was condensed with (*S*)-(+)- and (*R*)-(–)-MαNP acid using EDC and DMAP in $CH_2Cl_2$, yielding esters **A** and **B**, respectively (Figure 8-1). The signals of esters **A** and **B** were assigned using 2D NMR spectra (DQF COSY, HSQC, and HMBC). Negative $\Delta\delta$ (= $\delta$**B** – $\delta$**A**) values were observed at the methoxycarbonylmethyl group ($R_1$) of the alcohol moiety (Figure 8-2). Positive $\Delta\delta$ values were observed at the ethyl group ($R_2$). By contrast, the $\Delta\delta$ values of the MαNP moiety were between 0 and +0.02 ppm. These observations are consistent with the conformational hypothesis of the $^1$H NMR anisotropy method using MαNP acid.

Esters **A** and **B** were readily separable by normal-phase HPLC (Figure 8-3). These results suggest that the single-enantiomer MαNP acid could resolve methyl (*RS*)-3-hydroxypentanoate.

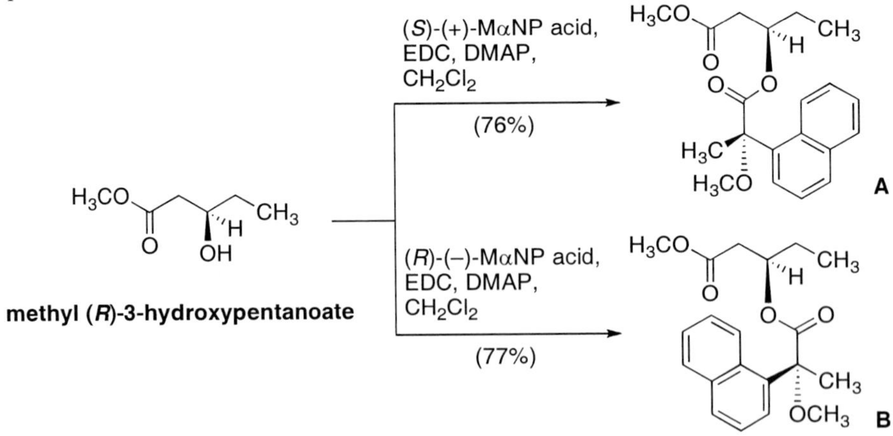


Figure 8-1. Preparation of the (*S*)-and (*R*)-MαNP esters of methyl (*R*)-3-hydroxypentanoate.

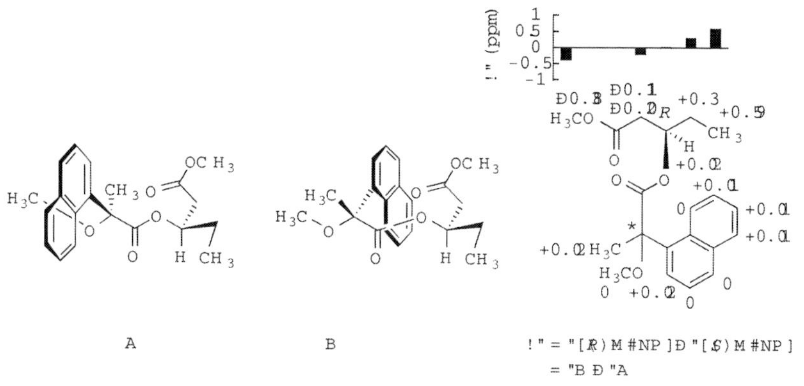

Figure 8-2. Preferred conformations of MαNP esters **A** and **B**, and their Δδ values (500 MHz, $CDCl_3$). The black bars represent the intensities of the Δδ values in the alcohol moiety.

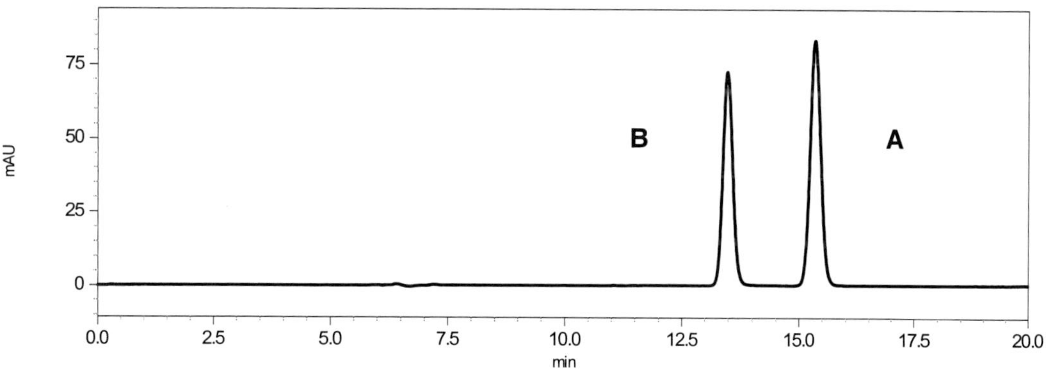


Figure 8-3. HPLC profile of diastereomeric esters formed from (*S*)- and (*R*)-MαNP acid and methyl (*R*)-3-hydroxypentanoate.
Column, Silica SG80 (250 mm × 4.6 mm I.D.); mobile phase, hexane:EtOAc = 3:1; flow rate, 0.5 mL/min; detection, UV 300 nm; $\alpha = 1.25$, $Rs = 4.60$; $t_0$, 1,3,5-tri-*tert*-butylbenzene; $\alpha = (t_2 - t_0)/(t_1 - t_0)$; $Rs = 2(t_2 - t_1)/(W_1 + W_2)$.

## 8.2. Secondary Alcohol at the C2 Position: Sulcatol [64]

Sulcatol (6-methyl-5-hepten-2-ol) is the aggregation pheromone of the ambrosia beetle, *Gnathotrichus sulcatus* [65,66]. The utility of MαNP acid for the preparation of single-enantiomer alcohol was exemplified by the enantioresolution of sulcatol. (*RS*)-Sulcatol was esterified with (*S*)-MαNP acid using EDC and DMAP in $CH_2Cl_2$, yielding (*S*)-MαNP esters **C** and **D** at 43 and 37%, respectively (Figure 8-4). The diastereomeric esters **C** and **D** were readily separable by normal-phase HPLC (Silica SG80, hexane:EtOAc = 9:1; Figure 8-5). Preparative HPLC, using a pre-packed glass column (Kusano C.I.G. HS, silica gel), afforded diastereomerically pure esters **C** and **D**, as the first and second eluted fractions, respectively.

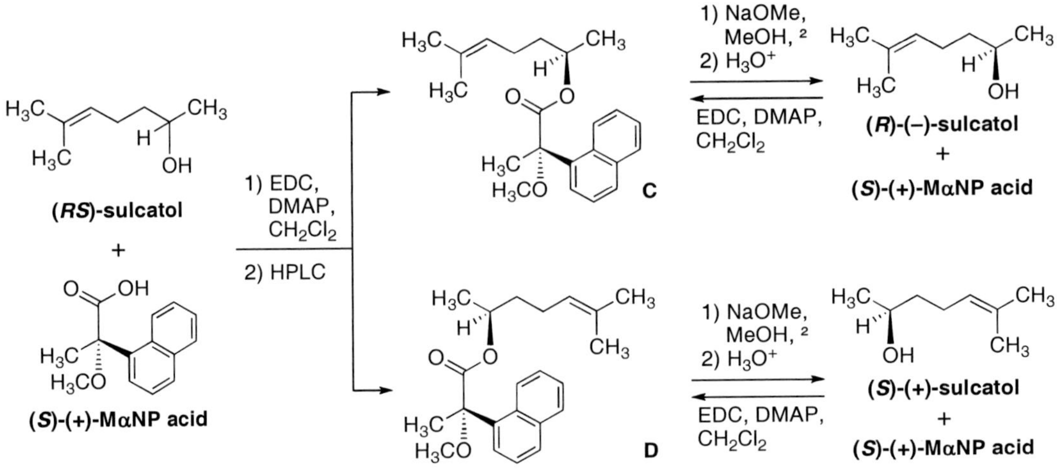


Figure 8-4. Preparation of single-enantiomer sulcatol.

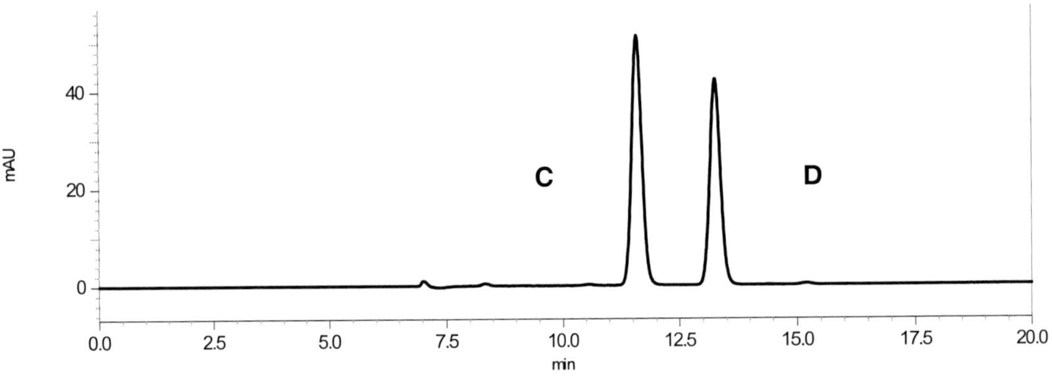


Figure 8-5. HPLC profile of diastereomeric esters formed from (*RS*)-sulcatol and (*S*)-(+)-MαNP acid. Column, Silica SG80 (250 mm × 4.6 mm I.D.); mobile phase, hexane:EtOAc = 9:1; flow rate, 0.5 mL/min; detection, UV 300 nm; $\alpha = 1.31$; $t_0$, 1,3,5-tri-*tert*-butylbenzene.

The $^1$H NMR signals of esters **C** and **D** were assigned using DQF COSY and NOESY spectra. A positive Δδ (= δ**D** – δ**C**) value (+0.27 ppm) was observed at the 1-methyl group of the alcohol moiety, and negative Δδ values were observed in the rest of the alcohol moiety (Figure 8-6); thus, the alcohol stereochemistry of esters **C** and **D** was assigned as *R* and *S*, respectively. Ester **D** and its enantiomer showed identical NMR chemical shifts; thus, Δδ (= δ[(*R*)-MαNP] – δ[(*S*)-MαNP]) for (*R*)-sulcatol is identical with δ**D** – δ**C**. This assignment was confirmed by the specific rotations of sulcatol: the solvolysis of esters **C** and **D** gave (*R*)-(–)- and (*S*)-(+)-sulcatol at 69 and 93%, respectively. Their specific rotations were consistent with previously reported values [65]. Enantiopure (*S*)-(+)-MαNP acid was recovered after solvolysis. To confirm the retention of stereochemistry, (*R*)-(–)- and (*S*)-(+)-sulcatol were esterified with the recovered (*S*)-(+)-MαNP acid, yielding esters **C** and **D**, respectively.

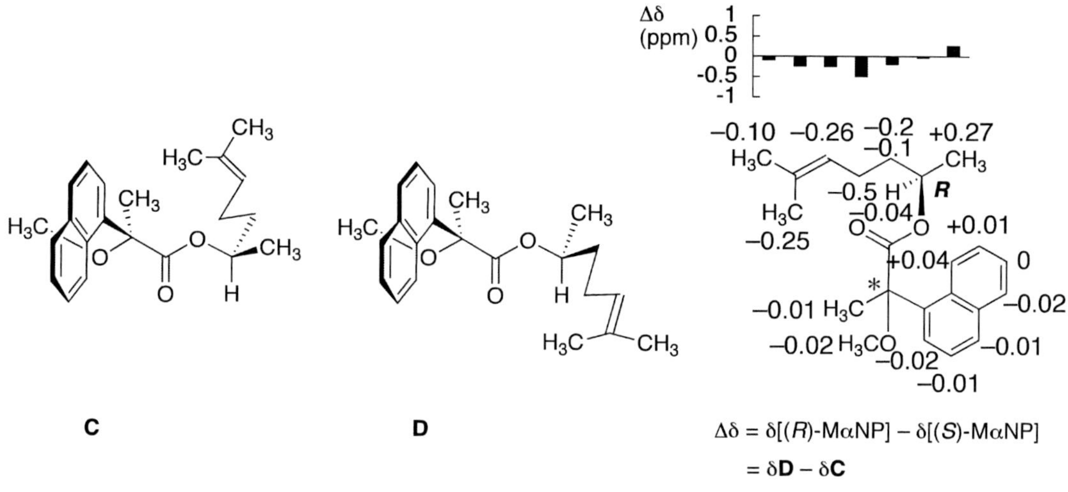

Figure 8-6. Structures of the (*S*)-MαNP esters **C** and **D**, and their Δδ values (800 MHz, $CDCl_3$). The black bars represent the intensities of the Δδ values in the alcohol moiety. The [1]H NMR chemical shifts of ester **D** were identical to those of its enantiomer.

## 8.3. Secondary Alcohol at the C3 Position: 3-Octanol and 1-Octen-3-ol [64]

MαNP acid exhibits superior ability for the enantioresolution of secondary alcohols at the 3-position [67]. Based on this, the enantioresolution of two pheromones, 3-octanol and 1-octen-3-ol, was conducted.

Both enantiomers of 3-octanol are ant pheromones. Fujiwhara and Mori synthesized enantiopure (*R*)-(–)- and (*S*)-(+)-3-octanol from methyl 3-hydroxypentanoate enantiomers [68]. (*RS*)-3-Octanol was condensed with (*S*)-(+)-MαNP acid yielding the diastereomeric esters **E** and **F** at 41 and 43%, respectively (Figure 8-7). The separation factor for these diastereomeric esters was 1.45 using a Silica SG80 column (Figure 8-8).

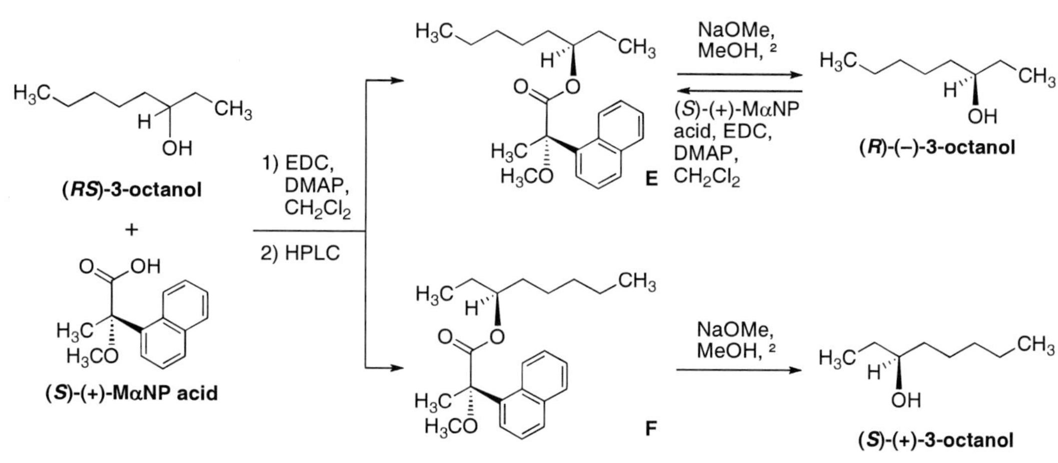

Figure 8-7. Preparation of single-enantiomer 3-octanol.

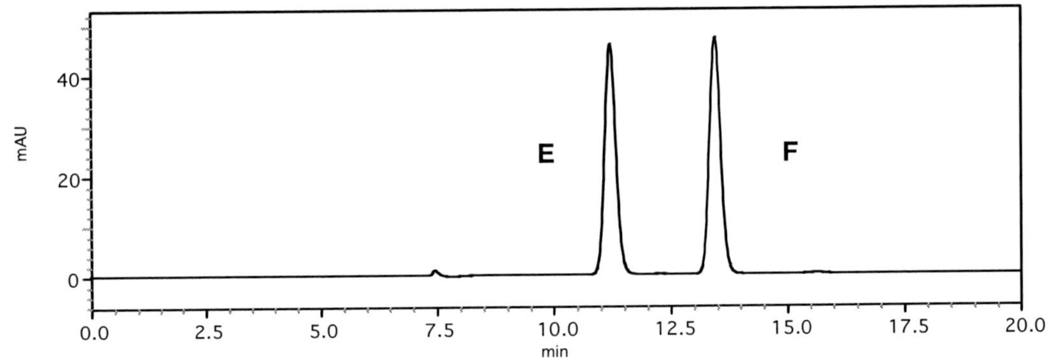


Figure 8-8. HPLC profile of diastereomeric esters formed from (*RS*)-3-octanol and (*S*)-(+)-MαNP acid. Column, Silica SG80 (250 mm × 4.6 mm I.D.); mobile phase, hexane:EtOAc = 9:1; flow rate, 0.5 mL/min; detection, UV 300 nm; $\alpha = 1.45$; $t_0$, 1,3,5-tri-*tert*-butylbenzene.

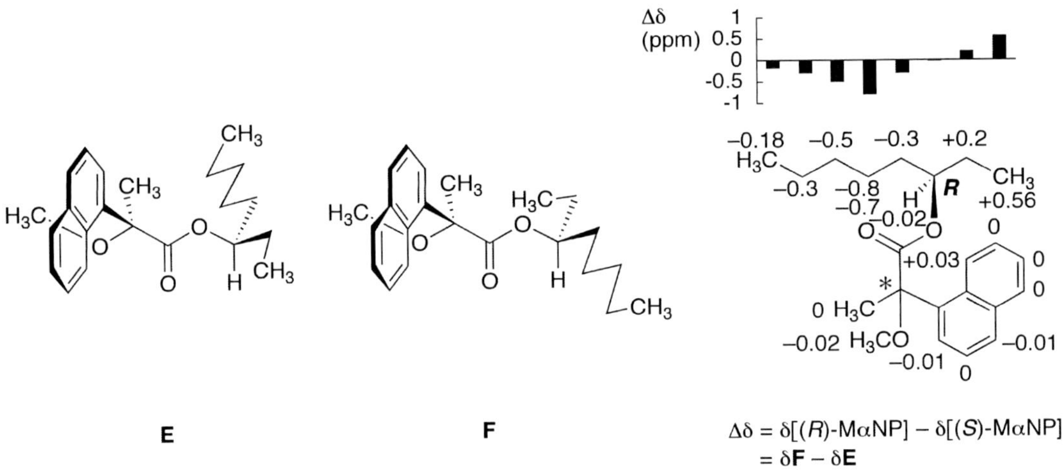


Figure 8-9. Structures of the (*S*)-MαNP esters **E** and **F**, and their Δδ values (600 MHz, $CDCl_3$). The black bars represent the intensities of the Δδ values in the alcohol moiety.

The NMR signals were assigned using DQF COSY, NOESY, and HMBC spectra. Positive Δδ (= δ**F** – δ**E**) values (+0.2 and +0.56 ppm) were observed at the 1-ethyl group. Negative Δδ values were observed in the rest of the alcohol moiety. Thus, the stereochemistry of the alcohol moieties of esters **E** and **F** was assigned as *R* and *S*, respectively (Figure 8-9). The solvolysis of esters **E** and **F** yielded (*R*)-(–)- and (*S*)-(+)-3-octanol at 63 and 54%, respectively. The stereochemistry of the single-enantiomer 3-octanol obtained was confirmed by comparing the specific rotation values with previously reported data [68]. (*R*)-(–)-3-Octanol was condensed with (*S*)-(+)-MαNP acid, yielding ester **E** only; thus, the prepared (*R*)-(–)-3-octanol was enantiopure.

1-Octen-3-ol (mushroom alcohol) is the compound responsible for the smell of mushrooms. Hall et al. identified 1-octen-3-ol from cattle odors as a potent olfactory stimulant and attractant of tsetse flies [69]. In addition, Pierce et al. reported 1-octen-3-ol as an aggregation pheromone for foreign grain beetle, *Ahasverus advena* [70]. Empirical NMR studies lead to the determination of the absolute configurations of both enantiomers as (*S*)-(+)- and (*R*)-(–)-1-octen-3-ol [71]. (*RS*)-1-Octen-3-ol was esterified with (*S*)-(+)-MαNP acid, yielding the diastereomeric esters **G** and **H** at 40 and 43%, respectively (Figure 8-10). The separation factor for esters **G** and **H** was 1.47 using a Silica SG80 column (Figure 8-11).

Their $^{1}$H NMR signals were assigned from the DQF COSY spectra. Positive Δδ (= δ**H** – δ**G**) values were observed at the 1-vinyl group, and negative Δδ values were observed in the rest of the alcohol moieties; thus, the absolute configurations of the alcohol moieties of esters **G** and **H** were assigned as *S* and *R*, respectively (Figure 8-12). The solvolysis of ester **G** yielded (*S*)-(+)-1-octen-3-ol at 77%. Conversely, the (*S*)-(+)-1-octen-3-ol was acylated with the recovered (*S*)-(+)-MαNP acid, yielding ester **G** only. The catalytic hydrogenation of ester **H** yielded ester **F**. The empirical NMR studies of 1-octen-3-ol were linked with the asymmetric synthesis of 3-octanol [68,71].

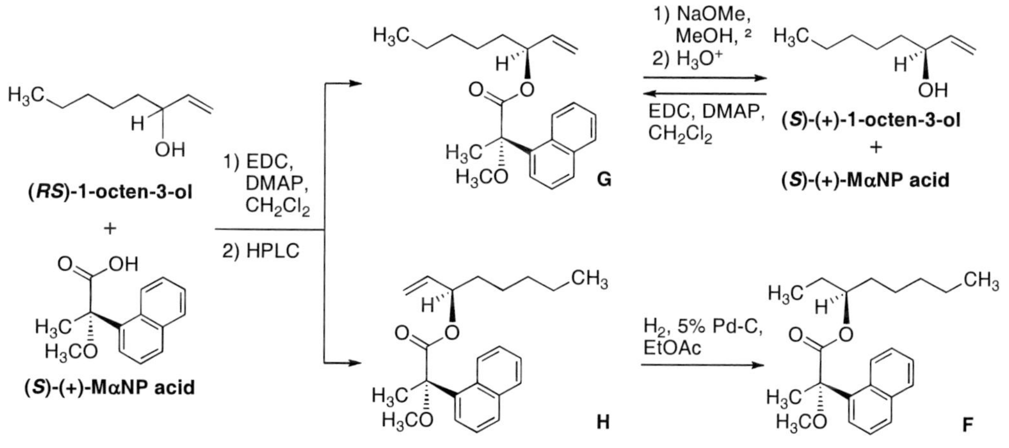


Figure 8-10. Preparation of (*S*)-(+)-1-octen-3-ol.

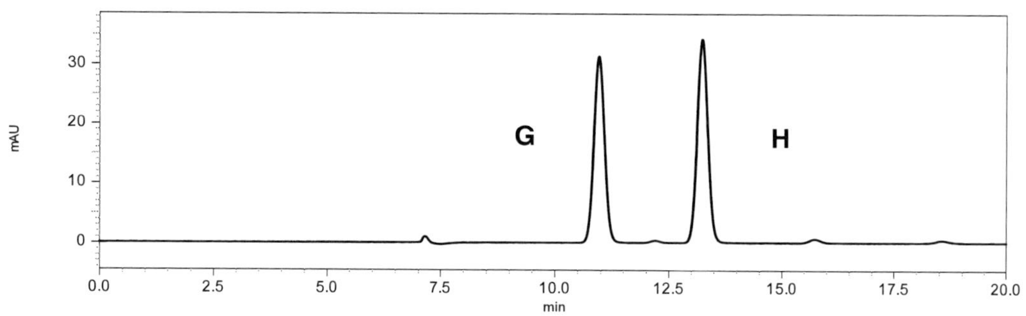


Figure 8-11. HPLC profile of diastereomeric esters formed from (*RS*)-1-octen-3-ol and (*S*)-(+)-MαNP acid. Column, Silica SG80 (250 mm × 4.6 mm I.D.); mobile phase, hexane:EtOAc = 92:8; flow rate, 0.5 mL/min; detection, UV 300 nm; $\alpha = 1.47$; $t_0$, 1,3,5-tri-*tert*-butylbenzene.

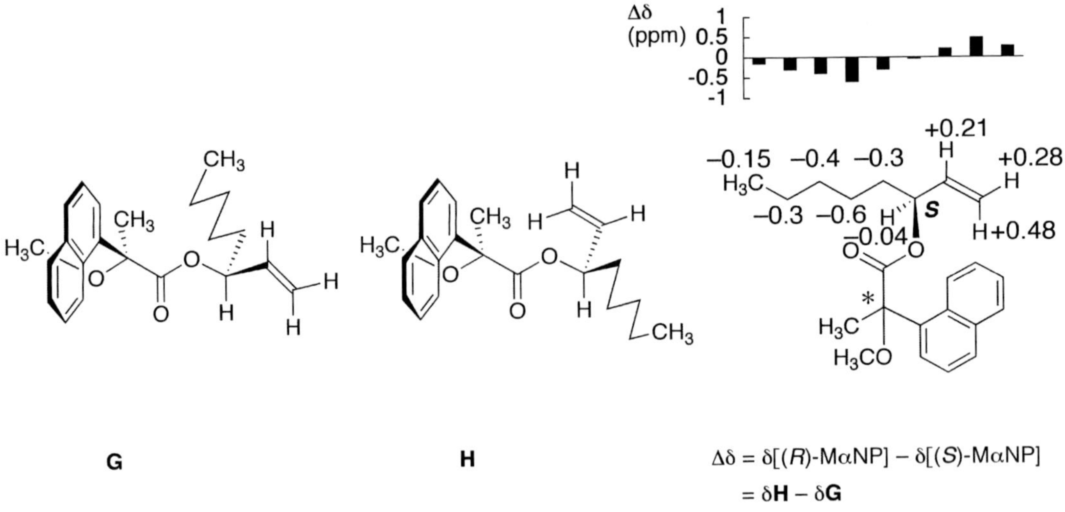


Figure 8-12. Structures of the (*S*)-MαNP esters **G** and **H**, and their Δδ values (500 MHz, $CDCl_3$). The black bars represent the intensities of the Δδ values in the alcohol moiety.

## 8.4. Secondary Alcohol at the C4 Position: 2-Methyl-4-heptanol [63]

Ramirez-Lucas et al. identified 2-methyl-4-heptanol from the volatile extract of the male West Indian sugarcane borer, *Metamasius hemipterus*, as an electroantennographically active compound [72]. Takenaka et al. synthesized (*R*)-(–)- and (*S*)-(+)-2-methyl-4-heptanol from (*S*)- and (*R*)-leucine, respectively [73].

(*RS*)-2-Methyl-4-heptanol was esterified with (*S*)-(+)-MαNP acid, yielding diastereomeric esters **I** and **J**, respectively (1:1 ratio; total yield, 86%; Figure 8-13). Esters **I** and **J** were separated by normal-phase HPLC using a Silica SG80 column ($\alpha$ = 1.07, *R*s = 1.42; Figure 8-14), but better separation was achieved with hexane/THF ($\alpha$ = 1.09, *R*s = 2.28).

The NMR signals of esters **I** and **J** were assigned using DQF COSY, HSQC, and HMBC spectra. Positive Δδ values were observed at the 1-propyl group, and negative Δδ values were observed in the rest of the alcohol moiety; thus, the stereochemistry of the alcohol moieties of esters **I** and **J** were assigned as *R* and *S*, respectively (Figure 8-15). This assignment was confirmed by solvolysis: esters **I** and **J** yielded (*R*)-(–)- and (*S*)-(+)-2-methyl-4-heptanol, respectively. Their specific rotations were consistent with the previous report [73]. Enantiopure (*S*)-(+)-MαNP acid was also recovered.

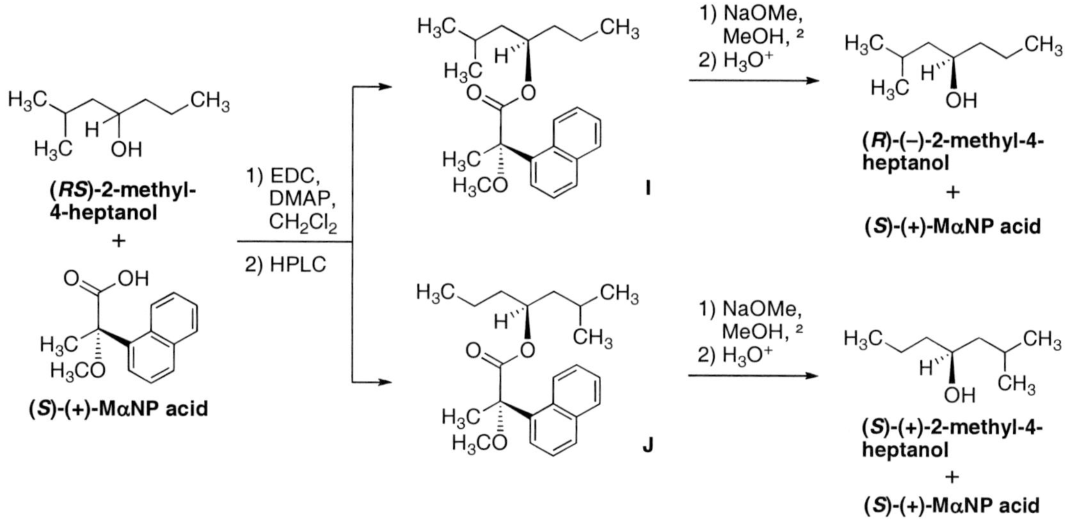


Figure 8-13. Preparation of single-enantiomer 2-methyl-4-heptanol.

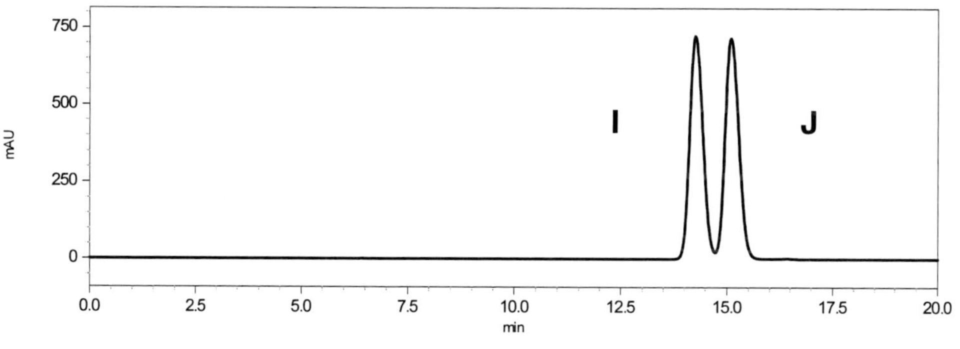


Figure 8-14. HPLC profile of diastereomeric esters formed from (*RS*)-2-methyl-4-heptanol and (*S*)-(+)-MαNP acid. Column, Silica SG80 (250 mm × 4.6 mm I.D.); mobile phase, hexane:EtOAc = 98:2; flow rate, 1 mL/min; detection, UV 300 nm; $\alpha = 1.07$, $Rs = 1.42$; $t_0$, 1,3,5-tri-*tert*-butylbenzene.

Then, (*R*)-(−)- and (*S*)-(+)-2-methyl-4-heptanol were esterified with (*R*)-MTPA, yielding esters **K** and **L**. The MTPA esters **K** and **L** were inseparable in normal-phase HPLC, and the absolute values of their chemical shift differences in the alcohol moiety (−0.16 to +0.17 ppm; Figure 8-16), determined based on COSY, HSQC, and HMBC spectra, were clearly smaller than those of the MαNP esters **I** and **J** (−0.88 to +0.7 ppm; Figure 8-15).

As discussed above, the proposed conformations of MαNP esters were confirmed by the enantioresolution of pheromones. Mori unambiguously determined the absolute configurations and specific rotations of pheromones and other bioregulators by asymmetric synthesis [1,74]. In addition, the simple structures of pheromones were useful for evaluating the properties of chiral resolving agents. The MαNP method is useful for the rapid preparation of single-enantiomer biofunctional molecules, and the determination of their stereochemistry by the $^1$H NMR anisotropy method.

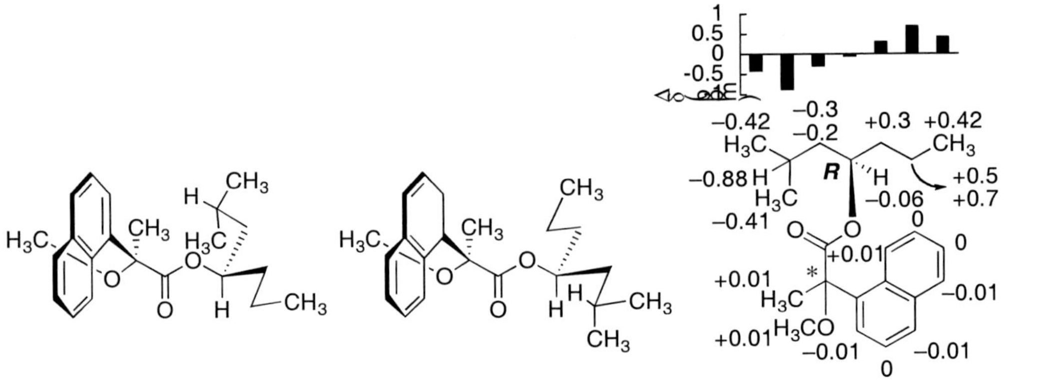

**I** **J**

$\Delta\delta = \delta[(R)\text{-M}\alpha\text{NP}] - \delta[(S)\text{-M}\alpha\text{NP}] = \delta\mathbf{J} - \delta\mathbf{I}$

Figure 8-15. Structures of the (*S*)-MαNP esters **I** and **J**, and their Δδ values (500 MHz, $CDCl_3$). The black bars represent the intensities of the Δδ values in the alcohol moiety.

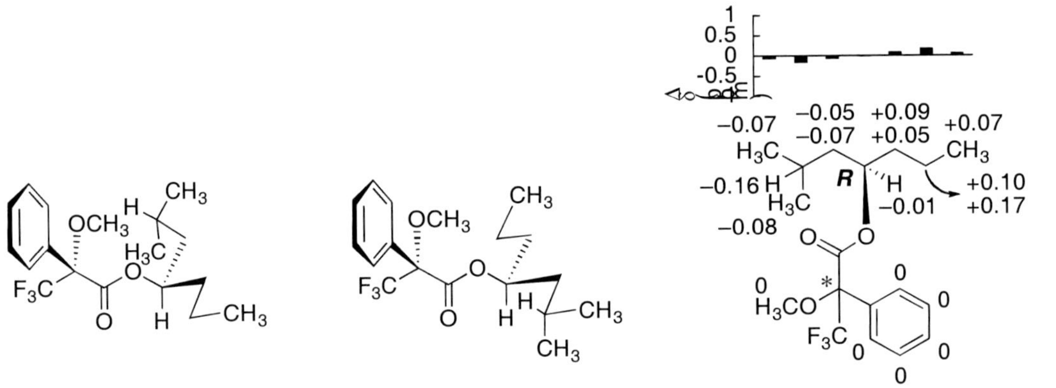

**K** **L**

$\Delta\delta = \delta(S)\text{-MTPA}] - \delta[(R)\text{-MTPA}] = \delta\mathbf{L} - \delta\mathbf{K}$

Figure 8-16. Structures of the (*R*)-MTPA esters **K** and **L**, and their Δδ values (800 MHz, $CDCl_3$). The black bars represent the intensities of the Δδ values in the alcohol moiety. The prochiral methylene protons of the alcohol moiety are not specified: the chemical shift differences between the higher and lower ones are represented.

## 9. M9PP Acid [52,62,64]

The chiral resolving agent M9PP acid exhibited similar properties to MαNP acid.

(*RS*)-M9PP acid was synthesized from methyl pyruvate in three steps (Figure 9-1), and then esterified with L-(−)-menthol, yielding esters **C** and **D** [62]. Esters **C** and **D** were readily separated by normal-phase HPLC (Figure 9-2).

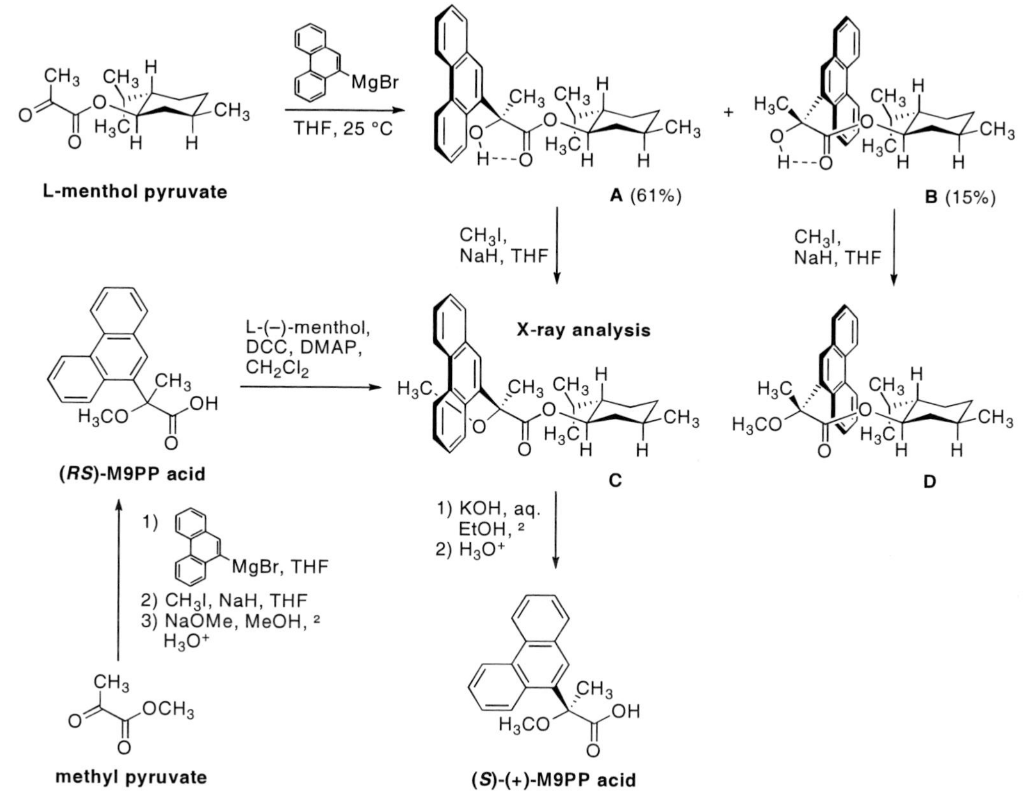


Figure 9-1. Synthesis of single-enantiomer M9PP acid.

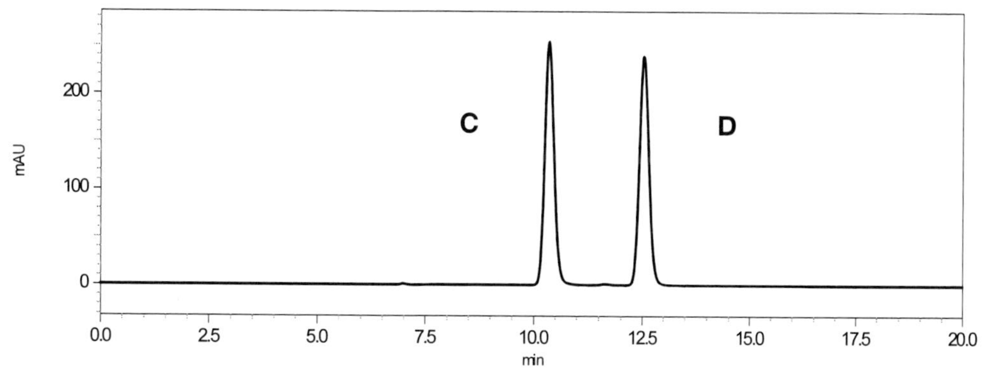


Figure 9-2. HPLC profile of L-menthol M9PP esters.
Column, Silica SG80 (250 mm × 4.6 mm I.D.); mobile phase, hexane:EtOAc = 9:1; flow rate, 0.5 mL/min; detection, UV 300 nm; $\alpha = 1.64$; $t_0$, 1,3,5-tri-*tert*-butylbenzene.

The first eluted ester **C** was crystallized from EtOAc with solvent removal in a draft chamber, and subjected to X-ray crystallography (Figure 9-3) [62]. The absolute configuration of the M9PP moiety of ester **C** was determined as *S*, using L-menthol moiety as an internal standard. Solvolysis of ester **C** gave (*S*)-(+)-M9PP acid. This crystal structure provided important information on conformation; specifically, the hydrogen atom at the α-position of the alcohol moiety and the carbonyl group were *syn*, the carbonyl group and the methoxy groups were *syn*, and the methyl group of the M9PP moiety and the hydrogen atom at the 10-position of the phenanthryl group were also *syn*.

Next, the α-hydroxy esters **A** and **B** were synthesized at a 4:1 ratio by stereoselective Grignard addition (Prelog's method) [52]. The methylation of esters **A** and **B** yielded the M9PP esters **C** and **D**, respectively. As expected, the Grignard reagent, 9-phenanthrylmagnesium bromide, selected the *Si*-face of the ketone carbonyl group.

The NMR signals of esters **A**, **B**, **C**, and **D** were assigned using 2D NMR spectra (DQF COSY, HSQC, and HMBC; Figure 9-4). Their chemical shift differences Δδ (= δ*R* – δ*S*) are shown in Figure 9-5. These Δδ values were similar to those for the 1-naphthyl analogs (i.e., HαNP and MαNP esters) [52].

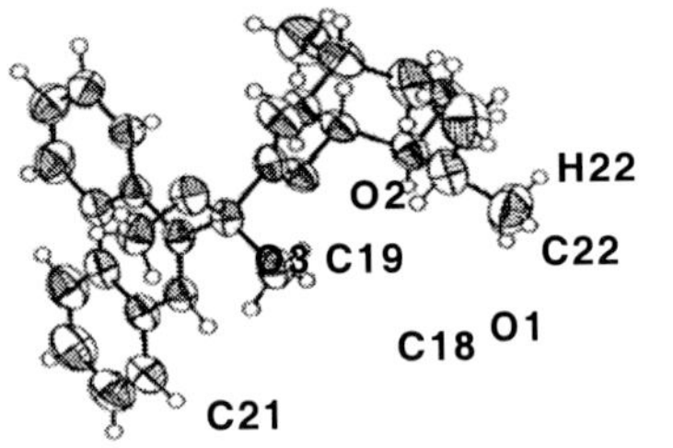


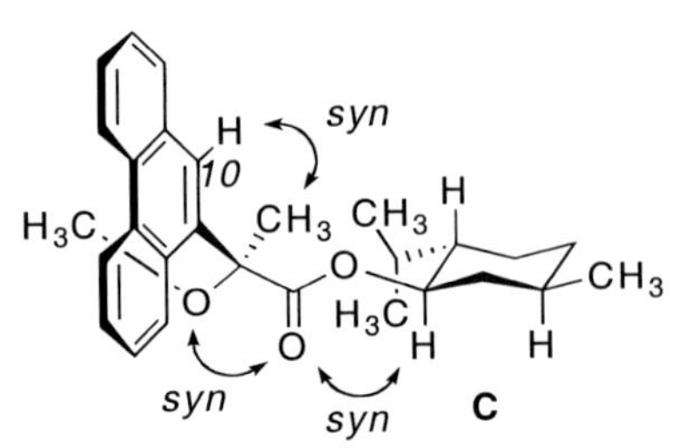


Dihedral angles

| | |
|---|---|
| H22-C22-O1-C19 | 6.6° |
| C22-O1-C19-O2 | –6.8° |
| O2-C19-C18-O3 | 11.8° |
| C19-C18-O3-C21 | 169.5° |

Figure 9-3. Stereochemistry of the (*S*)-M9PP ester **C** determined by X-ray crystallography. X-ray crystallography was performed by Prof. N. Harada at Tohoku University.

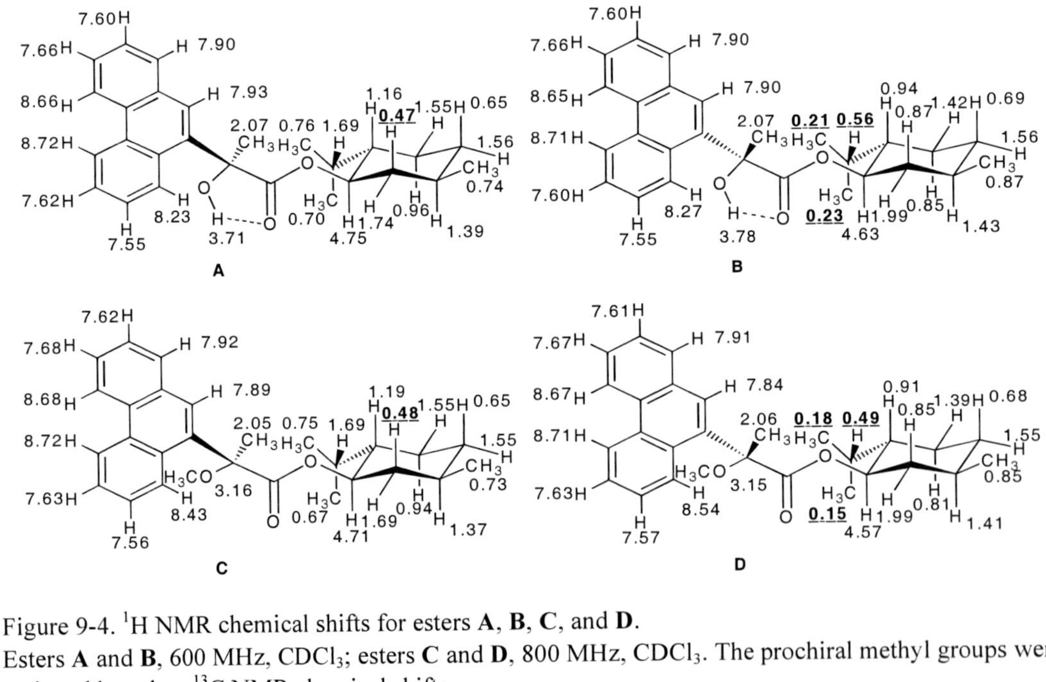


Figure 9-4. [1]H NMR chemical shifts for esters **A**, **B**, **C**, and **D**.
Esters **A** and **B**, 600 MHz, $CDCl_3$; esters **C** and **D**, 800 MHz, $CDCl_3$. The prochiral methyl groups were assigned based on [13]C NMR chemical shifts.

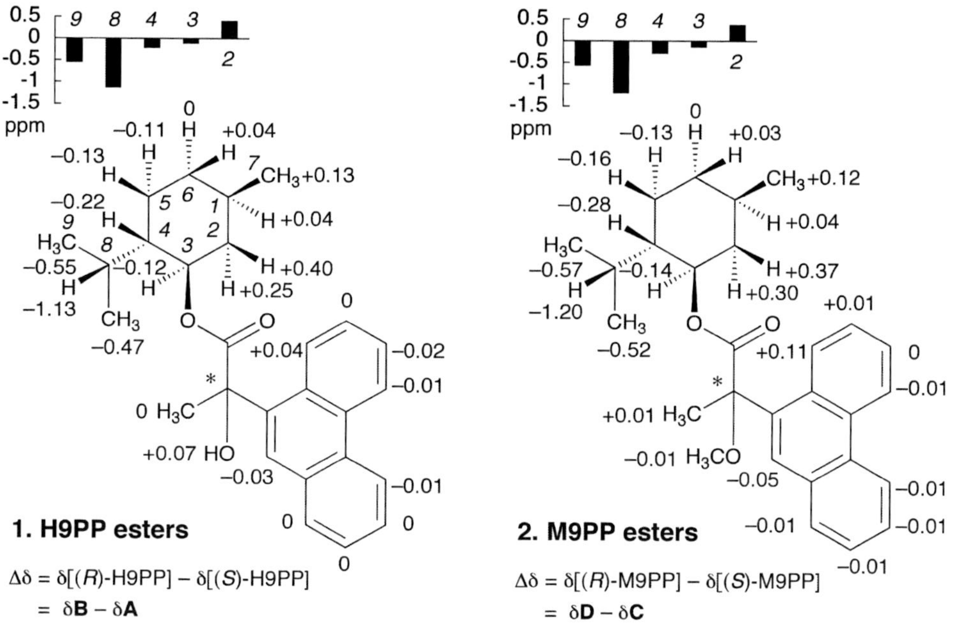


Figure 9-5. Δδ (= δ$R$ – δ$S$) values for L-menthol esters in $^1$H HMR (600 and 800 MHz, $CDCl_3$). The black bars represent the intensities of the Δδ values at the 2-, 3-, 4-, 8-, and 9-positions in the menthol moieties. Abbreviation: H9PP acid, 2-hydroxy-2-(9-phenanthryl)propanoic acid.

Similar to MαNP acid, M9PP acid can also be used for the enantioresolution of alcohols [64]. (*S*)-(+)-M9PP acid was used to resolve (*RS*)-sulcatol [52], (*RS*)-3-octanol [62], and (*RS*)-1-octen-3-ol (Figure 9-6) [64]. Esters **E**, **G**, and **I** were eluted first, followed by esters **F**, **H**, and **J** in normal-phase HPLC with hexane/EtOAc. The stereochemistry of the diastereomeric esters **E**–**J** was confirmed by the specific rotations of single-enantiomer alcohols recovered by solvolysis.

The separation factors for the diastereomeric esters **E**–**J** are summarized in Table 9-1. The chiral resolving ability of M9PP acid was slightly superior to that of MαNP acid in normal-phase HPLC [64]. This difference is important for difficult cases [e.g., (*RS*)-citronellol]. The Δδ (= δ[(*R*)-M9PP] – δ[(*S*)-M9PP]) values for the M9PP esters **E**–**J** were almost identical to those of MαNP esters (Figure 9-6).

The conformations of MαNP and M9PP esters shown in Figure 9-7 could explain these results. The aromatic hydrogen atoms at the 8-positions of 1-naphthyl and 9-phenanthryl moieties form the intramolecular bifurcated hydrogen bonds CH···O [51,75] with the carbonyl and methoxy oxygen atoms. Harada proposed this phenomenon from the crystalline-state conformational analysis of MαNP esters, observed by the X-ray crystallography [51]. The aromatic ring **X** of the M9PP moiety did not overlap with the alcohol moiety; therefore, the MαNP and M9PP acid exhibited similar resolving properties in HPLC and $^1$H NMR.

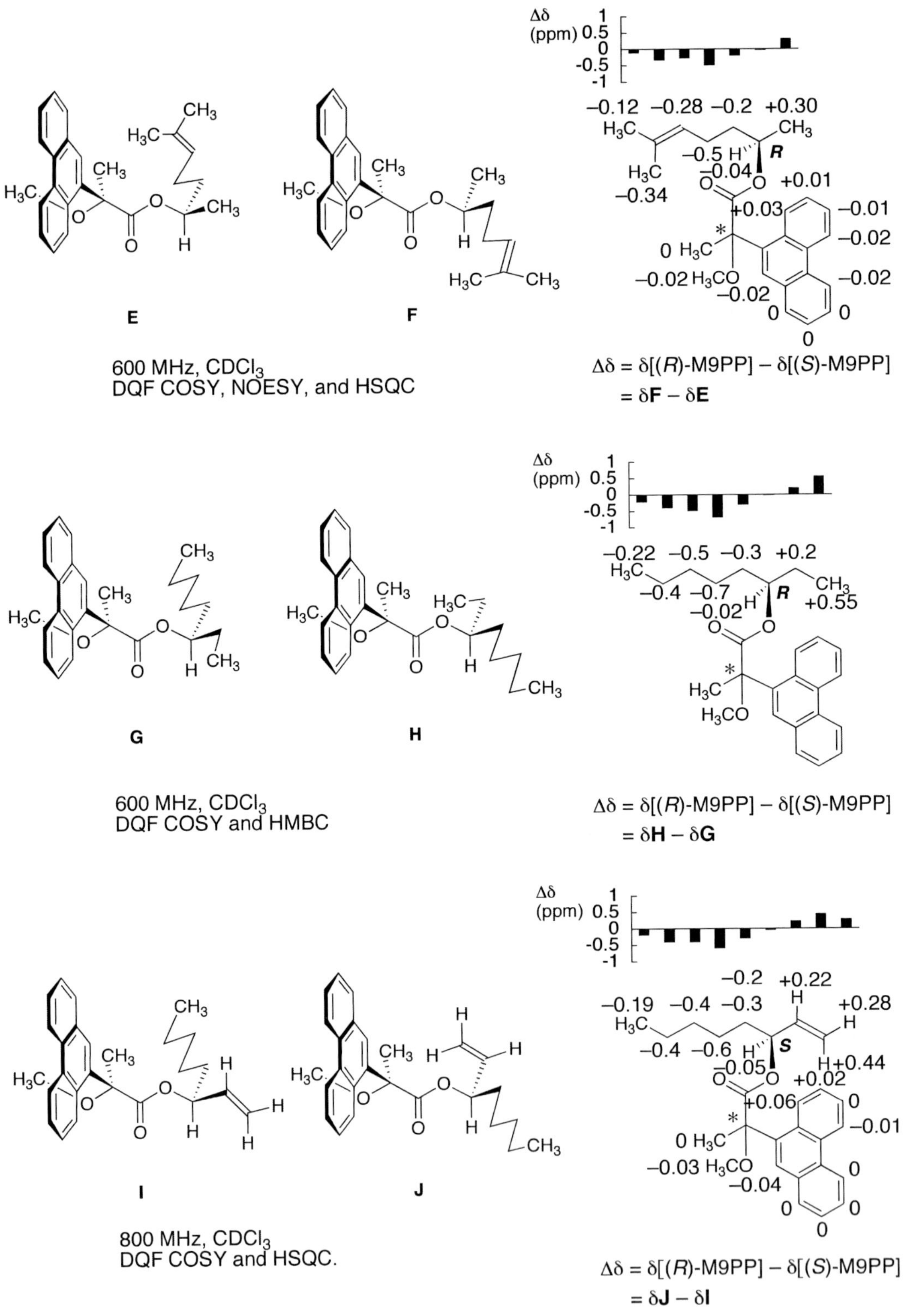


Figure 9-6. Structures of the (*S*)-M9PP esters **E**–**J** and their Δδ values.
The [1]H NMR chemical shifts of esters **F**, **H**, and **J** were identical to those of their enantiomers.

**Table 9-1. Separation factors for diastereomeric MαNP and M9PP esters**

| | sulcatol (6-mehyl-5-hepten-2-ol) | 3-octanol | 1-octen-3-ol |
|---|---|---|---|
| MαNP acid | $\alpha = 1.31$ | 1.45 | 1.47 |
| M9PP acid | $\alpha = 1.37$ | 1.57 | 1.52 |

Column, Silica SG80 (250 mm × 4.6 mm I.D.); mobile phase, hexane/EtOAc; detection, UV 300 nm; $t_0$, 1,3,5-tri-*tert*-butylbenzene.

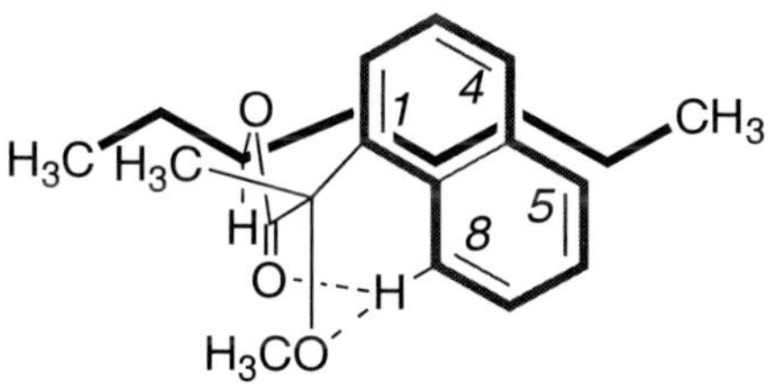

**(*S*)-MαNP ester**

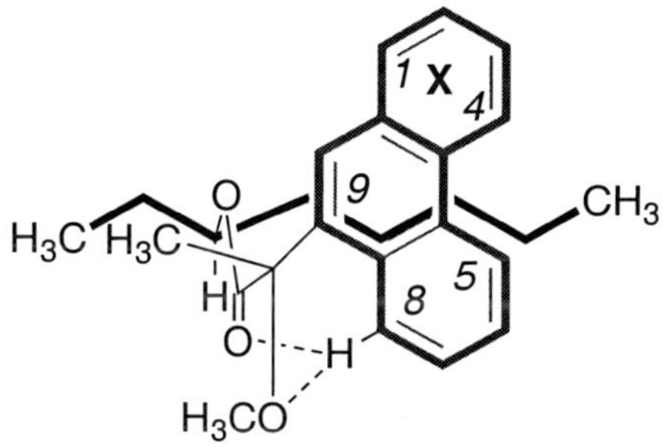

**(*S*)-M9PP ester**

Figure 9-7. Conformations of the (*S*)-MαNP and (*S*)-M9PP esters.

In terms of stability, MαNP acid is superior to M9PP acid. M9PP acid and its derivatives were slowly degraded and became brown-colored. However, our studies using M9PP acid have provided important information regarding the conformations and mechanisms of chiral resolving agents.

# 10. MβNP Acid

MβNP acid had very distinct properties from those of MαNP and M9PP acids in $^1$H NMR.

## 10.1. Prelog's Method [76]

The Grignard reaction of 2-naphthylmagnesium bromide and L-menthol pyruvate yielded the α-hydroxy esters **A** and **B** at 15 and 10%, respectively (Figure 10-1). Their NMR signals were assigned using 2D NMR spectra (COSY, CH COSY, HMBC, HOHAHA, and NOESY; Figure 10-2). Considering the intramolecular hydrogen bonds, the stereochemistry of esters **A** and **B** was assigned using the $^1$H NMR anisotropy method (Figure 10-2, 10-3). As discussed,

Grignard reagents select the *Si*-face of the ketone carbonyl group; the above assignment was confirmed by reaction selectivity.

Solvolysis of ester **A** yielded (*S*)-(+)-2-hydroxy-2-(2-naphthyl)propanoic acid [(*S*)-(+)-HβNP acid]. The methylation of esters **A** and **B** yielded esters **C** and **D**, respectively (Figure 10-1). The solvolysis of esters **C** and **D** yielded (*S*)-(+)- and (*R*)-(–)-MβNP acid, respectively. The $^1$H NMR chemical shifts of esters **C** and **D** are presented in Figure 10-2. The Δδ (= δ**D** – δ**C**) values suggested that the $^1$H NMR anisotropy method was not applicable (Figure 10-3). In contrast, 2NMA is reported to be a powerful $^1$H NMR anisotropic reagent [25,26]. The conformational difference between MβNP esters and 2NMA esters is unclear. The separation factor ($\alpha$) for MβNP esters **C** and **D** was 1.47 in normal-phase HPLC (Figure 10-4; cf., MαNP esters, $\alpha$ = 1.73). These results indicate that the chiral resolving ability of MβNP acid was inferior to that of MαNP and M9PP acids. However, it will be important to clarify the features of MβNP acid for the development of chiral resolving agents and understanding their chiral recognition mechanisms.

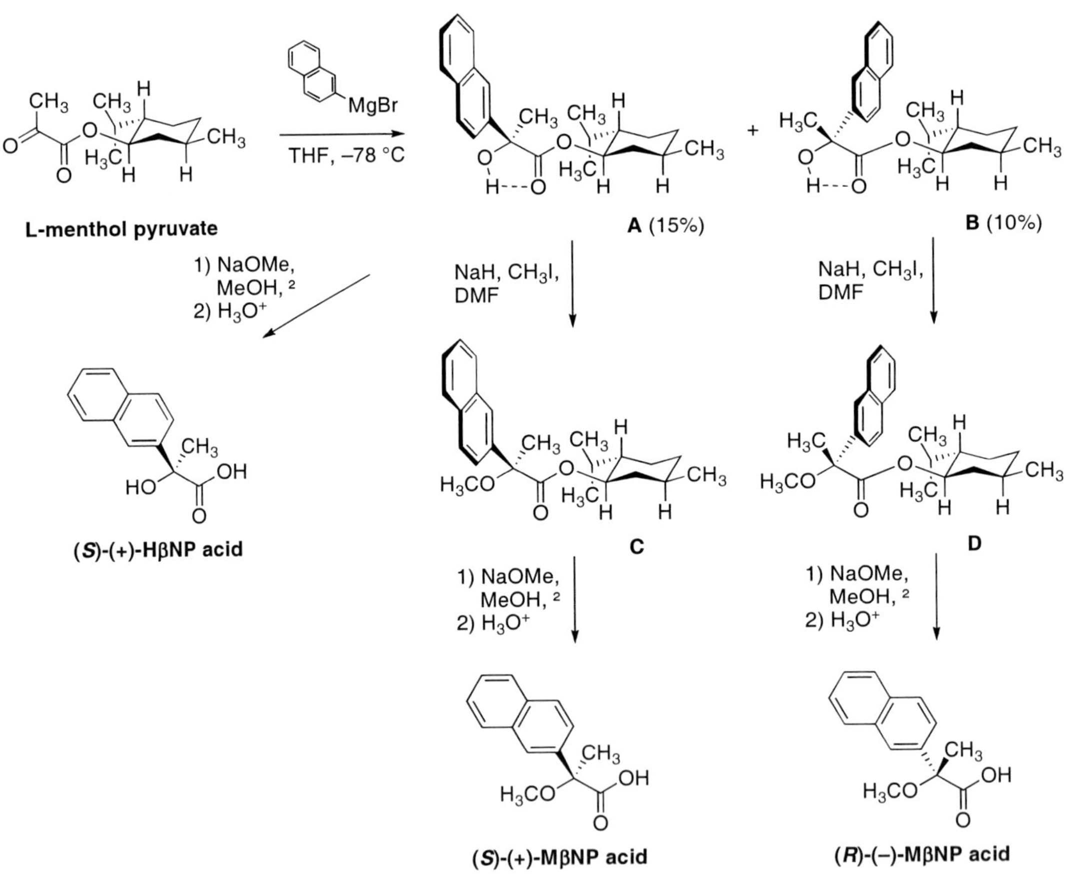


Figure 10-1. Synthesis of single-enantiomer MβNP acid.

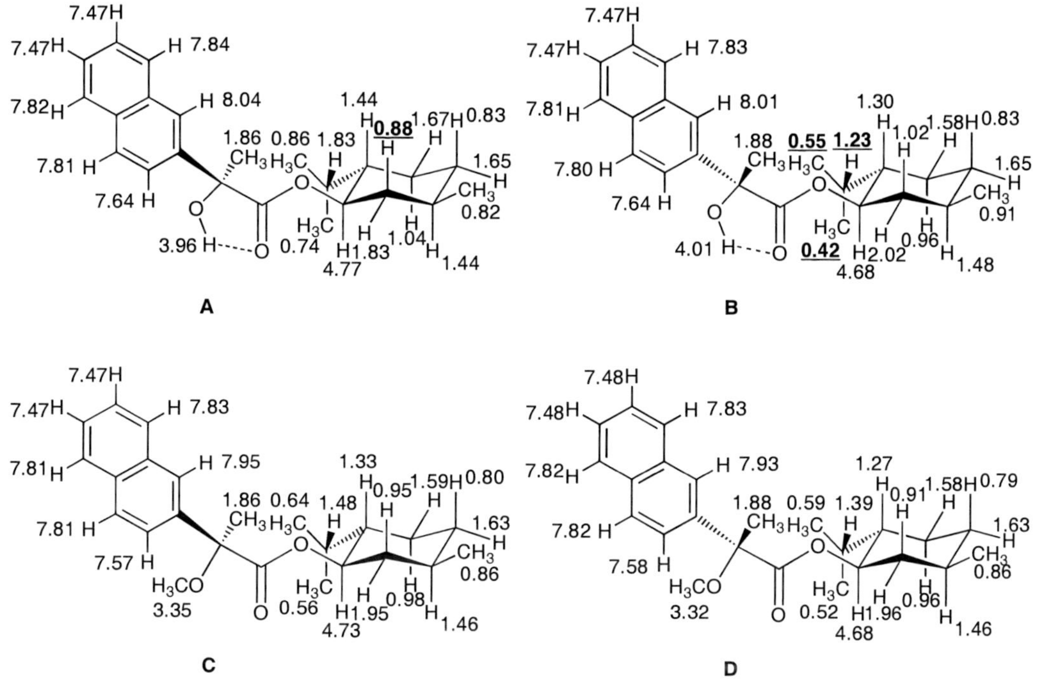


Figure 10-2. [1]H NMR chemical shifts for esters **A**, **B**, **C**, and **D** (600 MHz, $CDCl_3$). The prochiral methyl groups were assigned based on NOESY spectra. The conformations of esters **C** and **D** were not identified. The NMR spectra of esters **A**, **B**, **C**, and **D** were determined by Dr. S. Hiradate (NIAES).

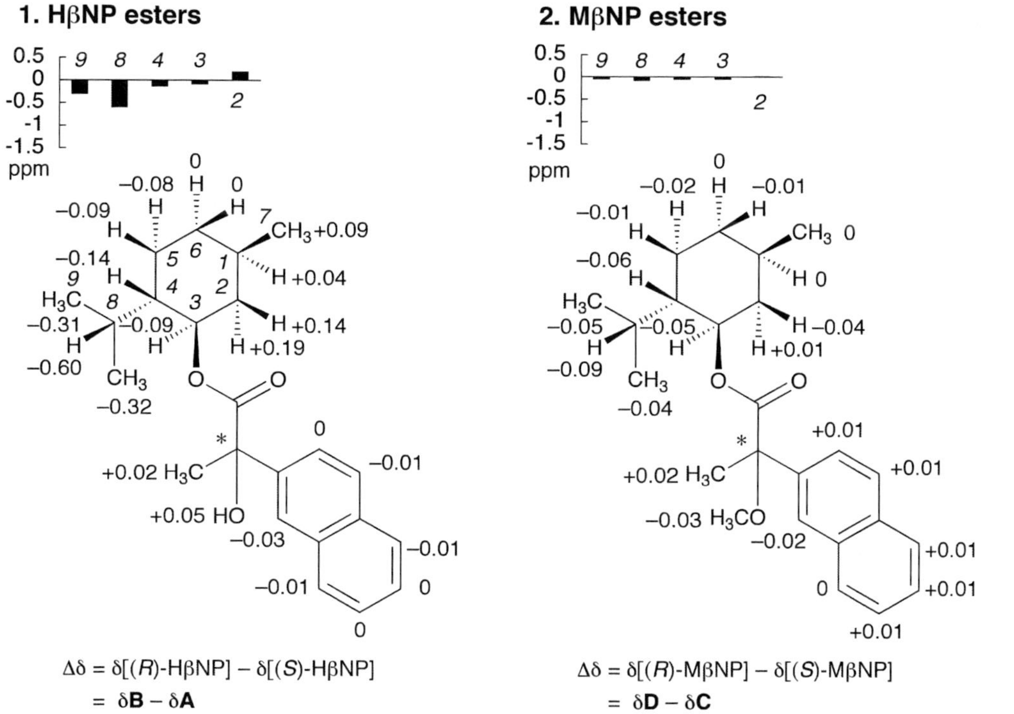


Figure 10-3. Δδ (= δ$R$ – δ$S$) values for L-menthol esters in [1]H HMR (600 MHz, $CDCl_3$).
The black bars represent the Δδ values at the 2-, 3-, 4-, 8-, and 9-positions in the menthol moieties.

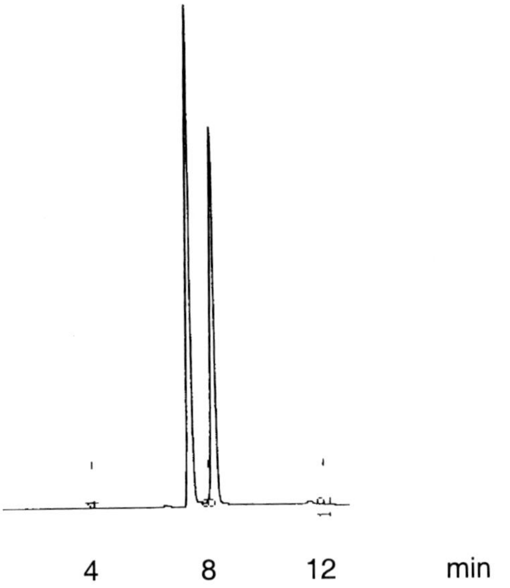


Figure 10-4. HPLC profile of L-menthol MβNP esters.
Column, Silica SG80 (250 mm × 4.6 mm I.D.); mobile phase, hexane:EtOH = 199:1; flow rate, 0.5 mL/min; detection, UV 260 nm; $\alpha = 1.47$; $t_0$, octadecylbenzene.

## 10.2. The PGME Method [55, 77]

In 2000, Yabuuchi and Kusumi developed a method that could be used to determine the stereochemistry of α-methoxy carboxylic acids using phenylglycine methyl ester (PGME; Figure 10-5) [78]. In (*S*)-PGME amide **X**, the α-methoxy group and NH were *syn*, and the NH and ester carbonyl group were *syn*. The same was true for (*R*)-PGME amide **Y**. The $R_2$ substituent of amide **X** and the $R_1$ substituent of amide **Y** are shielded by the phenyl groups. Thus, the absolute configurations of the α-methoxycarboxylic acids could be determined from the Δδ (= δ[(*S*)-PGME] – δ[*R*)-PGME]) values. Using the PGME method, we confirmed the absolute configuration of MβNP acid.

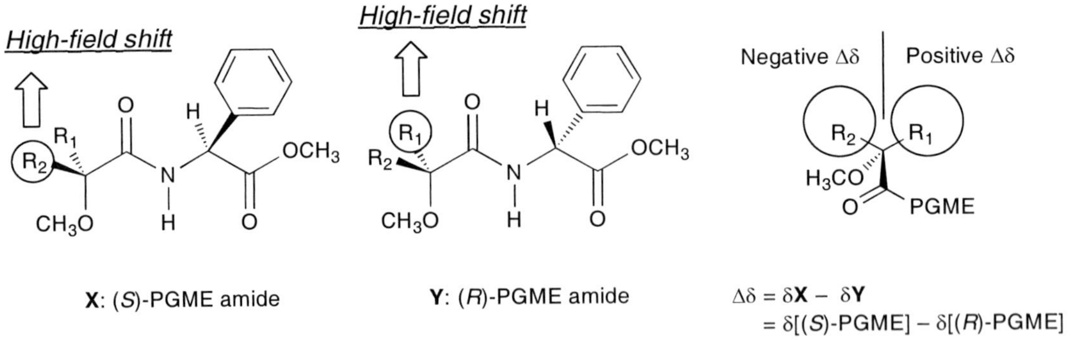


Figure 10-5. Preferred conformations of (*S*)- and (*R*)-PGME amides and their chemical shift differences in $^1$H NMR.

First, the PGME method was applied to (*S*)-(+)-MαNP acid. (*S*)-(+)-MαNP acid was condensed with (*S*)- and (*R*)-PGME, yielding amides **E** and **F**, respectively (Figure 10-6). Their NMR signals were assigned using 2D NMR spectra (DQF COSY, HSQC, and HMBC). Negative Δδ (= δ**E** – δ**F**) values were observed at the 1-naphthyl group. In contrast, a positive Δδ value (+0.09 ppm) was observed at the methyl group of the MαNP moiety. This result suggests that the PGME method is applicable to carboxylic acid possessing an aromatic group.

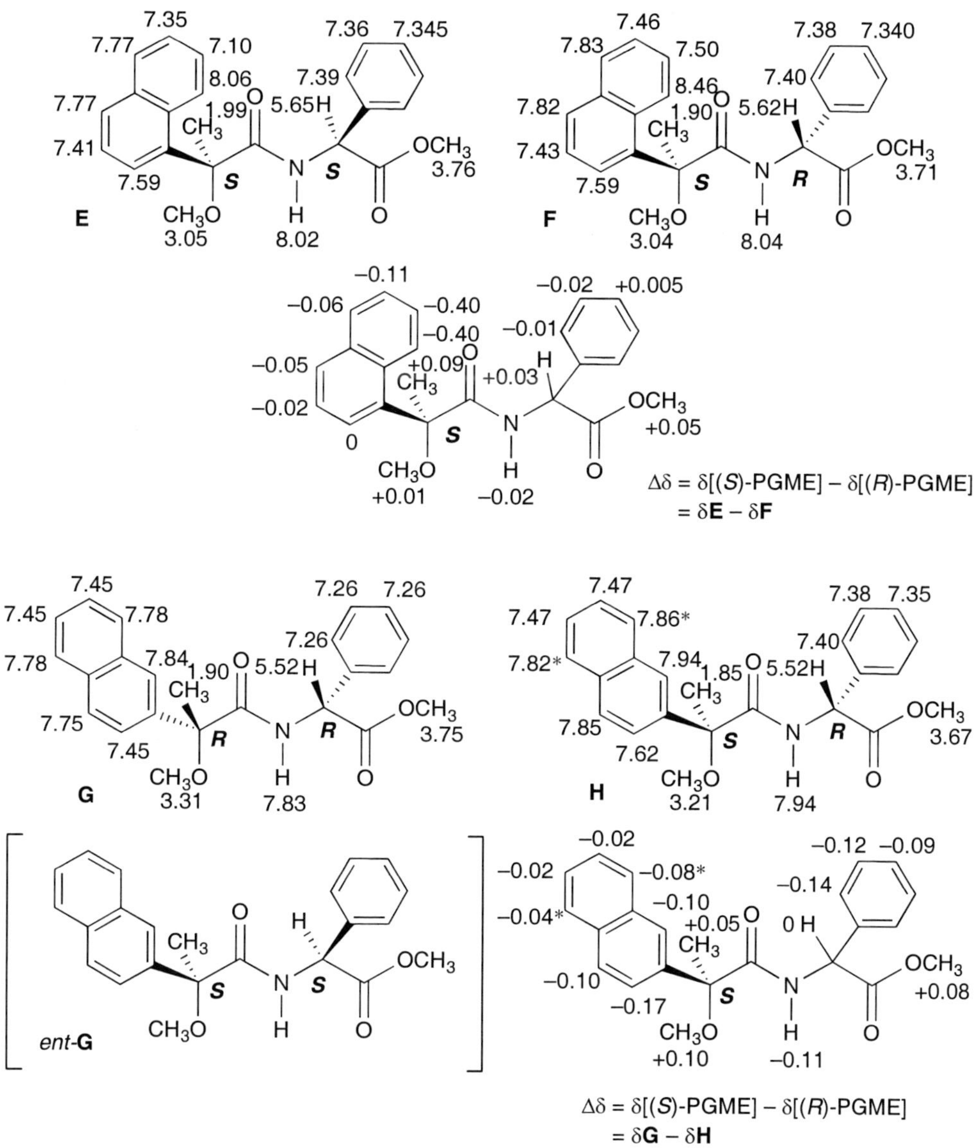


Figure 10-6. Application of PGME method for MαNP and MβNP acids (600 MHz, $CDCl_3$). Asterisks (*) indicate exchangeable signals.

Next, the PGME method was applied to MβNP acid. (*R*)-(–)- and (*S*)-(+)-MβNP acid were condensed with (*R*)-PGME, yielding amides **G** and **H**, respectively. Amide **G** and its enantiomer *ent*-**G** had identical NMR spectra; thus, the Δδ (= δ[(*S*)-PGME] – δ[(*R*)-PGME] = δ(*ent*-**G**) – δ**H**) values for (*S*)-(+)-MβNP acid were identical to δ**G** – δ**H**. The NMR signals of amides **G** and **H** were assigned using COSY, CH COSY, and HMBC spectra. Negative Δδ values were observed at the 2-naphthyl group, and a positive Δδ value (+0.05 ppm) was observed at the methyl group of the MβNP moiety. These results indicate that the stereochemical assignment of MβNP acid was correct.

In 2008, Sekiguchi et al. confirmed the absolute configuration of (*R*)-(–)-MβNP acid by X-ray crystallography [79].

## Conclusion

The stereochemistry of a sex pheromone isolated from the tea tussock moth and hydroxy-fatty acids isolated from the resinous secretion of the lac insect was determined via syntheses and NMR. The novel chiral resolving agents, MαNP acid, MβNP acid, and M9PP acid, were synthesized by asymmetric induction using enantiopure L-(–)-menthol. Grignard addition selected the *Si* face of the pyruvate ketone carbonyl group. The absolute configurations of the chiral resolving agents and the conformations of their esters were determined by X-ray crystallography and NMR. MαNP acid exhibited a superior ability for the enantioresolution of alcohols. The diastereomeric MαNP esters were readily separable by normal-phase HPLC, and the solvolysis of these esters yielded single-enantiomer alcohols. M9PP acid exhibited a slightly better resolving ability than MαNP acid. The $^{1}$H NMR anisotropy method using MαNP acid or M9PP acid was useful in determining the stereochemistry of diastereomeric esters. Because of its stability and handling properties, MαNP acid is a promising agent for future stereochemical studies.

## Acknowledgments

We are grateful to Former Professor N. Harada of Tohoku University, and the members of his laboratory. We also thank Dr. S. Hiradate (NIAES), Ms. I. Maeda (NFRI), Dr. K. Tanaka (NIFTS), and Mr. K. Esaki (NIHS) for NMR measurements and useful discussions. A.I. is grateful to Dr. H. Kato (NIAS), Dr. Y. Magoshi (NIAS), Dr. Y. Goto (NIAS), and Dr. T. Hata (NIAS) for their encouragement.

The assignment of absolute configurations by NMR is a research area developed largely by Prof. H. S. Mosher (Stanford University), Prof. B. M. Trost (Stanford University), Prof. Emeritus T. Kusumi (Tokushima University), Prof. R. Riguera (Universidad de Santiago de Compostela), and Dr. I. Ohtani.

The asymmetric synthesis of insect pheromones is a research area developed largely by Prof. Emeritus K. Mori (the University of Tokyo). We could unambiguously confirm the stereochemistry from his specific rotation data.

## REFERENCES

[1] Mori, K. *Chemical Synthesis of Bioactive Natural Products – How to Synthesize Biofunctional Molecules*; Shokabo: Tokyo, Japan, 1995 (in Japanese).

[2] Takahashi, S. *Insect Behavior*; Bioscience Series; Kagakudojin: Kyoto, Japan, 1987 (in Japanese).

[3] Wakamura, S.; Yasuda, T.; Ichikawa, A.; Fukumoto, T.; Mochizuki, F. *Appl. Entomol. Zool.* 1994, *29*, 403–411.

[4] Ogata, K. *Jpn. J. Sanit. Zool.* 1958, *9*, 116–129 (in Japanese with English abstract).

[5] Nelson, D. R.; Fatland, C. L. *Insect Biochem. Mol. Biol.* 1992, *22*, 99–110.

[6] Nelson, D. R.; Fatland, C. L. *Insect Biochem. Mol. Biol.* 1992, *22*, 111–123.

[7] Ichikawa, A.; Yasuda, T.; Wakamura, S. *J. Chem. Ecol.* 1995, *21*, 627–634.

[8] Wakamura, S.; Ichikawa, A.; Yasuda, T.; Arakaki, N.; Fukumoto, T. *Appl. Entomol. Zool.* 1996, *31*, 623–625.

[9] Windholz, M.; Budavari, S.; Blumetti, B. F.; Otterbein, E. S. *The Merck Index, Tenth Edition*; Merck & Co., INC.: Rahway, NJ, 1983, p332.

[10] Ishitsuka, M. O.; Kusumi, T.; Ichikawa, A.; Kakisawa, H. *Phytochemistry* 1990, *29*, 2605–2610.

[11] Zheng, G.-C.; Ichikawa, A.; Ishitsuka, M. O.; Kusumi, T.; Yamamoto, H.; Kakisawa, H. *J. Org. Chem.* 1990, *55*, 3677–3679.

[12] Kusumi, T.; Igari, M.; Ishitsuka, M. O.; Ichikawa, A.; Nakayama, N.; Kakisawa, H. *J. Org. Chem.* 1990, *55*, 6286–6289.

[13] Iguchi, K.; Kitade, M.; Yamada, Y.; Ichikawa, A.; Ohtani, I.; Kusumi, T.; Kakisawa, H. *Chem. Lett.* 1991, 319–322.

[14] Mori, K.; Harada, H.; Zagatti, P.; Cork, A.; Hall, D. R. *Liebigs Ann. Chem.* 1991, 259–267.

[15] Horiike, M.; Tanouchi, M.; Hirano, C. *Agric. Biol. Chem.* 1978, *42*, 1963–1965.

[16] Hoffman, J. M., JR.; Schlessinger, R. H. *J. Chem. Soc., Chem. Commun.* 1971, 1245–1246.

[17] Zhao, C.-H., Millar, J. G., Pan, K.-H.; Xu, C.-S. *J. Chem. Ecol.* 1998, *24*, 1347–1353.

[18] Bergot, B. J.; Anderson, R. J.; Schooley, D. A.; Henrick, C. A. *J. Chromatogr.* 1978, *155*, 97–105.

[19] Tani, K.; Yamagata, T; Akutagawa, S.; Kumobayashi, H.; Taketomi, T.; Takaya, H.; Miyashita, A.; Noyori, R.; Otsuka, S. *J. Am. Chem. Soc.* 1984, *106*, 5208–5217.

[20] Inoue, S.; Takaya, H.; Tani, K.; Otsuka, S.; Sato, T.; Noyori, R. *J. Am. Chem. Soc.* 1990, *112*, 4897–4905.

[21] Takaya, H.; Ohta, T.; Sayo, N.; Kumobayashi, H.; Akutagawa, S.; Inoue, S.; Kasahara, I.; Noyori, R. *J. Am. Chem. Soc.* 1987, *109*, 1596–1597.

[22] Ichikawa, A.; Takahashi, H.; Ooi, T.; Kusumi, T. *Biosci. Biotechnol. Biochem.* 1997, *61*, 881–883.

[23] Ohtani, I.; Kusumi, T.; Kashman, Y.; Kakisawa, H. *J. Am. Chem. Soc.* 1991, *113*, 4092–4096.

[24] Kusumi, T. *J. Syn. Org. Chem. Jpn.* 1993, *51*, 462–470 (in Japanese with English abstract).

[25] Kusumi, T.; Takahashi, H.; Xu, P.; Fukushima, T.; Asakawa, Y.; Hashimoto, T.; Kan, Y.; Inouye, Y. *Tetrahedron Lett.* 1994, *35*, 4397–4400.

[26] Seco, J. M.; Latypov, Sh. K.; Quiñoá E.; Riguera R. *Tetrahedron Lett.* 1994, *35*, 2921–2924.

[27] Takahashi, H. *Master's thesis, Tokushima University*, 1996.

[28] Arita, S.; Yabuuchi, T.; Kusumi, T. *Chirality* 2003, *15*, 609–614.

[29] Ichikawa, A. *Enantiomer* 1997, *2*, 327–331.

[30] Ichikawa, A. *Enantiomer* 1998, *3*, 255–261.

[31] Ichikawa, A. *Chirality* 1999, *11*, 338–342.

[32] Trost, B. M.; Belletire, J. L.; Godleski, S.; McDougal, P. G.; Balkovec, J. M.; Baldwin, J. J.; Christy, M. E.; Ponticello, G. S.; Varga, S. L.; Springer, J. P. *J. Org. Chem.* 1986, *51*, 2370–2374.

[33] Menger, F. M.; Mandell, L. *Electronic Interpretation of Organic Chemistry*; Plenum Press: New York, NY, 1980.

[34] Kuroda, R. *Homochiral Biological World*; Chu-ko Shinsho; Chuokoron-shinsha, Inc.: Tokyo, Japan, 1992, pp149–154 (in Japanese).

[35] Fujii, N. *Biol. Pharm. Bull.* 2005, *28*, 1585–1589.

[36] Nishimura, K.; Hashimoto, Y.; Iwasaki, S. *Chem. Pharm. Bull.* 1994, *42*, 1157–1159.

[37] Hashimoto, Y. *Bioorg. Med. Chem.* 2002, *10*, 461–479.

[38] Takeuchi Y.; Shiragami T.; Kimura K.; Suzuki E.; and Shibata N. *Org. Lett.* 1999, *1*, 1571–1573.

[39] Seco, J. M.; Quiñoá, E.; Riguera, R. *Chem. Rev.* 2004, *104*, 17–117.

[40] Goto, J.; Hasegawa, M.; Nakamura, S.; Shimada, K.; Nambara, T. *Chem. Pharm. Bull.* 1977, *25*, 847–849.

[41] Goto, J.; Hasegawa, M.; Nakamura, S.; Shimada, K.; Nambara, T. *J. Chromatogr.* 1978, *152*, 413–419.

[42] Ichikawa, A. Application of αMNPA in Enantioreric Resolution of Citronellol. *Book of Abstracts, The 42nd Symposium on the Chemistry of Terpenes, Essential Oils, and Aromatics (TEAC42)*, 1998, pp48–50 (in Japanese with English abstract).

[43] Ichikawa, A. Absolute configuration of chiral resolving agent 2-methoxy-2-(1-naphthyl)propanoic acid. *Book of Abstracts, The 76th Spring Meeting of the Chemical Society of Japan (Yokohama)*, 1999, Vol. II, p1343 (in Japanese).

[44] Ichikawa, A.; Hiradate, S.; Sugio, A.; Kuwahara, S.; Watanabe, M.; Harada, N. *Tetrahedron: Asymmetry* 1999, *10*, 4075–4078.

[45] Harada, N.; Watanabe, M.; Kuwahara, S.; Sugio, A.; Kasai, Y.; Ichikawa, A. *Tetrahedron: Asymmetry* 2000, *11*, 1249–1253.

[46] Chikamatsu, H.; Naemura, K. Determination of the absolute configuration of organic compounds by physical and chemical methods. In *Resolution of Optical Isomers;* The Chemical Society of Japan, Eds.; Kikan Kagaku Sosetsu; Japan scientific societies press: Tokyo, Japan, 1989, No. 6, pp198–209 (in Japanese with English abstract).

[47] Prelog, V. *Helv. Chim. Acta* 1953, *36*, 308–319.

[48] Ichikawa, A. *Chirality* 1999, *11*, 70–74.

[49] Latypov, Sh. K.; Seco, J. M.; Quiñoá, E.; Riguera, R. *J. Org. Chem.* 1995, *60*, 504–515.

[50] Kuwahara, S.; Fujita, K.; Watanabe, M.; Harada, N.; Ishida, T. *Enantiomer* 1999, *4*, 141–145.

[51] Kuwahara, S.; Naito, J.; Yamamoto, Y.; Kasai, Y.; Fujita, T.; Noro, K.; Shimanuki, K.; Akagi, M.; Watanabe, M.; Matsumoto, T.; Watanabe, M.; Ichikawa, A.; Harada, N. *E. J. Org. Chem.* 2007, 1827–1840.
[52] Ichikawa, A.; Ono, H.; Harada, N. *Chirality* 2004, *16*, 559–567.
[53] Corey, E. J.; Ensley, H. E. *J. Am. Chem. Soc.* 1975, *97*, 6908–6909.
[54] Whitesell, J. K.; Deyo, D.; Bhattacharya, A. *J. Chem. Soc., Chem. Commun.* 1983, 802.
[55] Ichikawa, A.; Ono, H.; Asaoka, K.; Harada, N. Enantioresolution using arylpropionic acids. *Book of Abstracts, Symposium on Molecular Chirality 2002 (MC2002)* 2002, pp103–106 (in Japanese with English abstract).
[56] Yoshida, K.; Osawa, S.; Monde, K.; Watanabe, M.; Harada, N. *Enantiomer* 2002, *7*, 23–32.
[57] Imaizumi, K.; Terasima, H.; Akasaka, K.; Ohrui, H. *Anal. Sci.* 2003, *19*, 1243–1249.
[58] Ohrui, H. *Proc. Jpn. Acad., Ser. B* 2007, *83*, 127–135.
[59] Kasai, Y.; Naito, J.; Kuwahara, S.; Watanabe, M.; Ichikawa, A.; Harada, N. *J. Syn. Org. Chem. Jpn.* 2004, *62*, 1114–1127.
[60] Kasai, Y.; Taji, H.; Fujita, T.; Yamamoto, Y.; Akagi, M.; Sugio, A.; Kuwahara, S.; Watanabe, M.; Harada, N.; Ichikawa, A.; Schurig, V. *Chirality* 2004, *16*, 569–585.
[61] Kasai, Y.; Sugio, A.; Sekiguchi, S.; Kuwahara, S.; Matsumoto, T.; Watanabe, M.; Icihkawa, A.; Harada, N. *E. J. Org. Chem.* 2007, 1811–1826.
[62] Ichikawa, A.; Ono, H.; Harada, N. *Tetrahedron: Asymmetry* 2003, *14*, 1593–1597.
[63] Ichikawa, A.; Ono, H. *J. Chromatogr. A* 2006, *1117*, 38–46.
[64] Ichikawa, A.; Ono, H. *Tetrahedron: Asymmetry* 2005, *16*, 2559–2568.
[65] Mori, K. *Tetrahedron* 1975, *31*, 3011–3012.
[66] Borden, J. H.; Chong, L.; McLean, J. A.; Slessor, K. N.; Mori, K. *Science* 1976, *192*, 894–896.
[67] Taji, H.; Kasai, Y.; Sugio, A.; Kuwahara, S.; Watanabe, M.; Harada, N.; Ichikawa, A. *Chirality* 2002, *14*, 81–84.
[68] Fujiwhara, M.; Mori, K. *Agric. Biol. Chem.* 1986, *50*, 2925–2927.
[69] Hall, D. R.; Beevor, P. S.; Cork, A.; Nesbitt, B. F.; Vale, G. A. *Insect Sci. Applic.* 1984, *5*, 335–339.
[70] Pierce, A. M.; Pierce, H. D., Jr.; Oehlschlager, A. C.; Borden, J. H. *J. Chem. Ecol.* 1991, *17*, 567–580.
[71] Mosandl, A.; Heusinger, G.; Gessner, M. *J. Agric. Food. Chem.* 1986, *34*, 119–122.
[72] Ramirez-Lucas, P.; Malosse, C.; Ducrot, P. H.; Lettere, M.; Zagatti, P. *Bioorg. Med. Chem.* 1996, *4*, 323–330.
[73] Takenaka, M.; Takikawa, H.; Mori, K. *Liebigs Ann. Chem.* 1996, 1963–1964.
[74] Mori, K. *Biosci. Biotechnol. Biochem.* 1996, *60*, 1925–1932.
[75] Desiraju, G. R. *Acc. Chem. Res.* 2002, *35*, 565–573.
[76] Ichikawa, A.; Hiradate, S.; Sugio, A.; Kuwahara, S.; Watanabe, M.; Harada, N. *Tetrahedron: Asymmetry* 2000, *11*, 2669–2675.
[77] Ichikawa, A.; Ono, H.; Hiradate, S.; Watanabe, M.; Harada, N. *Tetrahedron: Asymmetry* 2002, *13*, 1167–1172.
[78] Yabuuchi, T.; Kusumi, T. *J. Org. Chem.* 2000, *65*, 397–404.
[79] Sekiguchi, S.; Naito, J.; Taji, H.; Kasai, Y.; Sugio, A.; Kuwahara, S.; Watanabe, M.; Harada, N. *Chirality* 2008, *20*, 251–264.

[80] Harada, N. *Chirality* 2008, *20*, 691–723.
[81] Ichikawa, A.; Ono, H. *Biosci. Biotechnol. Biochem.* in press.

In: Stereochemistry Research Trends
Editor: M. A. Horvat and J. H. Golob

ISBN 978-1-60456-825-7


*Chapter 3*

# STEREOCHEMISTRY OF AFLASTATIN A AND BLASTICIDIN A: APPLICATION OF *J*-BASED, ACETONIDE AND NMR DATABASE METHODS

***Shohei Sakuda***
Department of Applied Biological Chemistry, University of Tokyo, Japan
***Nobuaki Matsumori***
Department of Chemistry, Osaka University, Japan

## ABSTRACT

Aflastatin A and blasticidin A are aflatoxin production inhibitors produced by *Streptomyces* sp. They have similar unique structures, which are tetramic acid derivatives with a highly oxygenated long alkyl chain containing a lot of chiral centers. In this article, our studies on stereochemistry of aflastatin A and blasticidin A are reviewed. Several fragment molecules were prepared from each of aflastatin A and blasticidin A. Relative or absolute configurations of the fragment molecules were assigned by a variety of methods including *J*-based, acetonide, MTPA and NMR database ones and by preparing model compounds. Basic concept and application of *J*-based method are also reviewed since the method was effectively used for assigning the relative stereochemistry of the acyclic moieties of aflastatin A and blasticidin A.

## INTRODUCTION

Aflatoxins are mycotoxins produced by fungi belonging to *Aspergillus* species, especially *Aspegillus parasiticus* and *Aspergillus flavus*. Aflatoxins show strong toxicity causing serious acute hepatitis in humans, such as the outbreak of acute aflatoxicosis occurred in Kenya in 2004 [1]. Aflatoxin $B_1$, a major component of aflatoxins, has the strongest carcinogenic activity in known natural products [2]. Several epidemiological studies in localities with a high incidence of liver cancer have strongly suggested the association of aflatoxins with hepatocellular carcinoma [3]. Aflatoxin-producing fungi are cosmopolitan organisms which can contaminate a wide range of natural substrates including cereal grains, oil seeds, cottonseed, etc. This aflatoxin contamination in crops is a serious problem from the viewpoint of not only public health but also economical loss [4,5]. It is, however, difficult to resolve the problem due to lack of an effective method to control aflatoxin production.

Use of antifungal agents may be an effective method to prevent aflatoxin contamination. But, strong fungicides are sometimes toxic to mammals, and rapid spread of resistant strains may occur as known in the case of use of antibiotics. Since aflatoxins are produced by fungi as secondary metabolites, which are not necessary for the fungal growth, specific inhibitors for aflatoxin production with no fungicidal activity may be useful to prevent foods and feeds from aflatoxin contamination without incurring rapid spread of resistant strains.

Until now, some natural products and synthetic pesticides [6,7] have been shown to have aflatoxin production inhibitory activity. Among them, aflastatin A [8,9,10], blasticidin A [11] and dioctatin A [12] are known as specific inhibitors of microbial origin (Figure 1). These compounds are all *Streptomyces* metabolites. Aflastatin A is the first compound found by a screening search for useful aflatoxin production inhibitors among microbial metabolites. Blasticidin A was isolated as an antibiotic in 1955 [13] and recently rediscovered as an aflatoxin production inhibitor. Dioctatin A was isolated as a human dipeptidyl peptidase inhibitor and shown to have a specific inhibitory activity toward aflatoxin production very recently. Among the three inhibitors, aflastatin A (AsA) and blasticidin A (BcA) are congeners. They have similar unique structures, which are tetramic acid derivatives with a highly oxygenated long alkyl chain. Our works on their isolation and planar structures have been reviewed previously [14]. They can strongly inhibit aflatoxin production by *A. parasiticus* at low concentrations such as 0.5 μM without inhibiting the fungal growth at the same concentration, but their molecular target for inhibition of aflatoxin production is not clarified yet [15]. Both compounds reduce the mRNA level of *aflR* gene encoding a key regulatory protein, which is absolutely necessary for starting expression of aflatoxin biosynthetic genes [16,17]. Therefore, their molecular target may be present in an unknown early regulatory system with an important role to connect primary metabolism to secondary one. Studies on the mode of action of these inhibitors are thought to be very significant not only for developing new effective aflatoxin production inhibitors but also to better characterize regulatory mechanisms of secondary metabolite production in fungi.

A lot of chiral centers are present in each molecule of AsA and BcA. Most of them are located in acyclic parts of the long alky side chain attaching to the tetramic acid moiety. Information on their stereochemistry is very important for studies at a molecular level such as their mode of action or interaction with a target molecule, but crystals of AsA, BcA or their derivatives suitable for X-ray analysis had not been obtained. Therefore, we tried to determine the absolute configuration of AsA and BcA chemically.

In this chapter, we review the determination of the absolute configuration of AsA and BcA [18,19,20]. Several fragment molecules were prepared from each compound, and their relative or absolute configurations were assigned by applying a variety of methods including *J*-based [21], acetonide [22], MTPA [23] and NMR database [24] ones and also by preparing model compounds. Among the methods used, the *J*-based method, which was developed by Matsumori (one of the authors) *et al.*, is very useful for assigning the relative stereochemistry of the acyclic moieties. Therefore, the basic concept and recent application of the *J*-based method are also reviewed in the latter half of this chapter.

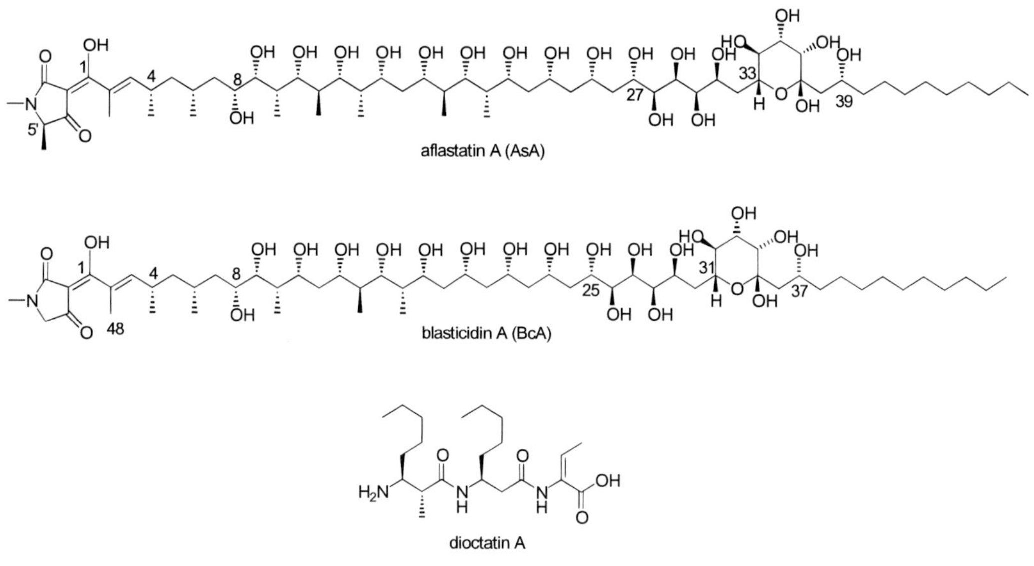

Figure 1. Structures of aflastatin A, blasticidin A and dioctatin A.

## ABSOLUTE CONFIGURATION OF BLASTICIDIN A

We started the study on the absolute configuration of AsA earlier than that of BcA [18], but correct assignment of the stereochemistry was first achieved on BcA [20]. Therefore, we describe the stereochemistry of BcA first and that of AsA later. The total absolute configuration of BcA was constructed by stepwise assignment of stereochemistry of each chiral center involved in the BcA molecule. We used a case-by-case method to determine each stereochemistry. Determination procedure we used is described in order of carbon number in BcA hereafter.

### Stereochemistry of C4 and C6

BcA afforded fragment *1* with $NaIO_4$ oxidation, followed by $NaBH_4$ reduction (Figure 2). Fragment *1* was further degraded to obtain 2,4-dimethyl-1,6-hexanediol dibenzoate *2* by $O_3$ oxidation and $LiAlH_4$ reduction. The small fragment molecule *2* was used for the assignment of the stereochemistry at C4 and C6 of BcA. To assign the absolute configuration of *2*, optically active (2*S*,4*S*)- and (2*R*,4*S*)-*2* were prepared as authentic samples from cycloheximide *3* according to the Scheme 1. Alkaline degradation of *1* afforded (2*S*,4*R*)- and (2*R*,4*R*)-2,4-dimethylcyclohexanone *4*, which could be converted to (2*S*,4*S*)- and (2*R*,4*S*)-*2*, respectively. NMR and HPLC analysis of natural and authentic *2* clarified natural *2* had *syn* stereochemistry. By comparing the CD spectrum of natural *2* with that of (2*R*,4*S*)-*2*, the absolute configuration of natural *2* was determined as (2*S*,4*R*), which afforded the absolute configurations at C4 and C6 of BcA.

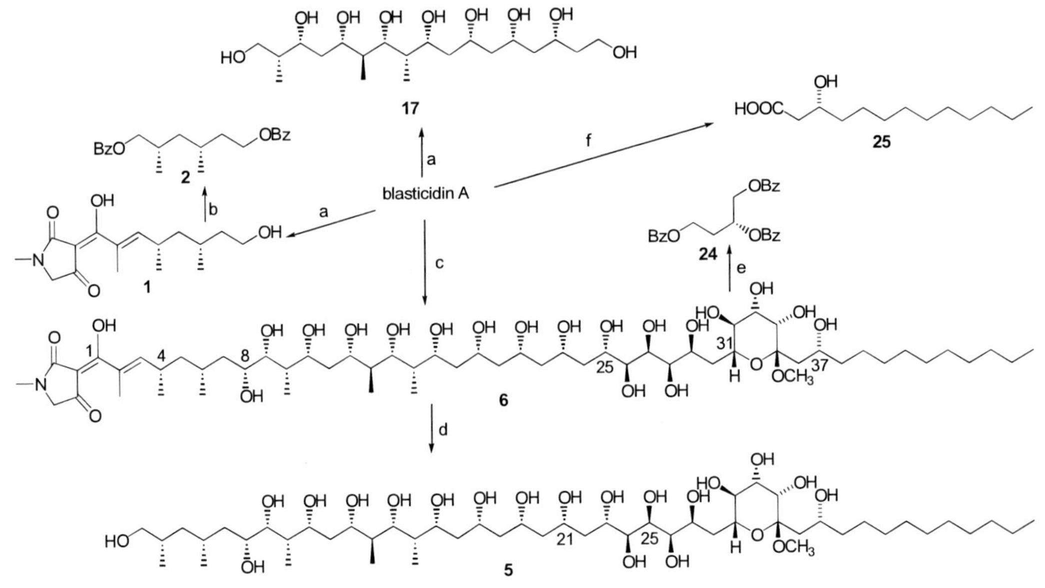

Figure 2. Procedures for the degradation of blasticidin A. (a) $NaIO_4$; $NaBH_4$. (b) $O_3$; $Me_2S$; $LiAlH_4$; BzCN. (c) HCl-MeOH. (d) $O_3$; $NaBH_4$. (e) $NaIO_4$; $NaBH_4$; HCl; BzCl. (f) $NaIO_4$.

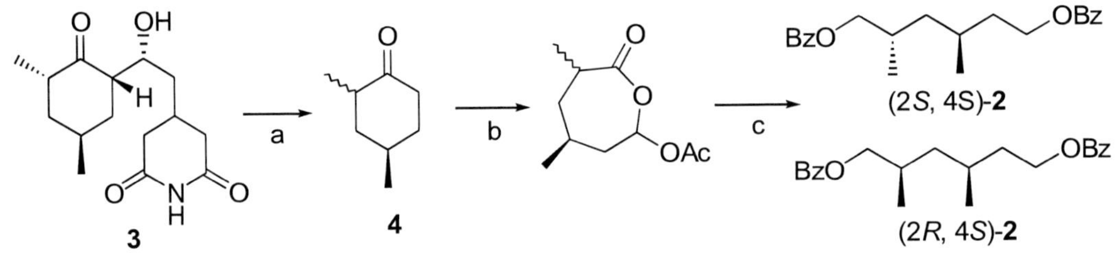

Scheme 1. Reagents: (a) 10 % NaOH; (b) $(AcO)_4Pb$, mCPBA; (c) $LiAlH_4$, BzCl.

## Stereochemistry of C8 and C9

Determination of the absolute configurations at C8 and C9 was so much trouble because a small fragment molecule containing the C8-C9 diol moiety could not be obtained. The absolute configuration at C6 of BcA was confirmed as mentioned above and that of C10-C11 moiety was also known as mentioned below. Therefore, determination of the relative stereochemistry either from C6 to C9 or from C8 to C10 was enough for assignment of the absolute configurations at C8 and C9. We first tried to elucidate the relative stereochemistry of the corresponding moiety of a large fragment *5*, which was obtained by $O_3$ degradation of methyl glycoside *6* of BcA [25], by the *J*-based method. However conformations around the bonds corresponding to C6-C7, C8-C9 and C9-C10 of BcA could not be determined by the obtained *J* values.

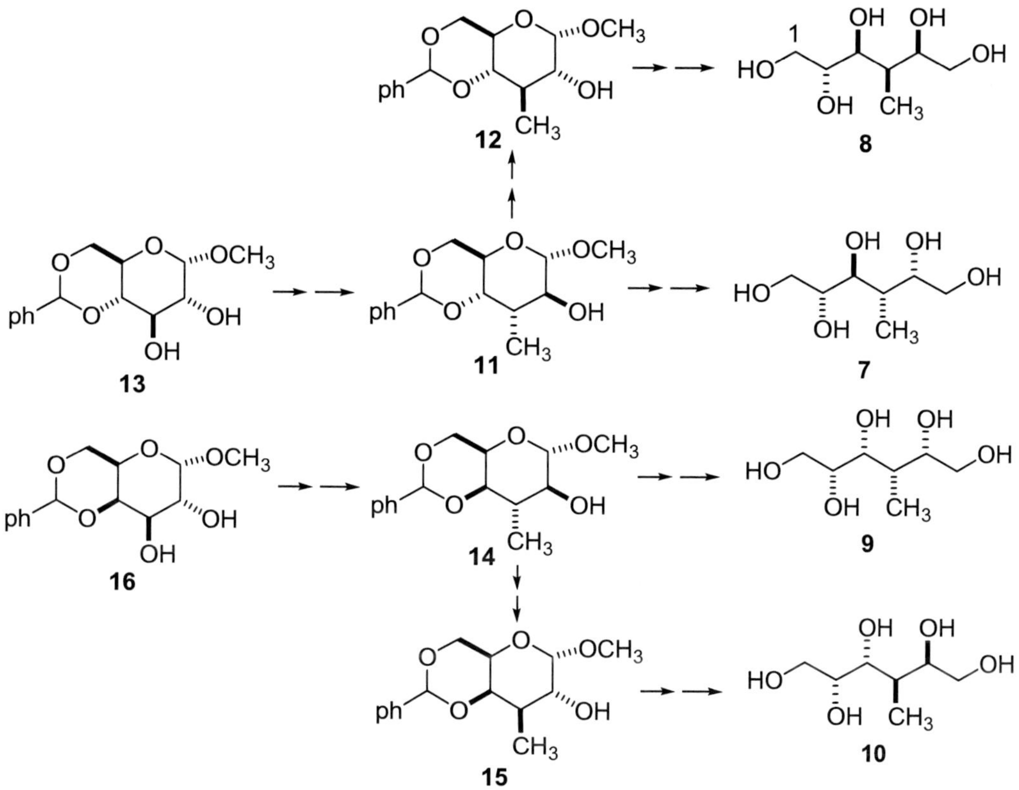


Scheme 2.

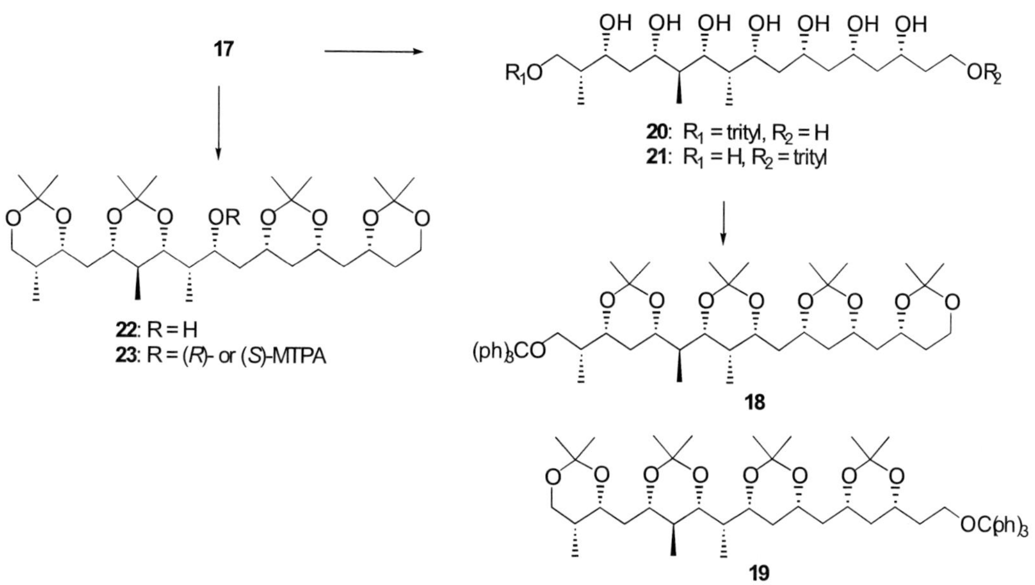


Scheme 3.

We next tried to determine the stereochemistry by comparison of the $^{3}J_{H,H}$ values around the C8-C11 moiety of BcA with those of four model stereoisomers *7*, *8*, *9* and *10*, which covered all possible stereochemistry for the moiety. These model compounds were prepared by the procedure shown in Scheme 2. Compounds *11* and *12* were prepared from methyl 4,6-*O*-benzylidene-α-D-glucopyranoside *13* according to the known method [26]. Similarly, compounds *14* and *15* were obtained from methyl 4,6-*O*-benzylidene-α-D-galactopyranoside *16*. Acid hydrolysis of *11*, *12*, *14* and *15*, followed by $NaBH_4$ reduction, afforded the model compounds, *7*, *8*, *9* and *10*, respectively.

The $^{3}J_{H\text{-}2,H\text{-}3}$ and $^{3}J_{H\text{-}3,H\text{-}4}$ values of these compounds were measured, and it was shown that only the values observed in compound *9* were comparable to those in the corresponding moiety of the fragment molecule *5*. Therefore, the relative configurations for both C8-C9 and C9-C10 were assigned as *threo*. Since the absolute stereochemistry at C10 was confirmed as mentioned next, the absolute configurations at C8 and C9 of BcA were determined [20].

## Stereochemistry of C10-C23

The stereochemistry of C10-C23 moiety of BcA was elucidated with a polyol fragment *17*. Fragment *17* was obtained by $NaIO_4$ oxidation of BcA, followed by $NaBH_4$ reduction (Figure 2). The fragment *17* had a typical structure consisting of 1,3-diol systems. Therefore, we applied the acetonide method to analyze its relative stereochemistry. Since *17* had nine hydroxyl groups, two tetraacetonide derivatives, *18* and *19*, of *17* were prepared from mono-tritylated-*17* (*20* and *21)* (Scheme 3). The $^{13}$C NMR spectra of *18* and *19* showed that each of the 6-membered acetonide rings in *18* and *19* was present in a chair form [22], indicating a *syn* relationship of each of the corresponding 1,3-diol systems. Analysis of the $^{3}J_{H,H}$ values observed in each of the three 6-membered rings with a methyl group in *18* and *19* clarified the orientation of each methyl group in the ring, which determined the relative configurations around the methyl groups. Thus, the relative stereochemistry of *17* was assigned.

When fragment *17* was acetonided directly, the tetraacetonide *22* with a secondary hydroxyl group was mainly obtained (Scheme 3). Compound *22* was converted to (*R*)- and (*S*)-MTPA esters *23*, and the absolute configuration at the esterified oxymethine was determined by analysis of $^{1}$H NMR spectra of the MTPA esters according to the MTPA method [23]. Thus, the absolute configurations at C10, C11, C13, C14, C15, C16, C17, C19, C21 and C23 of BcA were assigned [19].

## Stereochemistry of C25-C29

Assignment of the stereochemistry of the C25-C29 pentaol moiety was not easy due to the same reason encountered in that of the C8-C9 diol moiety above mentioned. The large fragment molecule *5* was used for the assignment of the stereochemistry similarly to the case of C8-C9. We tried to assign the absolute stereochemistry at the pentaol moiety based on the absolute configuration at C23 of BcA. Therefore, the relative configurations from C23 to C29 were elucidated by the *J*-based method. Analysis of the $^{3}J_{H,H}$ and $^{2,3}J_{C,H}$ values obtained in the corresponding moiety of fragment *5* could assign the relative configurations for C23-C25,

C27-C28 and C28-C29 of BcA as *syn*, *threo* and *threo*, respectively. However, the configurations for C25-C26 and C26-C27 of BcA could not be assigned due to ambiguous medium $J$ values around the corresponding bonds in *5*. The NMR database method developed by Kishi *et al.* [24] broke through this circumstance. In the method, a relative configuration for an acyclic pentaol moiety can be determined by the profile of $^3J_{H,H}$ values observed in the moiety. The pattern of the $^3J_{H,H}$ values in the pentaol moiety of *5* clearly showed *erythro/threo/threo/threo* configuration, which determined the configurations for C25-C26 and C26-C27 of BcA and confirmed those for C27-C28 and C28-C29. Thus, the relative stereochemistry from C23 to C29 of BcA was clarified, which afforded the absolute configurations at C25, C26, C27, C28 and C29 based on the stereochemistry at C23 [20].

## Stereochemistry of C31-C37

The absolute stereochemistry of the tetrahydropyran ring moiety of BcA was assigned based on the relative configurations around C31-C35 and the absolute configuration at C31. The absolute configuration at C31 was determined as follows. Methyl glycoside of BcA *6* was oxidized with $NaIO_4$, followed by $NaBH_4$ reduction, acid hydrolysis and benzoylation, to afford 1,2,4-butanetriol tribenzoate *24* (Figure 2), in which the configuration at C31 was maintained. The absolute configuration of the small fragment was assigned by comparison of its CD spectrum with that of an authentic sample. Since the relative configuration of the tetrahydropyran ring was verified by the $^3J_{H,H}$ values observed in the moiety, the absolute configuration from C31 to C35 was determined based on the configuration at C31.

A β-hydroxycarboxylic acid *25*, in which the absolute configuration at C37 was maintained, was obtained by $NaIO_4$ oxidation of BcA (Figure 2). The absolute stereochemistry of the acid was determined from its optical rotation value.

From the results described above, the absolute configuration of BcA was completely assigned. Figure 3 summarizes the methods used for the assignment.

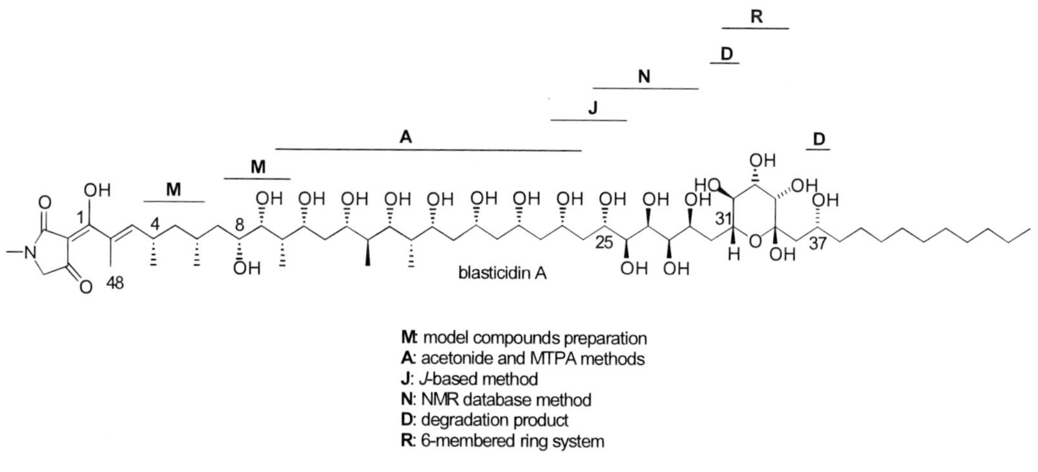


Figure 3. Methods used for assignment of stereochemistry of blasticidin A.

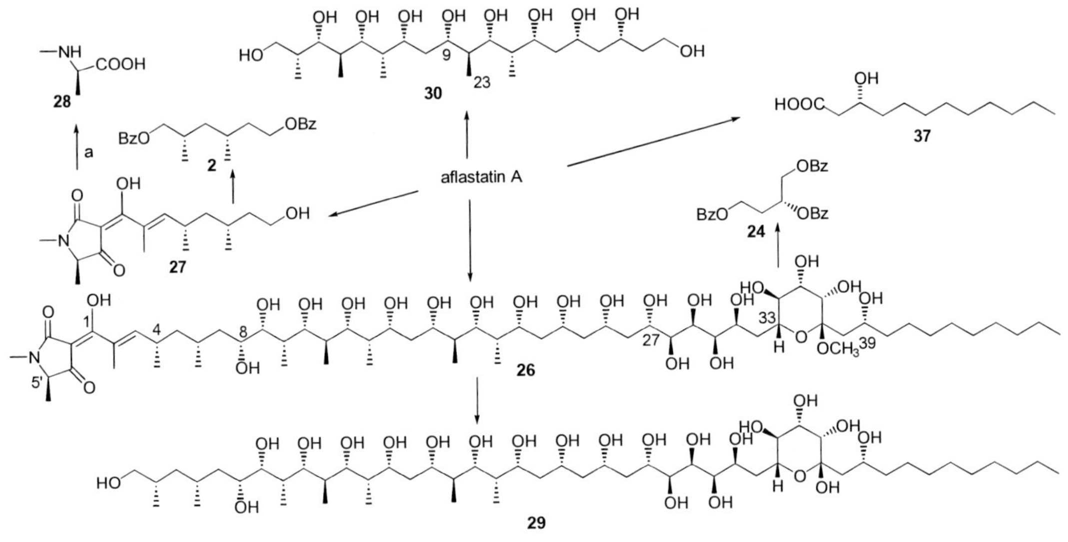

Figure 4. Procedures for the degradation of aflastatin A. (a) $NaIO_4$; HCl.

## Absolute Configuration of Aflastatin A

The planar structure of AsA is very similar to that of BcA. Differences between them are as follows. A methyl group in the tetramic acid moiety of AcA is not present in the corresponding part of BcA. The length of the carbon chain and the number or position of methyl or hydroxyl groups in the part of corresponding to fragment *5* are different from those in the counterpart of AsA. The length of the hydrocarbon end of AsA is shorter than that of BcA by one carbon unit. The strategy to determine the absolute stereochemistry of AsA was almost the same as that used in the case of BcA above mentioned. However, we tried to apply the *J*-based method to assign the relative configurations of the fragment molecules of AsA more frequently than the case of BcA. We succeeded in assignment of the relative stereochemistry of the polyol fragment *30* (Figure 4) by using only the *J*-based method, but we also encountered some cases in which application of the method was not easy. Determination procedure of the absolute stereochemistry of AsA is described below.

### Stereochemistry of C5', C4 and C6

AsA has a chiral center at C5' in the tetramic acid moiety. Fragment *27* (Figure 4) was used for determination of its absolute stereochemistry. Fragment *27* afforded *N*-methylalanine *28*, in which the configuration at C5' was maintained, by $NaIO_4$ oxidation, followed by hydrolysis with 6 M HCl. The absolute configuration of the amino acid was determined by a method with Marfey's reagent [27]. The absolute configurations at C4 and C6 of AsA were assigned by using the same method as that used for C4 and C6 of BcA.

## Stereochemistry of C8 and C9

We tried to assign the relative stereochemistry from C6 to C10 of AsA by the *J*-based method to determine the absolute configurations at C8 and C9 based on the configuration at C6. Large fragment molecule *29*, which was obtained from methyl glycoside *26,* was used for the analysis, and *anti*, *threo* and *erythro* configurations were proposed for C6-C8, C8-C9 and C9-C10, respectively [18]. To confirm the stereochemistry, we performed analysis of stereochemistry of the corresponding moiety of BcA in parallel because the NMR data at C6-C9 of AsA were very similar to those at C6-C9 of BcA. As a result, the stereochemistry of C8 and C9 of BcA was clearly determined as described above. Accordingly, the absolute configurations at C8 and C9 of AsA were revised to be identical with those of BcA [20].

## Stereochemistry of C10-C25

Stereochemistry of C10-C25 of AsA was analyzed with a polyol fragment *30* similarly to the case of BcA. We first assigned the relative stereochemistry of the fragment *30* by *J*-based method. Since a dominant conformation around each C-C bond involved in the fragment could be specified by the *J*-based method, a solution conformation of the polyol fragment *30* was also proposed [18]. To confirm the stereochemistry of C10-C25 of AsA, we prepared the pentaacetonide *31* and ditrityltetraacetonide *32* from frgament *30* and its ditrityl derivative *33*, respectively (Scheme 4). The relative stereochemistry of C10-C25 of AsA assigned by analysis of the NMR spectra of the two acetonides, *31* and *32*, was completely identical with that proposed by the *J*-based method. The mono-trityltetraacetonide *34* obtained from mono-trityl-*30* (*35*) was converted to the MTPA esters *36*. By applying the MTPA method to the esters *36*, the absolute configurations at all chiral centers involved in the fragment, corresponding to C10, C11, C12, C13, C14, C15, C17, C18, C19, C20, C21, C23 and C25 of AsA, were determined [19]. The chiral synthesis of the polyol fragment *30* was recently achieved by Evans *et al* [28].

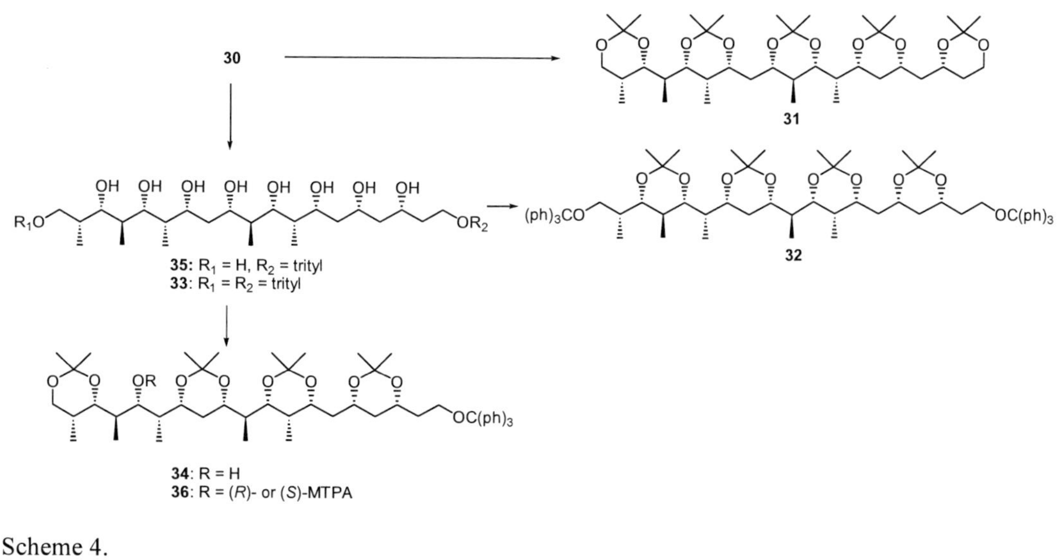


Scheme 4.

## Stereochemistry of C27-C39

The NMR data at C25-C31 of AsA coincided well with those at C23-C29 of BcA. Therefore, the relative stereochemistry of C25-C31 of AsA was the same as that of the corresponding moiety of BcA. We proposed all *threo* configurations for C27-C28, C28-C29, C29-C30 and C30-C31 of AsA from the results obtained by analysis with the *J*-based method [18]. As mentioned above, however, the NMR database method clearly showed that the configuration for C27-C28 of AsA was *erythro*. The absolute configurations at C33, C34, C35, C36, C37 and C39 of AsA were assigned from the degradation products, *24* and *37* (Figure 4), by the same methods as used in the case of BcA above mentioned.

Our study showed that the stereochemistry of AsA is very similar to that of BcA through the overall structure. We aggressively used the *J*-based method for assignment of the relative stereochemistry of AsA. Although sever signal overlappings in NMR spectra due to the highly repeated structure of AsA resultantly led to the wrong assignment in relative configurations for C6-C7, C9-C10 and C27-C28 of AsA by the *J*-based method, our trial left a lot of important information on usefulness of the *J*-based method.

# *J*-BASED METHOD

Since the *J*-based analysis method was reported in 1999 [21], the analysis has become a standard method widely used for the relative configuration determination of polysubstituted acyclic chains. In fact, as demonstrated above, the *J*-based configuration analysis method was effectively applied to the assignment of relative configurations in BcA and AsA. Since detailed description on the method is beyond the scope of this chapter and should be referred to the original paper and recent review [21, 29], we briefly overview the basic concept and application of the *J*-based configuration analysis in the following.

## Overview of *J*-based Configuration Analysis

The *J*-based configuration analysis is based on the dihedral angle dependence of spin-coupling constants. The spin-coupling constant is a useful NMR parameter for conformational studies of organic molecules. It is well known that proton-proton vicinal coupling constants ($^3J_{H,H}$) are correlated with their dihedral angles by the Karplus equation. Similarly, vicinal carbon-proton spin-coupling constants ($^3J_{C,H}$) follow a Karplus-type equation (Figure 5) [30], and thus can be utilized for stereochemical analysis as in the case of $^3J_{H,H}$. In addition, geminal carbon-proton coupling constants ($^2J_{C,H}$) also provide conformational information when the α-carbon bears an electronegative substituent [31]; i.e. when an oxygen functionality on a carbon atom is *gauche* to a proton, $^2J_{C,H}$ becomes large, and when it is *anti*, the value becomes small (Figure 5).

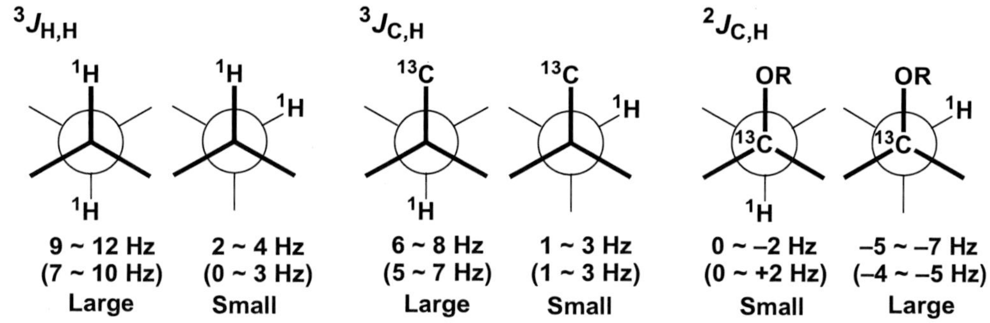


Figure 5. Dihedral angle dependence of coupling constants. The figures in parentheses represent the values of 1,2-dioxygenated systems.

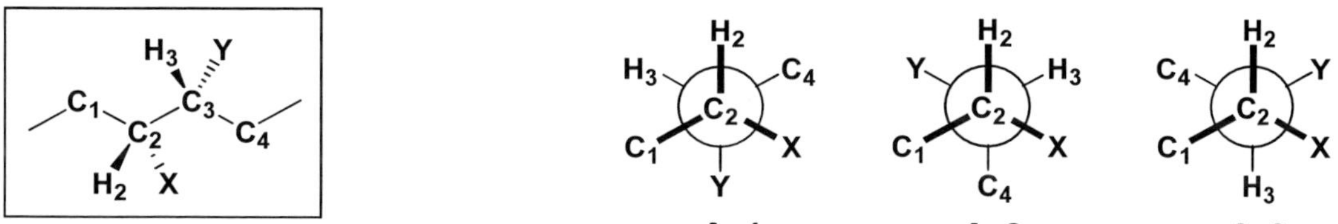


**A (*Threo*)**

| | | A-1 | A-2 | A-3 |
|---|---|---|---|---|
| | $^3J$ (H-2, H-3) | Small | Small | Large |
| | $^3J$ (H-2, C4) | Small | Large | Small |
| | $^3J$ (C1, H-3) | Small | Large | Small |
| X=Me, Y=OR | $^3J$ ($C_X$, H-3) | Large | Small | Small |
| | $^2J$ (C3, H-2) | Small | Large | Large |
| X=OR, Y=OR | $^2J$ (C2, H-3) | Small | Large | Large |
| | $^2J$ (C3, H-2) | Small | Large | Large |

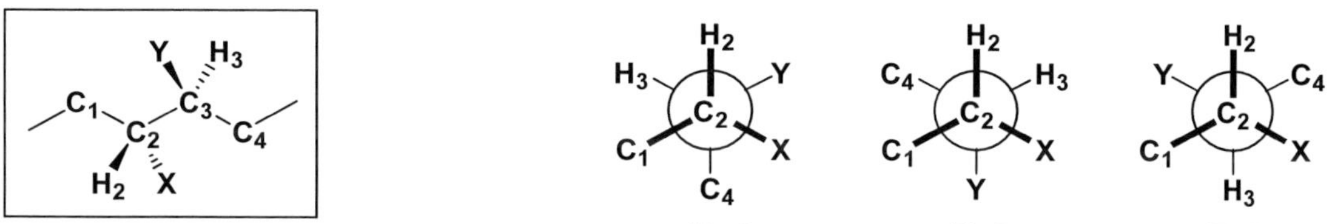


**B (*Erythro*)**

| | | B-1 | B-2 | B-3 |
|---|---|---|---|---|
| | $^3J$ (H-2, H-3) | Small | Small | Large |
| | $^3J$ (H-2, C4) | Large | Small | Small |
| | $^3J$ (C1, H-3) | Small | Large | Small |
| X=Me, Y=OR | $^3J$ ($C_X$, H-3) | Large | Small | Small |
| | $^2J$ (C3, H-2) | Large | Small | Large |
| X=OR, Y=OR | $^2J$ (C2, H-3) | Small | Large | Large |
| | $^2J$ (C3, H-2) | Large | Small | Large |

Figure 6. *J*-based configuration analysis for a 1,2-dimethine system. Expected magnitudes of coupling constants are shown for each rotamer. Different pattern of coupling constants can discriminate rotamers except H/H-*anti* rotamers A-3 and B-3, which can be distinguished on the basis of NOEs.

To determine the configuration of multiple asymmetric carbons in natural products, the diastereomeric relationship between two neighboring stereocenters must be assigned. In effect, 1,2-, and 1,3-asymmetric systems are commonly found in natural products, as seen in the structures of BcA and AsA. Therefore, the *J*-based method is focused on stereochemical correlations of 1,2-dimethines or 1,3-dimethines bearing hydroxy and/or methyl groups.

This method assumes that conformations can be represented by three staggered rotamers. This seems to be quite reasonable for acyclic carbon chains with hydroxy and methyl substitutions, which, unlike bulky substituents, usually cause no significant deviations from the *anti* or *gauche* orientation. To assign the stereochemical relationship for 1,2-dimethine system, a single conformer with a correct configuration has to be extracted from six staggered rotamers possible in *threo* and *erythro* diastereomers (Figure 6). Among those, four conformers, A-1, A-2, B-1, and B-2, can be identified using the magnitudes of $^3J_{H,H}$, $^2J_{C,H}$, and $^3J_{C,H}$. However, the two rotamers A-3 and B-3 with an H/H-*anti* orientation are still indistinguishable. For these *anti* conformers, NOE experiments should be applied to assign their configuration. In B-3 rotamer (Figure 6), no NOE (or ROE) should be observed between H-1 and H-4, whereas, in A-3 rotamer, H-1 and H-4 should come within the range of NOE. Using these criteria, all six conformers could be discriminated, with their relative configuration (*threo* or *erythro*) determined accordingly.

Here it should be noted that all of the coupling constants listed in Figure 6 are not necessary to identify a specific rotamer out of the six. In effect, it is not always possible to extract all the coupling constants from NMR spectra. Since each rotamer in Figure 6 can be defined by two pairs of *anti* arrangements, the coupling constants that specify two antiperiplanar relationships are enough to assign a single rotamer from the six. For example, A-2 in Figure 6 can be defined by *anti* orientations of H2/C4 and C1/H3; therefore, large $^3J_{C,H}$ values corresponding to H2/C4 and C1/H3 can specify this rotamer. Alternatively, since the *anti* arrangement of H2/C4 in rotamer A-2 is equivalent to two *gauche* arrangements of H2/H4 and H2/Y, their corresponding coupling constants can compensate for the lack of the coupling constant of H2/C4. In this way, finding two *anti* arrangements is a key to specify a single rotamer.

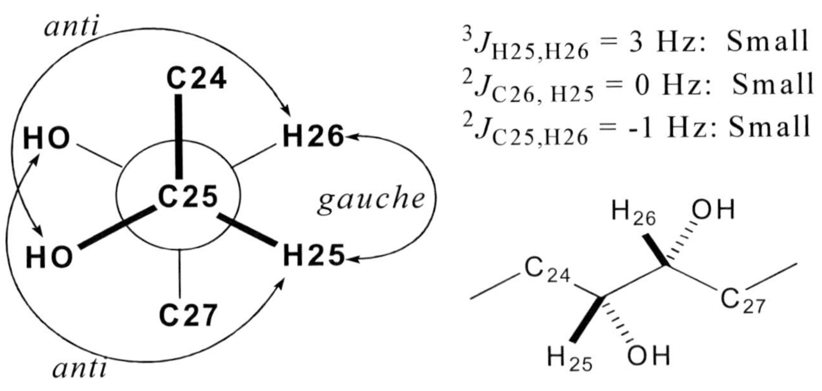


Figure 7. Stereochemical assignment for C25-C26 of BcA fragment *5* deduced from the *J*-based method. The observed coupling constants are typical for A-1 rotamer in Figure 5.

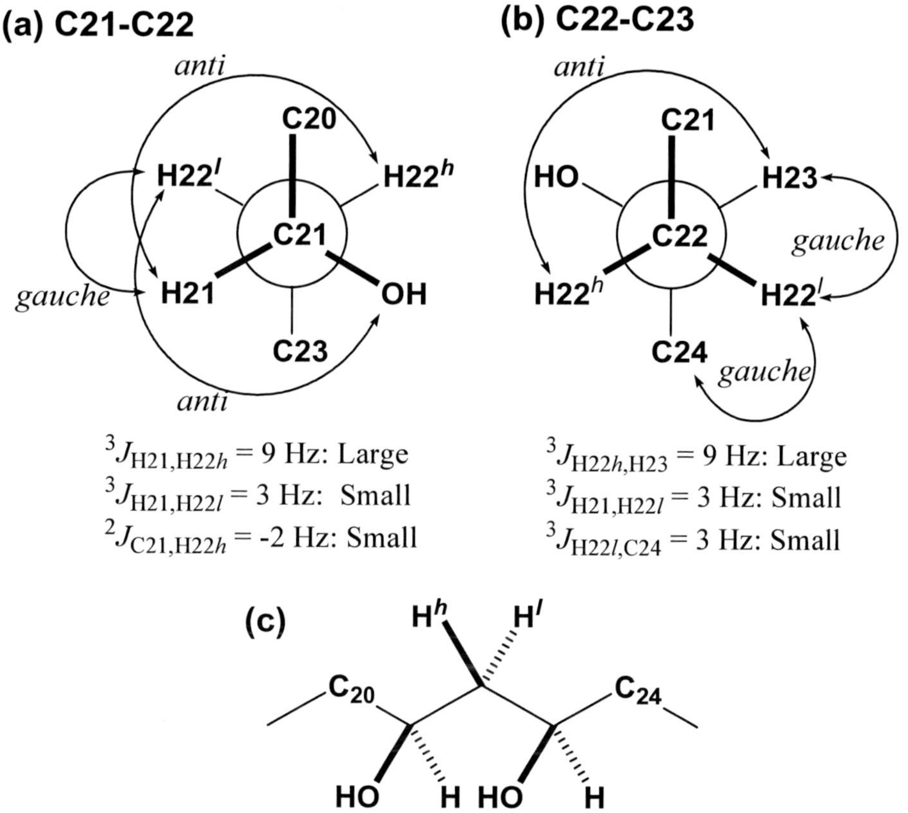


$^3J_{H21,H22h}$ = 9 Hz: Large
$^3J_{H21,H22l}$ = 3 Hz: Small
$^2J_{C21,H22h}$ = -2 Hz: Small

$^3J_{H22h,H23}$ = 9 Hz: Large
$^3J_{H21,H22l}$ = 3 Hz: Small
$^3J_{H22l,C24}$ = 3 Hz: Small

Figure 8. Stereochemical assignment for C21-C23 of BcA fragment *5* deduced from the *J*-based method. H22$^h$ and H22$^l$ represent diastereotopic methylene protons on C22 having chemical shifts at higher-field and lower-field, respectively.

Figure 7 shows an example of determining the relative stereochemistry in C25-C26 of BcA fragment *5* (Figure 2) [20]. The value of $^3J$(H25, H26) is typical of *gauche* orientation between H25 and H26. The values for $^2J$(C25, H26) and $^2J$(C26, H25) indicate that H26 is *anti* to OH on C25, and H25 is *anti* to OH on C26. These arrangements correspond to A-1 rotamer in Figure 5, thus unambiguously establishing the *threo* configuration for C25-C26 of fragment *5*. In this system, two *anti* orientations necessary for the assignment of the unique rotamer are seen in two vinical H/OH.

The analysis can be also extended to methylene-spaced stereocenters, provided that a stereospecific assignment of the methylene diastereotopic protons may be achieved. The method for assigning diastereotopic methylene protons with respect to the adjacent stereocenter is similar to that for the 1,2-dimethine system described above, when a pair of protons on a methylene is stereochemically labeled according to their chemical shifts. All possible rotamers for methine-methylene bond can be unambiguously identified using $^3J_{H,H}$ and $^{2,3}J_{C,H}$. With one methine-methylene relationship in hand, the same examination of another relationship leads to diastereomeric assignment of the 1,3-dimethine *via* the stereospecifically labeled methylene protons. An example is demonstrated by using the C21-C23 portion of BcA fragment *5* (Figure 2 and 8) [20]. The methylene protons on C22 give their $^1$H chemical shifts at 2.16 and 2.59 ppm, which are labeled as H22$^h$ (high-field proton) and H22$^l$ (low-field), respectively. For the C21-C22 bond, the observed coupling constants can identify the single rotamer shown in Figure 8a. In this rotamer, two *anti*-periplanar relations are shown in H21/H22$^h$ and OH/H22$^l$. Likewise, the rotamer for the C22-C23 bond was uniquely determined from *anti* orientation for H22$^h$/H23 and two *gauche* relations with respect to H22$^l$ (Figure 8b). Then the two elucidated rotamers shown in Figure 8a and 8b are connected at C22 position so that the H22$^h$ and H22$^l$ in the both rotamers are consistent, thus leading to establishment of the diastereomeric relation between C21 and C23 as depicted in Figure 8c.

If geminal proton signals are separated, this method can be applied to 1,4-methine systems separated by an ethylene group.

The aforementioned analysis assumes a single, highly populated (more than 85%), rotamer in a C-C bond. However, this is not always the case for acyclic carbon chains. When multiple rotamers exist, averaged coupling constants, which are described as medium $J$ values, are detected due to a fast conformational interconversion. In most cases, it seems reasonable to assume that the multiple conformer equilibrium is described by the interconversion between two staggered rotamers. Figure 9 represents an example for relative stereochemistry determination on the C9-C10 bond of AsA fragment *30* (Figure 4) [18]. The medium magnitude of $^3J$(H9, H10) (5.5 Hz) is accounted for by *gauche*/*anti* equilibrium of H9/H10 dihedral angles. The medium $^3J$(H9, C23) (3.5 Hz) is also indicative of equilibrium of *gauche*/*anti* between H9 and C23. The large value of $^2J$(H10, C9) (-5.6 Hz) further suggests two concurrent *gauche* arrangements of H10/OH. These data suggest the equilibrium of the two equilibrating rotamers shown in Figure 9, thus leading to the configurational assignment of the C9-C10 bond.

In case that three rotamers have comparable populations, which leads to an averaging of all $J$ values, the $J$-based method will not determine configuration. If a minor rotamer comprises less than 10% of the population, its contribution can be disregarded and the configuration can be assigned as described above.

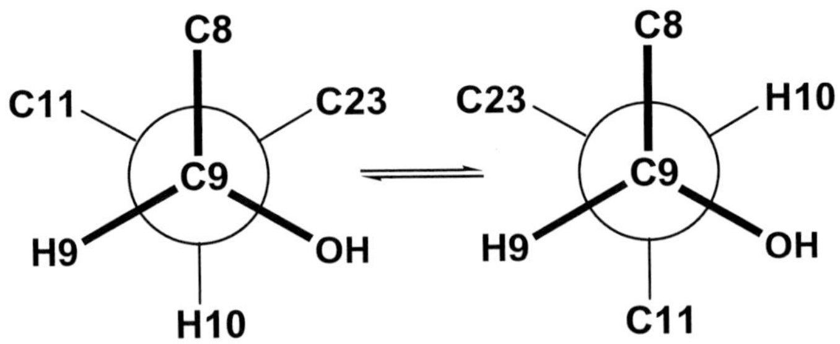


| | | | |
|---|---|---|---|
| $^3J_{H9,H10}$ = 5.5 Hz: Medium | *gauche* | H9/H10 | *anti* |
| $^3J_{H9,C23}$ = 3.5 Hz: Medium | *anti* | H9/C23 | *gauche* |
| $^2J_{H10,C9}$ = -5.6 Hz: Large | *gauche* | H10/OH | *gauche* |

Figure 9. Stereochemical assignment for C9-C10 of AsA fragment *30* deduced from the $J$-based method. A pair of equilibrating rotamers with the *erythro* configuration can account for the observed coupling constants.

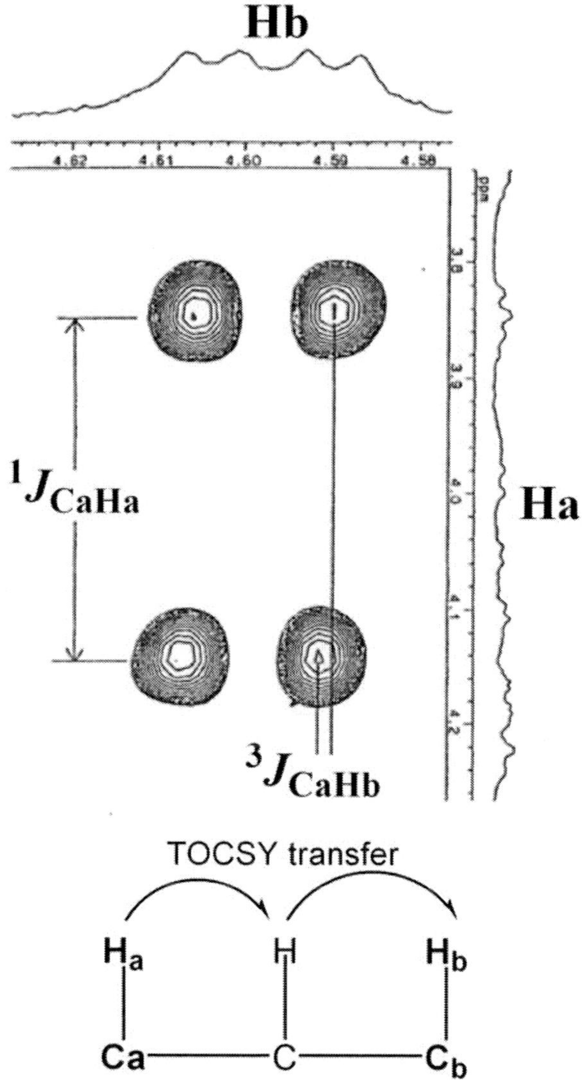


Figure 10. HELOC cross peak.

## Measurements of $^{2,3}J_{C,H}$

Now that the introduction of long-range carbon-proton coupling constants opens the way to the stereochemical assignments in acyclic natural products, measurements of $^{2,3}J_{C,H}$ are crucial to realize this methodology. Whereas the $^3J_{H,H}$ values can be straightforwardly measured by homonuclear NMR experiments such as E.COSY [32], since the 1990s several practical methods have been proposed for the measurement of $^{2,3}J_{C,H}$ [33] thanks to the introduction of inverse detection NMR techniques and the implementation of pulse-field-gradient (PFG) hardware. In the original paper reporting the *J*-based method [21], two methods were adopted: hetero half-filtered TOCSY (HETLOC) [34] and phase-sensitive HMBC (PS-HMBC) [35]. The HETLOC experiment gives a $^1$H-$^1$H 2D TOCSY-like spectrum, where each cross-peak is split by $^1J_{C,H}$ along $F_1$ direction and dislocated by $^{2,3}J_{C,H}$

along $F_2$ (Figure 10). Therefore, the cross-peak feature of HETLOC allows for efficient measurements of even small $^{2,3}J_{C,H}$ values. However, HETLOC has intrinsic drawbacks: the two protons must belong to the same spin system and must exhibit a TOCSY correlation peak. For this reason, it is impossible to use this experiment to measure coupling constants between a proton and a quaternary carbon or between a proton and a carbon belonging to a different spin system. To overcome this problem, PS-HMBC experiment [35] was also used. In the PS-HMBC experiments, the relative cross-peak intensities arising from the same proton depend on the magnitude of $^{2,3}J_{C,H}$, as given by the following equation:

$$I_{Ca,H} / I_{Cb,H} = \sin^2(\pi\, ^{2,3}J_{Ca,H}\, \Delta) / \sin^2(\pi\, ^{2,3}J_{Cb,H}\, \Delta)$$

where $I_{Ca,H}$ and $I_{Cb,H}$ are the intensities of cross-peaks due to Ca-H and Cb-H couplings, respectively, and $\Delta$ is a time interval in the pulse sequence [35]. Therefore, the values of $^{2,3}J_{Ca,H}$ can be obtained by the equation using the known $^{2,3}J_{Cb,H}$ derived from other methods (e.g., HETLOC) as reference. Because the correlations measured in the PS-HMBC are not read through homonuclear TOCSY correlations, as in the HETLOC spectrum, it is possible to estimate coupling constants between protons and quaternary carbons or between protons and carbons belonging to different spin systems.

Several pulse sequences including *J*-HMBC [36] were also presented for the estimation of long-range C-H coupling constants. The *J*-HMBC signal amplitude of spectra shows the characteristic sin ($\pi\, ^{2,3}J_{C,H}$ t)-dependence with increasing coupling evolution time t. By fitting a sine-curve to the experimental data or by performing a 3D-version of *J*-HMBC, accurate C-H coupling constants are obtained. This method is especially useful for estimating coupling constants between protons and unprotonated quarternary carbons.

## APPLICATION OF THE *J*-BASED METHOD

The *J*-based method was first applied to maitotoxin [37], the largest secondary metabolite known to date, dysiherbaine [38], and amphidinol 3 [39]. Since the method was systematized in 1999 [21], this method has been used for structure determination of wide variety of natural products [29]. During the course of its application, the method has been modified according to each compound, and been getting robust. For example, the application of the *J*-based method had extended from stereocenters bearing methyl and hydroxyl groups to those having amines or chlorides [29]. Besides, around the same time, Kishi *et al* reported universal NMR database method, which provided another tool for determining relative stereochemistry [24]. The principle of universal NMR database is based on the $^{13}$C chemical shift comparison of the compounds with unknown configuration with libraries of model compounds with known stereostructures. In effect, the method was also effectively applied for BcA as described previously [20]. Combined use of the *J*-based method and universal NMR database surely provides a more practical way for stereochemical analysis of natural products by resorting to NMR spectroscopy.

## References


[1] Probst, C.; Njapau, H.; Cotty, P. J. *Appl. Environ. Microbiol.* 2007, *73*, 2762-2764.

[2] Wong, J. J.; Hsieh, D. P. H. *Proc. Natl. Acad. Sci. USA* 1976, *73*, 2241-2244.

[3] Henry, S. H.; Bosch, G. X.; Troxell, T. C.; Bolger, P. M., *Science* 1999, *286*, 2453-2454.

[4] Bennett, J. W.; Klich, M. *Clin. Microbiol. Rev.* 2003, *16*, 497-516.

[5] Payne, G. A. In *Mycotoxins in Agriculture and Food Safety*; Sinha, K. K., Bhatnagar, D., Eds. Marcel Dekker: New York, 1998; pp 279-306.

[6] Zaika, L. L.; Buchanan, R. L. *J. Food. Prot.* 1987, *50*, 691-708.

[7] Wheeler, M. F.; Bhatnagar, D., *Pestic. Biochem. Physiol.* 1995, *52*, 109-115.

[8] Sakuda, S.; Ono, M.; Furihata, K.; Nakayama, J.; Suzuki, A.; Isogai, A. *J. Am. Chem. Soc.* 1996, *118*, 7855-7856.

[9] Ono, M.; Sakuda, S.; Suzuki, A.; Isogai, A, *J. Antibiotics* 1997, *50*, 111-118.

[10] Ono, M.; Sakuda, S.; Ikeda, H.; Furihata, K.; Nakayama, J.; Suzuki, A.; Isogai, A. *J. Antibiotics* 1998, *51*, 1019-1028.

[11] Sakuda, S.; Ono, M.; Ikeda, H.; Nakamura, T.; Inagaki, Y.; Kawachi, R.; Nakayama, J.; Suzuki, A.; Isogai, A.; Nagasawa, H. *J. Antibiotics* 2000, *53*, 1265-1271.

[12] Yoshinari, T.; Akiyama, T.; Nakamura, K.; Kondo, T.; Takahashi, Y.; Muraoka, Y.; Nonomura, Y.; Nagasawa, H.; Sakuda, S. *Microbiol.* 2007, *153*, 2774-2780.

[13] Kono, Y.; Takeuchi, S.; Yonehara, H. *J. Antibiot.* 1968, *21*, 433-438.

[14] Sakuda, S.; Ono, M.; Ikeda, H.; Sakurada, M.; Nakayama, J.; Suzuki, A.; Isogai, A. In *Biologically Active Natural Products: Agrochemicals*; Cutler, G., Cutler, S. J., Eds. CRC Press: Boca Raton, London, New York, Washington, D.C., 1999; pp 185-199.

[15] Sakuda, S. *Mycotoxins* 2002, *52*, 153-158.

[16] Kondo, T.; Sakurada, M.; Okamoto, S.; Ono, M.; Tsukigi, H.; Suzuki, A.; Nagasawa, H.; Sakuda, S. *J. Antibiotics* 2001, *54*, 650-657.

[17] Okamoto, S.; Sakurada, M.; Kubo, Y.; Tsuji, G.; Fujii, I.; Ebizuka, Y.; Ono, M.; Nagasawa, H.; Sakuda, S. *Microbiology* 2001, *147*, 2623-2628.

[18] Ikeda, H.; Matsumori, N.; Ono, M.; Suzuki, A.; Isogai, A.; Nagasawa, H.; Sakuda, S. *J. Org. Chem.* 2000, *65*, 438-444.

[19] Sakuda, S.; Ikeda, H.; Nakamura, T.; Nagasawa, H. *Biosci. Biotechnol. Biochem.* 2004, *68*, 407-412.

[20] Sakuda, S.; Matsumori, N.; Furihata, K.; Nagasawa, H. *Tetrahedron Lett.* 2007, *48*, 2527-2531.

[21] Matsumori, N.; Kaneko, D.; Murata, M.; Nakamura, H.; Tachibana, K. *J. Org. Chem.* 1999, *64*, 866-876.

[22] Rychnovsky, S. D.; Rogers, B.; Yang, G. *J. Org. Chem.*, 1993, *58*, 3511-3515.

[23] Ohtani, I.; Kusumi, T.; Kashman, Y.; Kakisawa, H. *J. Am. Chem. Soc.*, 1991, *113*, 4092-4096.

[24] Higashibayashi, S.; Czechtizky, W.; Kobayashi, Y.; Kishi, Y. *J. Am. Chem. Soc.* 2003, *125*, 14379-14393.

[25] Sakuda, S.; Ikeda, H.; Nakamura, T.; Kawachi, R.; Kondo, T.; Ono, M.; Sakurada, M.; Inagaki, H.; Ito, R.; Nagasawa, H. *J. Antibiotics* 2000, *53*, 1378-1384.

[26] Pougny, J. R.; Sinay, P. *J. Chem. Res. Synopses* 1982, 1.

[27] Marffey, P. *Carlsberg Res. Commun.* 1984, *49*, 591-596.

[28] Evans, D. A.; Trenkle, W. C.; Zhang, J.; Burch, J. D. *Org. Lett.* 2005, *7*, 3335-3338.

[29] Bifulco, G.; Dambruoso, P.; Gomez-Paloma, L.; Riccio, R. *Chem. Rev.* 2007, *107*, 3744-3779.

[30] (a) Lemieux, R. U.; Nagabhushan, T. L.; Paul, B. *Can. J. Chem.* 1972, *50*, 773-776. (b) Schwarcz, J. A.; Perlin, A. S. *Can. J. Chem.* 1972, *50*, 3667-3676. (c) Wasylishen, R.; Schaefer, T. *Can. J. Chem.* 1973, *51*, 1, 961-973. For a review, see: Hansen, P. E. *Prog. NMR Spectrosc.* 1981, *14,* 175-296.

[31] Schwarcz, J. A.; Cyr, N.; Perlin, A. S. *Can. J. Chem.* 1975, *53*, 1872-1875. (b) Parella, T.; Sanchez-Ferrando, F.; Virgili, A. *Magn. Reson. Chem.* 1994, *32*, 657-664. (c) Podlasek, C. A.; Wu, J.; Stripe, W. A.; Bondo, P. B.; Serianni, A. S. *J. Am. Chem. Soc.* 1995, *117*, 8635-8644.

[32] Griesinger, C.; Sørensen, O. W.; Ernst, R. R. *J. Am. Chem. Soc.* 1985, *107*, 6394-6396.

[33] Marquez, B. L.; Gerwick, W. H.; Williamson, R. T. *Magn. Reson. Chem.* 2001, *39*, 499-530.

[34] (a)Otting, G.; Wüthrich, K. *Quart. Rev. Biophys.* 1990, *23*, 39-96. (b) Wollborn, U.; Leibfritz, D. *J. Magn. Reson.* 1992, *98*, 142-146. (c) Kurz, M.; Schmieder, P.; Kessler, H. *Angew. Chem., Int. Ed. Engl.* 1991, *30*, 1329-1331.

[35] Zhu, G.; Bax, A. *J. Magn. Reson.* 1993, *104A*, 353-357. (b) Zhu, G.; Live, D.; Bax, A. *J. Am. Chem. Soc.* 1994, *116*, 8370-8371.

[36] Willker, W.; Leibfritz, D. *Magn. Reson. Chem.* 1995, *33*, 632-638.

[37] (a) Sasaki, M.; Matsumori, N.; Maruyama, T.; Nonomura, T.; Murata, M.; Tachibana, K.; Yasumoto, T. *Angew. Chem., Int. Ed. Engl.* 1996, *35*, 1672-1675. (b) Nonomura, T.; Sasaki, M.; Matsumori, N.; Murata, M.; Tachibana, K.; Yasumoto, T. *Angew. Chem., Int. Ed. Engl.* 1996, *35*, 1675-1678. (c) Matsumori, N.; Nonomura, T.; Sasaki, M.; Murata, M.; Tachibana, K.; Satake, M.; Yasumoto, T. *Tetrahedron Lett.* 1996, *37*, 1269-1272.

[38] Sakai, R.; Kamiya, H.; Murata, M.; Shimamoto, K. *J. Am. Chem. Soc.* 1997, *119*, 4112-4116.

[39] Murata, M.; Matsuoka, S.; Matsumori, N.; Paul, G. K.; Tachibana, K. *J. Am. Chem. Soc.* 1999, *121*, 870-871.

In: Stereochemistry Research Trends
Editor: M. A. Horvat and J. H. Golob

ISBN 978-1-60456-825-7


*Chapter 4*

# SYNTHESES, STRUCTURE, AND ELECTRONIC PROPERTIES OF ASSEMBLIES OF DIASTEREOMERS OF NICKEL(II) COMPLEXES WITH TETRACYANOMETALATES $[M(CN)_4]^{2-}$ (M = NI, PD, AND PT)

***Takashiro Akitsu****
Department of Chemistry, Faculty of Science and Technology, Keio University, 3-14-1 Hiyoshi, Kohoku-ku, Yokohama, Kanagawa 223-8522, Japan

## ABSTRACT

Nine new complexes of assemblies containing diastereomers of nickel(II) complexes incorporating 1,2-diaminocyclohexane ligands and tetracyanometalates, $[Ni(RR\text{-}chxn)_3][M(CN)_4]\cdot 3H_2O$ (M = Ni (*1*), Pt (*4*), Pd (*7*)), $[Ni(trans\text{-}chxn)_3][M(CN)_4]\cdot 3H_2O$ (M = Ni (*2*), Pt (*5*), Pd (*8*)), and $[Ni(cis\text{-}chxn)_3][M(CN)_4]\cdot 3H_2O$ (M = Ni (*3*), Pt (*6*), Pd (*9*)) (*RR*-chxn = cyclohexane-(1*R*,2*R*)-diamine, *trans*-chxn = racemic *trans*-cyclohexane-(1,2)-diamine, and *cis*-chxn = racemic *cis*-cyclohexane-(1,2)-diamine) have been synthesized and characterized by SQUID and IR, XPS, and electronic spectra. Crystal structure of *1* (space group $P2_12_12_1$) revealed that both $\Delta$- and $\Lambda$-$[Ni(RR\text{-}chxn)_3]^{2+}$ cations and isolated square planar $[Ni(CN)_4]^{2-}$ anions without cyano-bridges were contained in the assembly. The emission spectra of *4*, *5*, *6* ($\lambda_{ex}$ = 360 nm) exhibited that different intensity of a peak around 470 nm which depended on interaction of quenching between paramagnetic $[Ni(RR\text{-}chxn)_3]^{2+}$ moieties and $[Pt(CN)_4]^{2-}$ moieties. The CD spectra for *1*, *4*, and *7* showed not only d-d bands around 20000 $cm^{-1}$ of $[Ni(RR\text{-}chxn)_3]^{2+}$ moieties by vicnal effect but also charge transfer bands of $[M(CN)_4]^{2-}$ moieties induced by chiral ligands.

---

* E-mail: akitsu@chem.keio.ac.jp Present Address (concurrent): Department of Chemistry, Stanford University, Stanford, CA 94305, USA. E-mail: akitsu@stanford.edu

## INTRODUCTION

Because of appropriate superexchange pathway between paramagnetic centers, the design and synthesis of multifunctional molecular magnetic materials of cyanometalates have been the focus of increasing attention over the past decades [1, 2]. In particular, photomagnets [3-7] and chiral magnets [8-13] are still a main challenge in advanced functional materials chemistry as well as inorganic chemistry.

I have extensively studied new cyano-bridged bimetallic assemblies as a function of structural and electronic factors and new possibility of interaction between photofunctional and magnetic moieties. Recently, I found novel long axial coordination bonds of Cu-N = 3.120(8) Å due to pseudo Jahn-Teller elongation and semi-coordination with ionic nature in a one-dimensional zigzag coordination polymer, $[Cu(RR\text{-chxn})_2][Ni(CN)_4]\cdot 2H_2O$ [14]. On the other hand, I observed novel luminescence of $[Pt(CN)_4]^{2-}$ moieties in one-dimensional coordination polymers, $[Cu(N\text{-ethylethylenediamine})_2][Pt(CN)_4]$, without quenching by paramagnetic ions [15]. Although pseudo Jahn-Teller effect may result in so-called 'Jahn-Teller switching', crossover of magnetic orbitals of the $^2D$ term [16, 17], these characteristic cyano-bridged copper(II) complexes can be attributed to weak interaction between a copper(II) complex and tetracyanometalates. In this context, it is worth investigating assemblies of non-bonded cyanometalates and related paramagnetic chiral complexes to compare interactions between metal centers or induced by chirality, quenching of luminescence, and electronic properties by various spectroscopic methods.

Herein, I report the synthesis, crystal structure, and magnetic and spectroscopic properties (IR, XPS, luminescence, and electronic and CD spectra) of nine new complexes of assemblies containing diastereomers of nickel(II) complexes incorporating 1,2-diaminocyclohexane ligands and tetracyanometalates, $[Ni(RR\text{-chxn})_3][M(CN)_4]\cdot 3H_2O$ (M = Ni (*1*), Pt (*4*), Pd (*7*)), $[Ni(trans\text{-chxn})_3][M(CN)_4]\cdot 3H_2O$ (M = Ni (*2*), Pt (*5*), Pd (*8*)), and $[Ni(cis\text{-chxn})_3][M(CN)_4]\cdot 3H_2O$ (M = Ni (*3*), Pt (*6*), Pd (*9*)) (*RR*-chxn = cyclohexane-(1*R*,2*R*)-diamine, *trans*-chxn = racemic *trans*-cyclohexane-(1,2)-diamine, and *cis*-chxn = racemic *cis*-cyclohexane-(1,2)-diamine). I investigate systematically tuning of electronic properties depending on both diastereomers of chiral ligands, namely *stereochemistry*, and substitution of transition metal in tetracyanometalates.

## EXPERIMENTAL SECTION

### Materials

All reagents and solvents (Wako and Aldrich) were commercially available and were used as purchased without further purification.

The precursor complexes, $[Ni(RR\text{-chxn})_3](ClO_4)_2$, $[Ni(trans\text{-chxn})_3](ClO_4)_2$, and $[Ni(cis\text{-chxn})_3](ClO_4)_2$ (*RR*-chxn = cyclohexane-(1*R*,2*R*)-diamine, *trans*-chxn = racemic *trans*-cyclohexane-(1,2)-diamine, and *cis*-chxn = racemic *cis*-cyclohexane-(1,2)-diamine), were prepared by mixing an ethanol solutions of $Ni(ClO_4)\cdot 6H_2O$ (0.515 g, 2.00 mmol) and the corresponding chxn ligands (0.457 g, 4.00 mmol) and filtration of the resulting red violet precipitates.

## Preparations

*Preparation of [Ni(RR-chxn)$_3$][Ni(CN)$_4$]·3H$_2$O (1).* Slow diffusion of acetone solution (20 mL) of [Ni(*RR*-chxn)$_3$](ClO$_4$)$_2$ (0.40 mg, 0.68 mmol) onto aqueous solution (10 mL) of $K_2$[Ni(CN)$_4$] (0.164 g, 0.68 mmol) gave rise to reddish violet or pink precipitates at 298 K for several days. Yield 0.3835 g (93.0 %). Anal. Found; C 43.61; H 5.99; N 23.11. Calc. for $C_{22}H_{36}N_{10}Ni_2O_3$: C 43.77; H 5.82; N 23.02. IR (KBr, cm$^{-1}$): 1010(s), 1118(s), 1202(w), 1236(w), 1361(w), 1400(w), 1445(s), 1604(s), 2118(s) (cyanide), 2858(s), 2932(s), 3175(m), 3271(s), 3327(s).

*Preparation of [Ni(trans-chxn)$_3$][Ni(CN)$_4$]·3H$_2$O (2).* Slow diffusion of acetone solution (20 mL) of [Ni(*trans*-chxn)$_3$](ClO$_4$)$_2$ (0.40 mg, 0.68 mmol) onto aqueous solution (10 mL) of $K_2$[Ni(CN)$_4$] (0.164 g, 0.68 mmol) gave rise to reddish violet or pink precipitates at 298 K for several days. Yield 0.1567 g (33.0 %). Anal. Found; C 43.61; H 5.99; N 23.11. Calc. for $C_{22}H_{36}N_{10}Ni_2O_3$: C 43.79; H 5.79; N 22.93. IR (KBr, cm$^{-1}$): 1017(s), 1121(s), 1230(w), 1444(m), 1577(s), 2113(s) (cyanide), 2858(s), 2926(s), 3175(m), 3273(s), 3325(s), 3446(s).

*Preparation of [Ni(cis-chxn)$_3$][Ni(CN)$_4$]·3H$_2$O (3).* Slow diffusion of acetone solution (20 mL) of [Ni(*cis*-chxn)$_3$](ClO$_4$)$_2$ (0.40 mg, 0.68 mmol) onto aqueous solution (10 mL) of $K_2$[Ni(CN)$_4$] (0.164 g, 0.68 mmol) gave rise to reddish violet or pink precipitates at 298 K for several days. Yield 0.2459 g (59.7 %). Anal. Found; C 43.61; H 5.99; N 23.11. Calc. for $C_{22}H_{36}N_{10}Ni_2O_3$: C 43.69; H 5.83; N 22.85. IR (KBr, cm$^{-1}$): 976(s), 1016(s), 1050(s), 1089(s), 1157(m), 1197(m), 1253(w), 1383(m), 1445(m), 1592(s), 2118(s) (cyanide), 2154(s), 2853(s), 2931(s), 3169(s), 3265(s), 3322(s).

*Preparation of [Ni(RR-chxn)$_3$][Pt(CN)$_4$]·3H$_2$O (4).* Slow diffusion of acetone solution (20 mL) of [Ni(*RR*-chxn)$_3$](ClO$_4$)$_2$ (0.40 mg, 0.68 mmol) onto aqueous solution (10 mL) of $K_2$[Pt(CN)$_4$] (0.293 g, 0.68 mmol) gave rise to reddish violet or pink precipitates at 298 K for several days. Yield 0.2911 g (57.8 %). Anal. Found; C 35.59; H 4.89; N 18.87. Calc. for $C_{22}H_{36}N_{10}NiO_3Pt$: C 35.55; H 4.90; N 18.83. IR (KBr, cm$^{-1}$): 1016(s), 1118(s), 1230(w), 1298(w), 1361(w), 1400(w), 1445(m), 1598(s), 2129(s) (cyanide), 2858(s), 2920(s), 3163(s), 3270(s), 3293(s), 3321(s).

*Preparation of [Ni(trans-chxn)$_3$][Pt(CN)$_4$]·3H$_2$O (5).* Slow diffusion of acetone solution (20 mL) of [Ni(*trans*-chxn)$_3$](ClO$_4$)$_2$ (0.40 mg, 0.68 mmol) onto aqueous solution (10 mL) of $K_2$[Pt(CN)$_4$] (0.293 g, 0.68 mmol) gave rise to reddish violet or pink precipitates at 298 K for several days. Yield 0.1088 g (21.6 %). Anal. Found; C 35.59; H 4.89; N 18.87. Calc. for $C_{22}H_{36}N_{10}NiO_3Pt$: C 35.74; H 4.95; N 18.82. IR (KBr, cm$^{-1}$): 1019(s), 1118(s), 1235(w), 1452(m), 1579(s), 1649(s), 2125(s) (cyanide), 2860(s), 2936(s), 3166(m), 3261(s), 3322(s), 3477(s).

*Preparation of [Ni(cis-chxn)$_3$][Pt(CN)$_4$]·3H$_2$O (6).* Slow diffusion of acetone solution (20 mL) of [Ni(*cis*-chxn)$_3$](ClO$_4$)$_2$ (0.40 mg, 0.68 mmol) onto aqueous solution (10 mL) of $K_2$[Pt(CN)$_4$] (0.293 g, 0.68 mmol) gave rise to reddish violet or pink precipitates at 298 K for several days. Yield 0.3908 g (79.6 %). Anal. Found; C 35.59; H 4.89; N 18.87. Calc. for $C_{22}H_{36}N_{10}NiO_3Pt$: C 35.69; H 4.99; N 18.81. IR (KBr, cm$^{-1}$): 976(s), 1016(s), 1044(s), 1072(s), 1089(s), 1123(s), 1157(s), 1197(m), 1259(w), 1394(s), 1445(s), 1575(s), 2129(s) (cyanide), 2864(s), 2937(s), 3169(m), 3270(s), 3321(s).

*Preparation of [Ni(RR-chxn)$_3$][Pd(CN)$_4$]·3H$_2$O (7).* Slow diffusion of acetone solution (20 mL) of [Ni(*RR*-chxn)$_3$](ClO$_4$)$_2$ (0.40 mg, 0.68 mmol) onto aqueous solution (10 mL) of $K_2$[Pd(CN)$_4$] (0.197 g, 0.68 mmol) gave rise to reddish violet or pink precipitates at 298 K for several days. Yield 0.3749 g (84.6 %). Anal. Found; C 40.42; H 5.55; N 21.43. Calc. for $C_{22}H_{36}N_{10}NiO_3Pd$: C 40.20; H 5.81; N 21.22. IR (KBr, $cm^{-1}$): 1022(s), 1115(s), 1202(m), 1231(m), 1298(m), 1349(m), 1394(s), 1439(s), 1456(s), 1603(s), 2135(s) (cyanide), 2858(s), 2921(s), 2954(s), 3158(m), 3282(m).

*Preparation of [Ni(trans-chxn)$_3$][Pd(CN)$_4$]·3H$_2$O (8).* Slow diffusion of acetone solution (20 mL) of [Ni(*RR*-chxn)$_3$](ClO$_4$)$_2$ (0.40 mg, 0.68 mmol) onto aqueous solution (10 mL) of $K_2$[Pd(CN)$_4$] (0.197 g, 0.68 mmol) gave rise to reddish violet or pink precipitates at 298 K for several days. Yield 0.1514 g (34.1 %). Anal. Found; C 40.42; H 5.55; N 21.43. Calc. for $C_{22}H_{36}N_{10}NiO_3Pd$: C 40.16; H 5.66; N 21.35. IR (KBr, $cm^{-1}$): 1010(s), 1113(s), 1233(w), 1404(w), 1445(m), 1581(s), 2123(s), 2858(s), 2129(s) (cyanide), 3183(m), 3275(s), 3326(s), 3452(s).

*Preparation of [Ni(cis-chxn)$_3$][Pd(CN)$_4$]·3H$_2$O (9).* Slow diffusion of acetone solution (20 mL) of [Ni(*RR*-chxn)$_3$](ClO$_4$)$_2$ (0.40 mg, 0.68 mmol) onto aqueous solution (10 mL) of $K_2$[Pd(CN)$_4$] (0.197 g, 0.68 mmol) gave rise to reddish violet or pink precipitates at 298 K for several days. Yield 0.2751 g (62.0 %). Anal. Found; C 40.42; H 5.55; N 21.43. Calc. for $C_{22}H_{36}N_{10}NiO_3Pd$: C 40.19; H 5.79; N 21.31. IR (KBr, $cm^{-1}$): 982(s), 1016(s), 1044(s), 1090(m), 1118(m), 1158(w), 1193(w), 1262(w), 1387(m), 1449(s), 1592(s), 2129(s) (cyanide), 2859(s), 2933(s), 3270(s), 3320(s), 3435(s).

## Physical Measurements

Elemental analyses (C, H, N) were carried out an Elementar Vario EL analyzer at Keio University. Infrared spectra were recorded as KBr pellets on a JASCO FT-IR 660 plus spectrophotometer in the range of 4000-400 $cm^{-1}$ at 298 K. Diffuse reflectance electronic spectra were measured on a JASCO V-560 spectrophotometer equipped with an integrating sphere in the range of 850-220 nm at 298 K. Luminescence excitation and emission spectra of the solid compounds were recorded by a JASCO FP-6500 spectrofluorometer using a sample holder developed for powders at 298 K. X-ray photoelectron spectra (XPS) were recorded with a JEOL JPS-9000MX at 298 K. Powdered samples were pressed as pellets and put under UHV to reach the $10^{-8}$ Pa range. The nonmonochromatized Mg K$\alpha$ source was used at 10 kV and 10 mA, as a flood gun to compensate for the nonconductive samples. The binding energy of the spectra was calibrated in relation to the C 1s binging energy (284.0 eV), which was applied as an internal standard. The XPS data have been deposited to the database of *Surface Science Spectra* (American Vacuum Society). The magnetic properties of *1*, *2*, and *3* were investigated with a Quantum Design MPMS-XL superconducting quantum interference device magnetometer (SQUID) at an applied field 5000 Oe in a temperature range 5-300 K at 5000 Oe. Powder samples (20.17, 19.52, and 17.78 mg for *1-3*, respectively) were measured in a pharmaceutical cellulose capsule. The apparatus signals and the diamagnetic corrections were evaluated from Pascal's constants.

## X-ray Crystallography

Reddish violet or pink single crystals of *1* (0.30 x 0.30 x 0.30 mm) were glued on top of a glass fiber and coated with a thin layer of epoxy resin to measure the diffraction data. The X-ray intensities were measured at 297 K with graphite-monochromated Mo K$\alpha$ radiation ($\lambda$ = 0.71073 Å) on a Rigaku four-circle diffractometer AFC-7R. The structures were solved by direct methods using SIR 92 [18] and expanded by Fourier techniques. The structures were refined on $F^2$ anidotropically for non-hydrogen atoms by full-matrix least-square methods with SHELXL97 [19] on a teXsan program package [20] on an SGI Indy workstation. Empirical absorption corrections were applied based on $\psi$ scans. No significant decay in the intensity of three standard reflections was observed throughout the data collection. All the non-hydrogen atoms were refined anisotropically. The hydrogen atoms were located at geometrically calculated position with C-H = 0.950 Å and refined isotropically. Since hydrogen atoms could not be observed in the difference Fourier map, I could not introduce hydrogen atoms of water molecules connected to O1, O2, and O3 atoms. Although residual electron density was also found in the difference Fourier maps, the models containing these peaks as water molecules resulted in increase of R-values. Therefore, I could not assign them and I could not discuss hydrogen bonds closely. Unfortunately, I could not obtain suitable single crystals for *2-9*.

Crystallographic data for *1*. $C_{22}H_{36}N_{10}Ni_2O_3$, orthorhombic, $P2_12_12_1$, $a$ = 16.223(3) Å, $b$ = 16.392(4) Å, $c$ = 11.855(3) Å, and $V$ = 3152.6(12) Å$^3$, and $Z$ = 4, $M$ = 606.03, $D_c$ = 1.277 Mgm$^{-3}$, $R_1$ = 0.0726, $wR_2$= 0.3233 (4035 reflections), $S$ = 1.189, $T$ = 297(2) K, $\mu$= 1.232 mm$^{-1}$. Flack value is equal to 0.00(7).

# Results and Discussion

## Crystal Structure of 1

The asymmetric unit of *1* consists of a $[Ni(RRchxn)_3]^{2+}$ cation and an independent $[Ni(CN)_4]^{2-}$ anion and three molecules of water of crystallization. An ORTEP drawing of *1* is shown in Figure 1 with selected bond distances and angles in its caption. The $[Ni(RRchxn)_3]^{2+}$ cation adopts a slightly distorted octahedral coordination geometry with C2-Ni1-C1 = 178.7(7), C4-Ni1-C3 = 179.0(10), and N6-Ni2-N8 = 171.5(5)° incorporating three chiral *RR*chxn ligands through amino groups. The cation has pseudo $D_3$ symmetry with an approximately threefold rotation axis, which is in agreement with paramagnetic nickel(II) center of $^3A_{2g}$ ground state. Because there are $\Delta,\Lambda$-enantiomers of $[Ni(RRchxn)_3]^{2+}$ cations with the same conformation in the crystal, the only chiral source was 1- and 2- carbon atoms in the chiral bidentate ligands. The cyclohexane rings of the ligands appear to have a chair conformation. Most of the bond lengths and angles of the $[Ni(RRchxn)_3]^{2+}$ cation are in usual values [21-24].

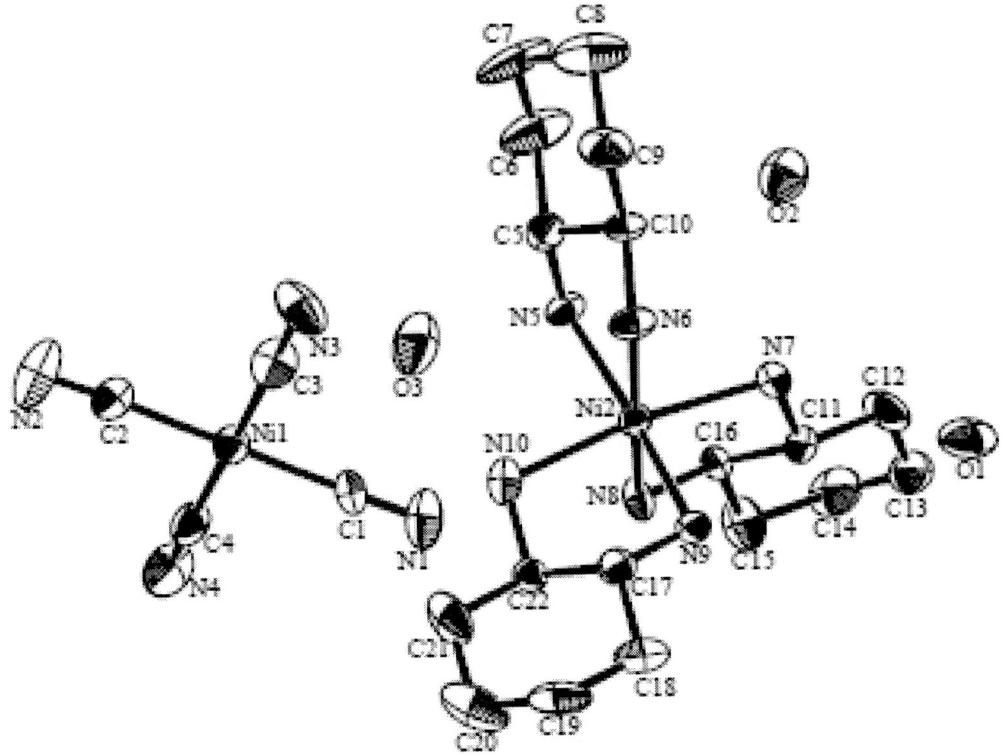


Figure 1. ORTEP drawing of the asymmetric unit of *1* showing the atom-labeling scheme. Displacement ellipsoids are drawn at the 30 % probability level. Hydrogen atoms are omitted for clarity. Selected bond distances (Å) and angles (°) for *1*: Ni1-C1 = 1.861(15), Ni1-C2 = 1.859(17), Ni1-C3 = 1.867(16), Ni1-C4 = 1.834(17), Ni2-N5 2.129(11), Ni2-N6 = 2.118(12), Ni2-N7 = 2.134(12), Ni2-N8 = 2.123(12), Ni2-N9 = 2.127(11), Ni2-N10 = 2.128(13), N1-C1 = 1.12(2), N2-C2 = 1.13(2), N3-C3 = 1.14(2), N4-C4 = 1.14(2), N5-C5 = 1.479(19), N6-C10 = 1.516(18), N7-C11 = 1.47(2), N8-C16 = 1.475(19), N9-C17 = 1.48(2), N10-C22 = 1.51(2), C4-Ni1-C2 = 90.7(7), C4-Ni1-C1 = 88.0(7), C2-Ni1-C1 = 178.7(7), C4-Ni1-C3 = 179.0(10), C2-Ni1-C3 = 90.0(8), C1-Ni1-C3 = 91.3(7), N6-Ni2-N8 = 171.5(5), N6-Ni2-N9 = 92.8(4), N8-Ni2-N9 = 94.9(5), N6-Ni2-N10 = 92.0(5), N8-Ni2-N10 = 92.6(5), N9-Ni2-N10 = 81.9(5), N6-Ni2-N5 = 81.9(4), N8-Ni2-N5 = 90.7(5), N9-Ni2-N5 = 172.5(5), N10-Ni2-N5 = 92.9(5), N6-Ni2-N7 = 94.5(5), N8-Ni2-N7 = 81.7(5), N9-Ni2-N7 = 91.7(5), N10-Ni2-N7 = 171.2(5), N5-Ni2-N7 = 93.9(5).

On the other hand, the $[Ni(CN)_4]^{2-}$ anion adopts a four-coordinated square planar coordination geometry without bridging cyano ligands. The observed bond lengths and angles exhibit usual values, though *1* is not one-dimensional structure [25-28] or two-dimensional or more complicated cyano-bridged structures [29-31] but discrete cocrystal [33-39], and both nickel atoms of *1* are not at the center of symmetry.

Possible hydrogen bonds are as follows (parent atoms, Å): O(1)···N(7) = 3.33(1), O(2)···N(7) = 3.35(2), O(3)···N(3) = 3.03(2), O(3)···N(6) = 3.25(2), O(3)···N(10) = 3.10(2), N(1)···N(8) = 3.08(1), N(1)···N(10) = 3.22(1), O(1)···$N(4)^{i}$ = 3.03(2), O(1)···$N(2)^{ii}$ = 2.77(2), O(2)···$O(3)^{iii}$ = 2.93(2), N(4)···$N(6)^{iv}$ = 3.23(1), N(4)···$N(9)^{iv}$ 3.04(1), N(8)···$N(3)^{ii}$ = 3.34(2) [Symmetry codes: (i) x, y, z+1, (ii) x+1/2,-y+3/2,-z+1, (iii) -x, y+1/2,-z+3/2, (iv) x, y, z-1].

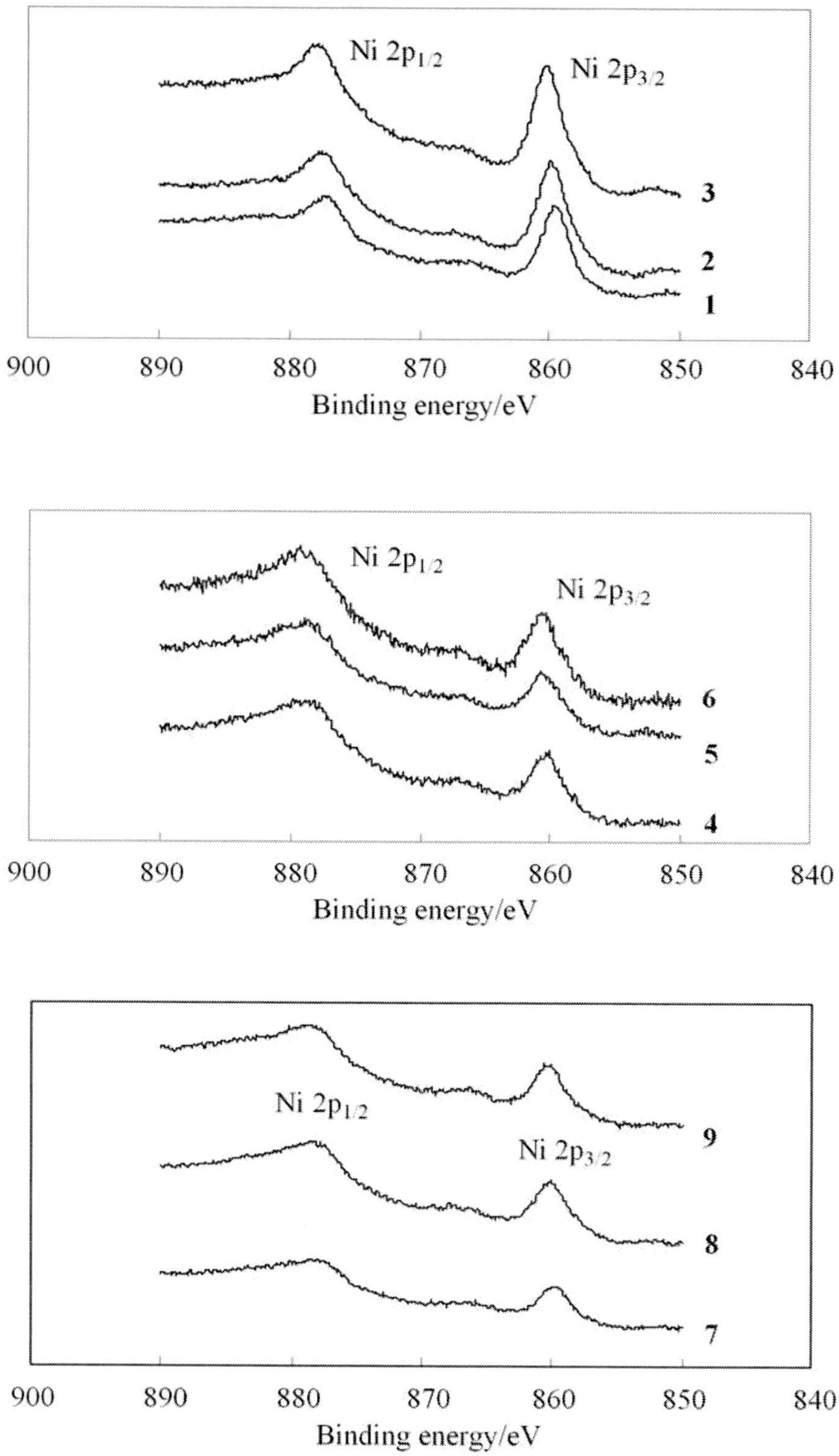


Figure 2. XPS spectra of Ni $2p_{1/2}$ and Ni $2p_{3/2}$ peaks for *1-9* at 298 K.

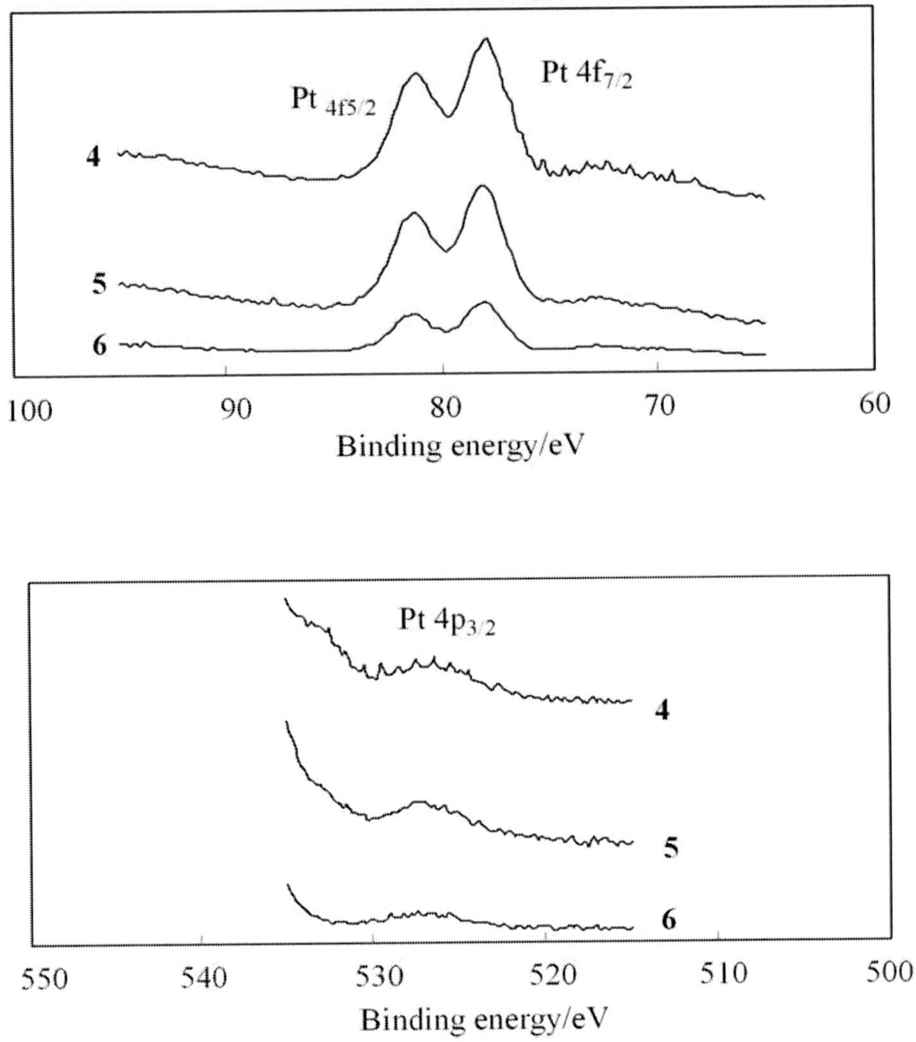


Figure 3. XPS spectra of Pt $3d_{5/2}$, Pt $3d_{5/2}$, and Pt $4p_{3/2}$ peaks for *4-6* at 298 K.

## XPS

The wide survey XPS of *1-9* was measured in the range 0 - 1000 eV at 298 K (not shown). The C1s, N1s, and O1s peaks were observed at 284, 398, and 530 eV, respectively. The spectral features of *1-9* indicated that organic ligands did not exhibit detectable shift of peaks or satellite peaks and all co-crystals did not exhibit changes of valence states. They are similar to those of cyano-bridged complexes such as [Cu(*RR*-chxn)$_2$][Ni(CN)$_4$]$\cdot$2H$_2$O and [Cu(*RR*-chxn)$_2$][Pt(CN)$_4$]$\cdot$2H$_2$O [14].

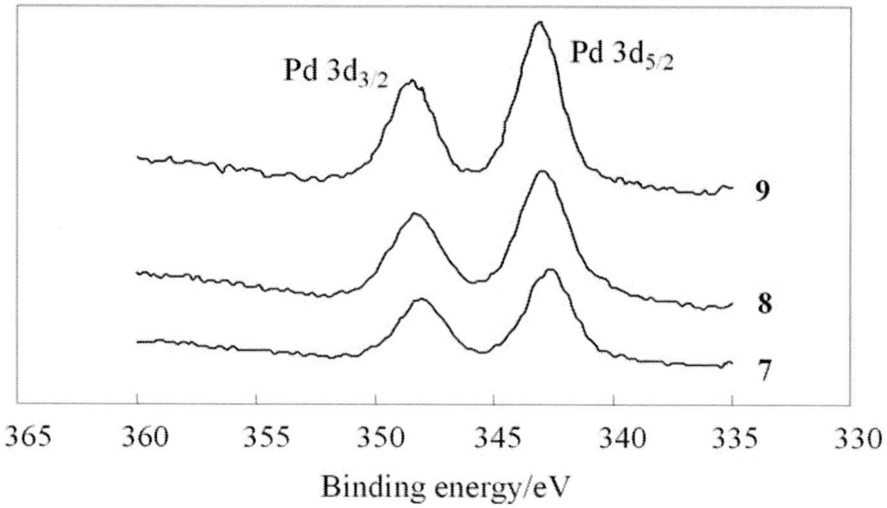


Figure 4. XPS spectra of Pd $3d_{3/2}$ and Pd $3d_{5/2}$ peaks for *7-9* at 298 K.

I plotted the intensity of XPS spectra (Figures 2-4) in arbitrary unit scale. The Ni $2p_{1/2}$ and Ni $2p_{3/2}$ peaks are observed at about 860 and 878 eV, respectively (Figure 2). There are little differences of binding energies between complexes containing only $[Ni(chxn)_3]^{2+}$ moiety (*4-9*) and both $[Ni(chxn)_3]^{2+}$ and $[Ni(CN)_4]^{2-}$ moieties (*1-3*), which also support double salts without cyano-bridges for *1-9*. However, the Pt $4p_{3/2}$, $4f_{5/2}$, and $4f_{7/2}$ peaks for *4-6* were observed at about 527, 81, 78, respectively, showing quite slight shifts against 525.6, 79.8, and 76.6 eV for$[Cu(RR\text{-}chxn)_2][Pt(CN)_4]\cdot 2H_2O$ and 525.8, 80.9, and 77.5 eV for $K_2[Pt(CN)_4]$ [14] (Figure 3). The $Pd3d_{3/2}$ and $Pd3d_{5/2}$ peaks are appeared at about 348 and 343 eV for *7-9* (Figure 4). As the crystal structure of *1*, there are no direct bonding between cations and anions, though negative chemical shifts in XPS can be expected for the system containing more electronegative atoms. The results of XPS are in good agreement with the structural features and do not show changes of valence states by chirality of ligands or metal ions in counter ions.

## IR Spectra

The relevant bands of IR spectra are listed in the experimental section. As for *1-9*, four cyanide groups in $[M(CN)_4]^{2-}$ moieties are non-bridging ones. IR spectra show strong cyanide stretching peaks without splitting at about 2100-2150 $cm^{-1}$ for *1-9*. Though the spectral features are in agreement with common values of 2000-2100 $cm^{-1}$ for general cyanide complexes [39], it is unclear why gradual shift by metal substitution for these double salts, namely $[Ni(CN)_4]^{2-}$ at 2128 $cm^{-1}$, $[Pd(CN)_4]^{2-}$ at 2143 $cm^{-1}$, $[Pt(CN)_4]^{2-}$ at 2150 $cm^{-1}$. Presumably, it can be assumed that fashions of hydrogen bonds and other weak intermolecular interactions contribute to the band shifts.

Unfortunately, I could not detect changes of IR spectra (especially cyanide stretching peak) after UV or visible light illumination at room temperature for *1-9* in the solid state.

## Emission Spectra

Room temperature solid-state emission spectra of cyano-bridged complexes are complicated by quenching and may be useful evidence for electronic interactions between (paramagnetic) ions and $[Pt(CN)_4]^{2-}$ moieties. The emission spectra ($\lambda_{ex}$ = 360 nm) at 298 K exhibited a strong peak at 475 nm for *5*, a weak peak at 475 nm for *4*, while almost no peak around 475 nm for *6*. The results of emission spectra of *4-6* exhibited that different intensity of a peak around 470 nm which depended on interaction of quenching between paramagnetic $[Ni(chxn)_3]^{2+}$ moieties and emissive $[Pt(CN)_4]^{2-}$ moieties. To our best knowledge, this is the first example that observed different emission quenching for non-cyano-bridging systems by varying chirality of ligands.

On the other hand, the corresponding peaks were observed as a peak around 446 nm for $K_2[Pt(CN)_4]$ and a broad band with peaks at 439 and 417 nm for a one-dimensional cyano-bridged coordination polymer, [Cu(*N*-ethylethylenediamine)$_2$][Pt(CN)$_4$] [15]. However, one-dimensional zigzag cyano-bridged coordination polymer, [Cu(*RR*-chxn)$_2$][Pt(CN)$_4$]$\cdot$2H$_2$O with long Cu-N semi-coordination bond did not exhibit emission peak under the same conditions [14].

Furthermore, although monocuclear europium(III) complexes are emissive with $^5D_0 \rightarrow {}^7F_2$ and $^5D_0 \rightarrow {}^7F_1$ luminescence generally, it is not emissive in the heterobimetallic magnetically-coupled systems commonly because of quenching. In Eu(DMF)$_4$(H$_2$O)$_3$Co(CN)$_6\cdot$H$_2$O (DMF = *N,N'*-dimethylformamide), cyano-bridged metal ions may potentially indicated superexchange interaction, but the europium(III) ion is emissive because of diamagnetic ground state of cobalt(III) ion mainly. However, superexchange interaction between diamagnetic ions and paramagnetic ions and its contribution for quenching of luminescence is complicated for cyano-bridged systems [40].

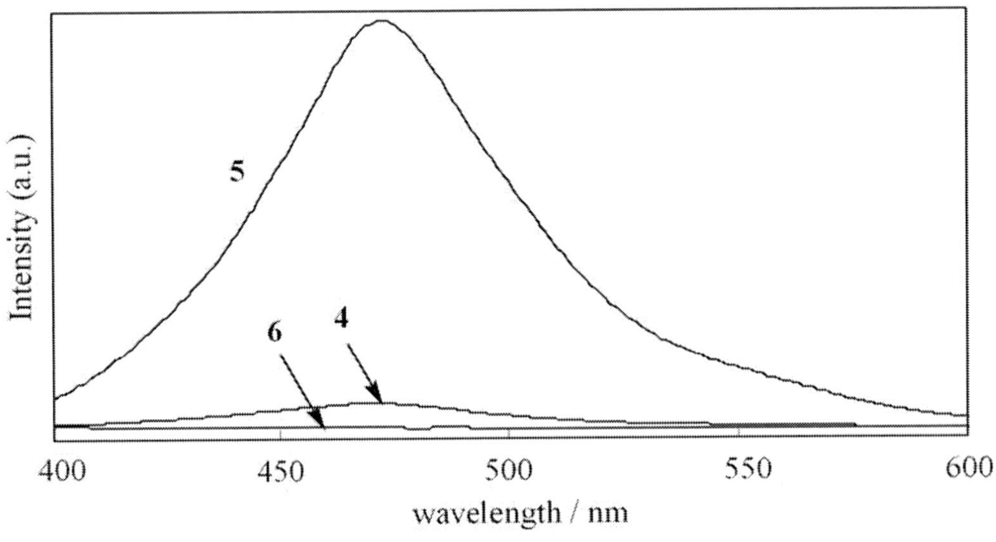


Figure 5. Solid state emission spectra ($\lambda_{ex}$ = 360 nm) of *4-6* at 298K.

## Electronic and CD Spectra

Diffuse reflectance electronic spectra showed strong d-d band of nickel(II) moieties and charge transfer bands of cyanide ligands. The d-d bands (mainly $^3A_{2g}$ -> $^3T_{1g}$) are observed at 18500, 18300, 18000, 18500, 18200, 18000, 18300, 18300, and 17900 $cm^{-1}$ for *1-9*, respectively. The assignment of d-d transitions has been established for $[Ni(ethylenediamine)_3]^{2+}$ as follows [41]: $^3T_{2g}$ 11700 $cm^{-1}$, $^3T_{1g}$ 18350 $cm^{-1}$, $^3T_{1g}(P)$ 29000 $cm^{-1}$ from the $^3A_{2g}$ ground state.

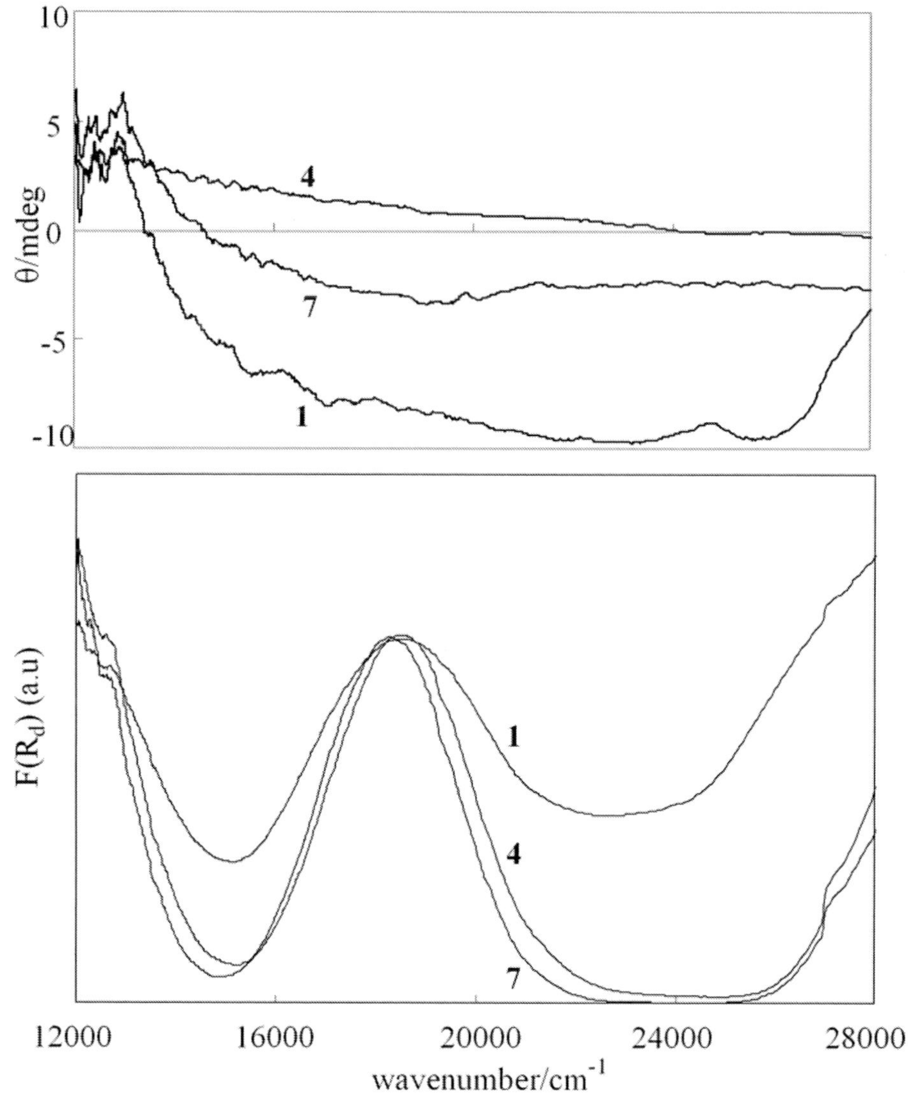


Figure 6. Diffuse reflectance electronic spectra and CD spectra in KBr pellets for *1*, *4*, and *7* at 298 K in the region of d-d bands.

As for intermolecular interaction of square planar $[M(CN)_4]^{2-}$ complexes, so-called "solid state effect" has been known and investigated for their interesting electronic perturbation along the stacking axis. It causes the out-of-plane-polarized $a_{1g}$ ($d_{z2}$) $a_{2u}$ ($p_z$, *) transition to red shift and intensity as the stack closer in stacked planes of $[Pt(CN)_4]^{2-}$ in crystals prepared from mixtures of various $[M(CN)_4]^{2-}$ complexes. And single out-of-plane-polarized $a_{1g}$ ($d_{z2}$) $a_{2u}$ ($p_z$, *) transition was found whose intensity and energy were proportional to the composition of mixture [42, 43]. The results of ZINDO calculations for simulating electronic spectra with assignment of predominant peak components for *1* by using molecular models of isolated $[Ni(CN)_4]^{2-}$, $[Ni(RRchxn)_3]^{2+}$ complexes and $[Ni(RRchxn)_3][Ni(CN)_4]$ double salts based on crystal structures are summarized in supplementary data (*vide infra*).

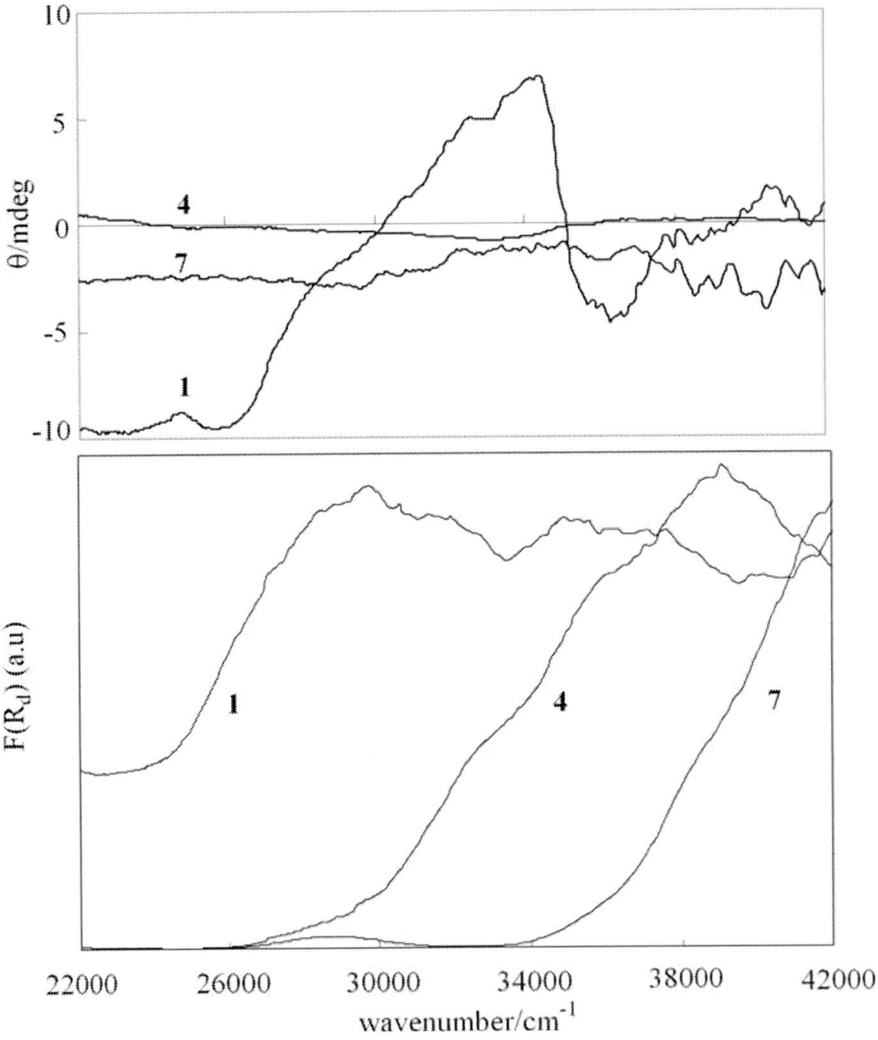


Figure 7. Diffuse reflectance electronic spectra and CD spectra in KBr pellets for *1*, *4*, and *7* at 298 K in the region of charge transfer bands.

As for chiral crystals *1*, *4*, and *7*, solid state CD spectra will provide chiral vicinal effects by the ligands of $[Ni(RRchxn)_3]^{2+}$ cations and induced effect detected as achiral $[M(CN)_4]^{2-}$ anions. The visible region mainly composed of d-d bands of nickel(II) ions are shown in Figure 6. The band width around 19000 $cm^{-1}$ of *1* is larger than that of *4* and *7*, which are attribute that *1* has two different nickel(II) complexes. The corresponding CD spectra are weak and indicate similar fashions to the diffuse reflectance spectra, which is support that the predominant origin of the CD spectra is vicinal effects of chiral ligands of $[Ni(RRchxn)_3]^{2+}$ moiety. On the other hand, charge transfer region exhibits drastic differences depending on central metal ions of $[M(CN)_4]^{2-}$ moieties. Diffuse reflectance spectra show gradual shift around 30000, 39000, and 42000 $cm^{-1}$ for *1*, *4*, and *7*, respectively. However, the corresponding CD spectrum of *1* is characteristic, namely strong positive peak around 33000 $cm^{-1}$ and weak negative peak around 35000 $cm^{-1}$. While the spectra of *4* and *7* exhibit merely weak Cotton effect in the corresponding wavenumber region. The features suggest that interaction between $[Ni(RRchxn)_3]^{2+}$ cation and $[M(CN)_4]^{2-}$ anion may be weaken as the larger ion radii of $[M(CN)_4]^{2-}$. The range of electronic and magnetic dipole transitions in molecular assemblies are dependent on slight distance of components, which is confirmed directly as far as I investigated.

## Magnetic Properties

The plots of $\chi_M T$ vs $T$ for *1-3* are shown in Figure 8, where the $\chi_M$ values are molar susceptibility for the $[Ni(chxn)_3][Ni(CN)_4]$ units. The temperature variations of the magnetic susceptibility at $H = 0.5$ T in the temperature range 5-300 K. The magnetic data are in accord with their spin states, $[Ni(chxn)_3]^{2+}$ (S = 1) and $[Ni(CN)_4]^{2-}$ (S = 0), as expected based on electronic spectra. However, magnetic interactions between complexes can not be observed as the crystal structure of isolated co-crystals of *1* in which no cyano-bridges are found.

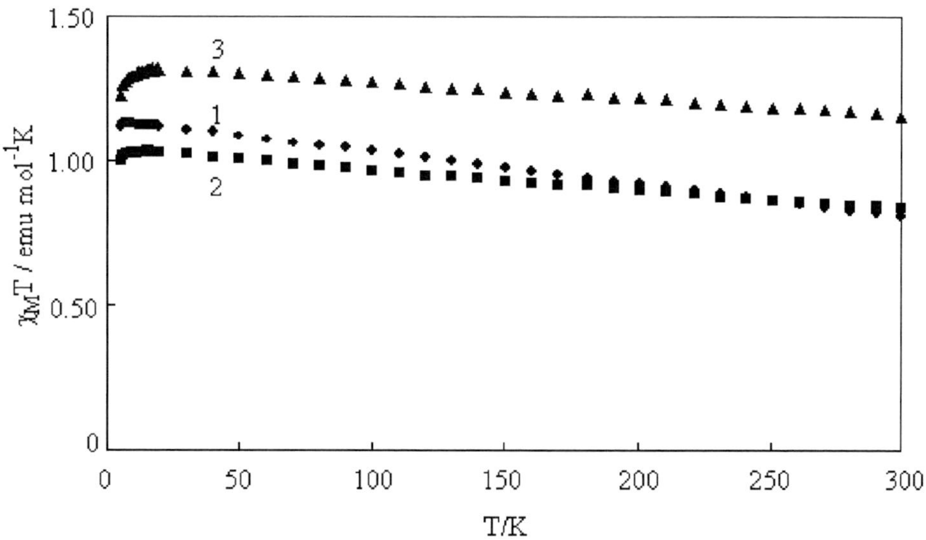


Figure 8. Plots of the $\chi_M T$ vs $T$ in the range of 5–300 K for *1-3* at 0.5 T.

## CONCLUSION

In summary, I prepared and characterized nine new complexes and investigate systematically tuning of electronic properties depending on diastereomers of chiral ligands and substitution of transition metal in tetracyanometalates. I could obtained spectroscopic evidence of intermolecular interactions between $[Ni(RR\text{-chxn})_3]^{2+}$ moieties and $[Pt(CN)_4]^{2-}$ moieties. The emission spectra of *4*, *5*, *6* ($\lambda_{ex}$ = 360 nm) exhibited that different intensity of a peak around 470 nm which depended on interaction of quenching between paramagnetic $[Ni(RR\text{-chxn})_3]^{2+}$ moieties and $[Pt(CN)_4]^{2-}$ moieties. The CD spectra for *1*, *4*, and *7* showed not only d-d bands around 20000 $cm^{-1}$ of $[Ni(RR\text{-chxn})_3]^{2+}$ moieties by vicnal effect but also charge transfer bands of $[M(CN)_4]^{2-}$ moieties induced by chiral ligands. These new results will be useful for designing new molecular assemblies containing photofunctional and magnetic materials.

## SUPPLEMENTARY DATA

### Crystallographic Data

Crystallographic information will be available (CCDC 649723) from the Cambridge Crystallographic Data Centre, 12 Union Road, Cambridge CB2 1EZ, UK (e-mail: deposit@ccdc.cam.ac.uk).

### ZINDO Calculations

The results of ZINDO calculations for simulating electronic spectra with assignment of predominant peak components for *1* by using molecular models of isolated $[Ni(CN)_4]^{2-}$, $[Ni(RR\text{chxn})_3]^{2+}$ complexes and $[Ni(RR\text{chxn})_3][Ni(CN)_4]$ double salts based on crystal structures are summarized. Semi-empirical molecular orbital calculations were performed with a program ZINDO [44, 45] in a CAChe software package based on the crystal structures determined.

Musselman *et al.* reported the results of ZINDO calculations to establish detailed assignment of electronic transitions for an isolated $[Ni(CN)_4]^{2-}$ complex [42]. In addition, they discussed "solid state effect" by comparing between isolated $[Ni(CN)_4]^{2-}$ or $[Pt(CN)_4]^{2-}$ complexes and three $[Ni(CN)_4]^{2-}$ models for one-dimensional stacked structures [43]. According to these studies, I carried out calculations by using our crystal structural data in order to intermolecular interactions between $[Ni(CN)_4]^{2-}$ and $[Ni(RR\text{chxn})_3]^{2+}$ moieties.

At first, I dealt with a square planar $[Ni(CN)_4]^{2-}$ moiety, of which coordination geometry exhibit distortion slightly. The electronic spectrum (Figure S1) appear two strong bands at about 240 and 270 nm assigned to HOMO-1 (-2.180 eV) -> LUMO (8.456 eV) (Figure S4) and HOMO (1.987 eV) -> LUMO (8.456 eV) (Figure S5) transitions, respectively. The distribution of molecular orbitals associated with transitions are symmetric and on the atoms. The $[Ni(RR\text{chxn})_3]^{2+}$ moiety appeared expected electronic spectrum (Figure S2) with strong bands less than 300 nm, which are assigned to HOMO-6 (-17.466 eV) -> LUMO (-7.507 eV) (Figure S6) and HOMO-2 (-17.003 eV) -> LUMO (-7.507 eV) (Figure S7).

On the other hands, the double salt [Ni(*RR*chxn)$_3$][Ni(CN)$_4$] model provided different expected spectrum from simple superimposed spectra of the components. Figure S3 appeared distinct peaks at about 200, 280, 360, 450, and 640 nm. In contrast to the cases of each moiety, the components are attributed to transitions between molecular orbitals distributing intermolecular $[Ni(CN)_4]^{2-}$ and [Ni(*RR*chxn)$_3$]$^{2+}$ moieties. The peak at 200 nm is convolution of a shorter wavelength component of HOMO-3 (-7.392 eV) on the $[Ni(CN)_4]^{2-}$ -> LUMO+7 (2.077 eV) on [Ni(*RR*chxn)$_3$]$^{2+}$ moiety (Figure S8) and a longer wavelength component of HOMO-4 (-7.675 eV) on the $[Ni(CN)_4]^{2-}$ -> LUMO+1 (0.696 eV) on [Ni(*RR*chxn)$_3$]$^{2+}$ moiety (Figure S9). The peak at 280 and 640 nm is ascribed to HOMO-2 (-7.060 eV) on the $[Ni(CN)_4]^{2-}$ -> LUMO (-2.785 eV) on [Ni(*RR*chxn)$_3$]$^{2+}$ moiety (Figure S10).

Interestingly, the following bands are not observed in the corresponding spectra of the components. The peak at 360 nm is ascribed to HOMO-6 (-8.658 eV) on the $[Ni(CN)_4]^{2-}$ -> LUMO (-2.785 eV) on [Ni(*RR*chxn)$_3$]$^{2+}$ moiety (Figure S11). The peak at 450 nm is convolution of three predominant bands; a shorter wavelength component of HOMO-3 (-7.392 eV) on the $[Ni(CN)_4]^{2-}$ -> LUMO (-2.785 eV) on [Ni(*RR*chxn)$_3$]$^{2+}$ moiety (Figure S12) and a middle wavelength component of HOMO-4 (-7.675 eV) on the $[Ni(CN)_4]^{2-}$ -> LUMO (-2.785 eV) on [Ni(*RR*chxn)$_3$]$^{2+}$ moiety (Figure S13) and a longer wavelength component of HOMO-5 (-7.957 eV) on the $[Ni(CN)_4]^{2-}$ -> LUMO (-2.785 eV) on [Ni(*RR*chxn)$_3$]$^{2+}$ moiety (Figure S14).

Interaction between $[Ni(CN)_4]^{2-}$ and [Ni(*RR*chxn)$_3$]$^{2+}$ moieties can be suggested by heat of formation, 82.9, 223.0, and 313.16 kcal/mol for $[Ni(CN)_4]^{2-}$ and [Ni(*RR*chxn)$_3$]$^{2+}$ and [Ni(*RR*chxn)$_3$][Ni(CN)$_4$], respectively. Thus, the [Ni(*RR*chxn)$_3$][Ni(CN)$_4$] double salt is more stable in 7.7 kcal/mol than simple sum of $[Ni(CN)_4]^{2-}$ and [Ni(*RR*chxn)$_3$]$^{2+}$ moieties.

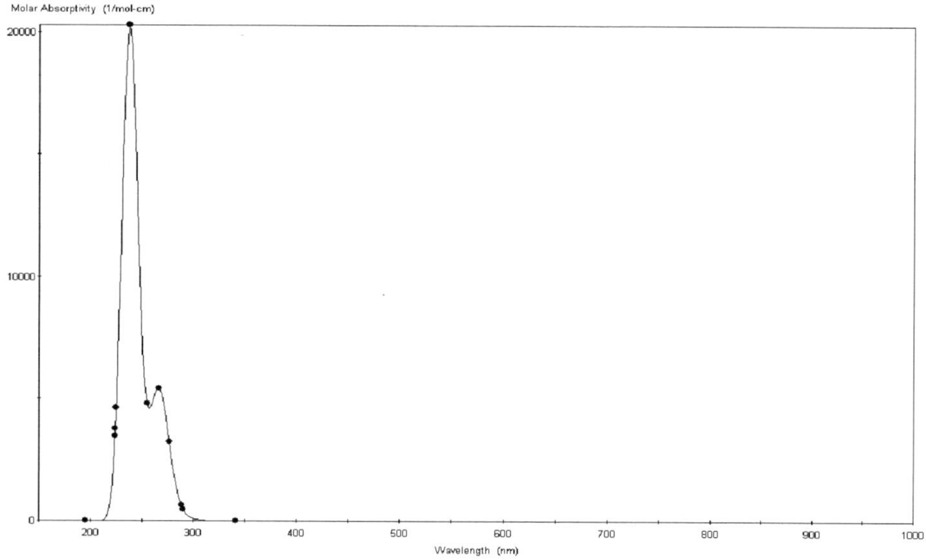


Figure S1. Calculated electronic spectra of a $[Ni(CN)_4]^{2-}$ complex in *1* by ZINDO.

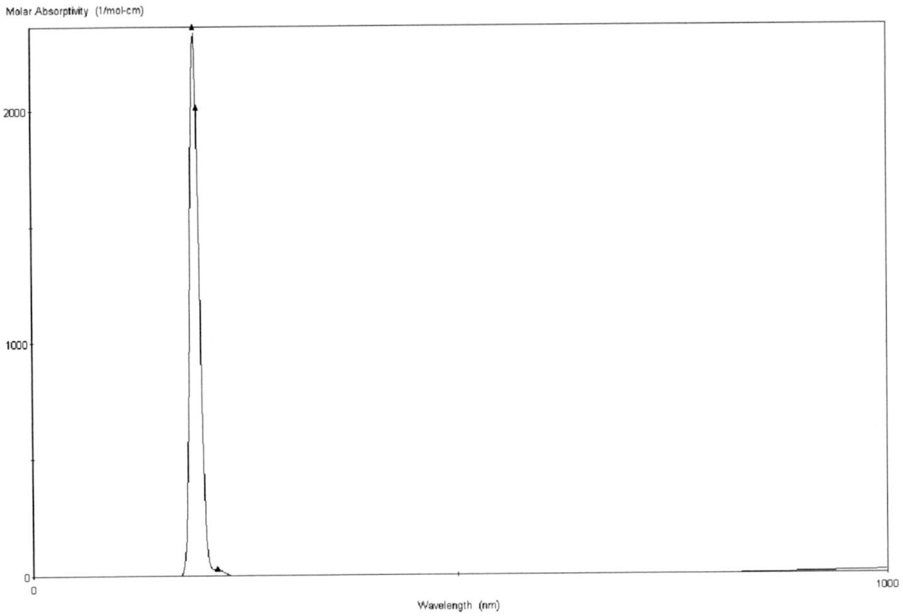


Figure S2. Calculated electronic spectra of a $[Ni(\mathit{RR}chxn)_3]^{2+}$ complex in *1* by ZINDO.

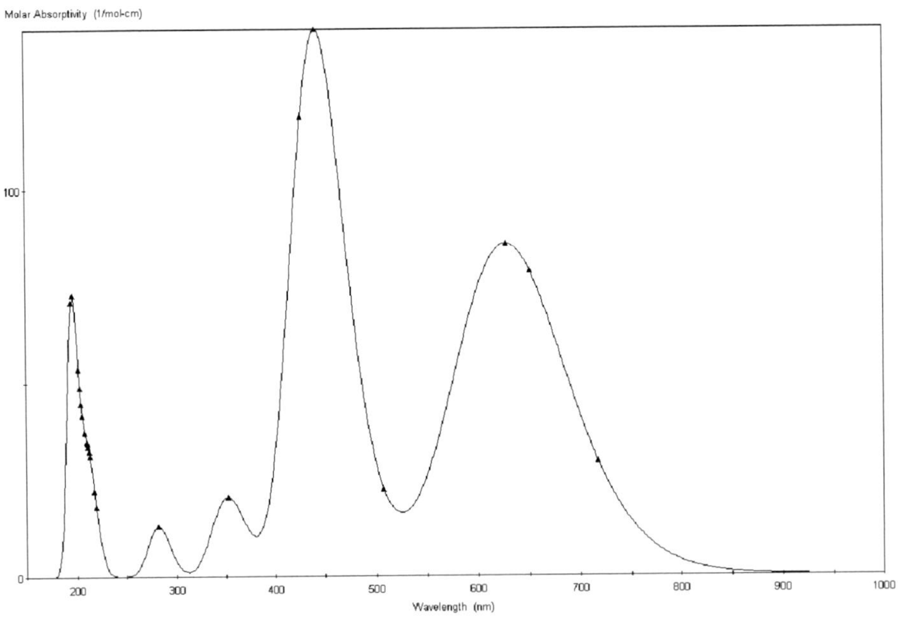


Figure S3. Calculated electronic spectra of a $[Ni(\mathit{RR}chxn)_3][Ni(CN)_4]$ double salt in *1* by ZINDO.

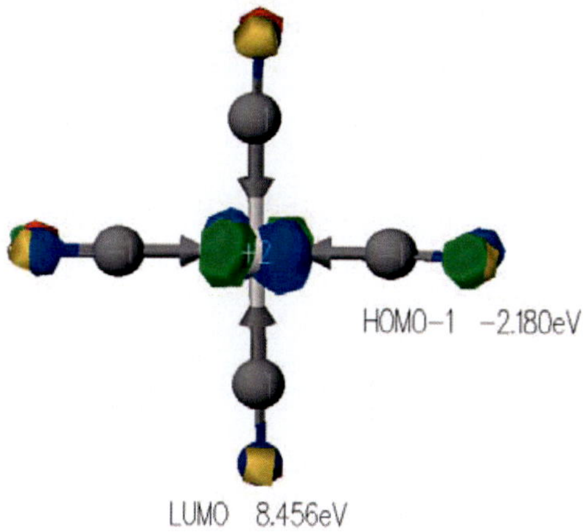


Figure S4. Calculated molecular orbitals associated with HOMO-1 -> LUMO transition.

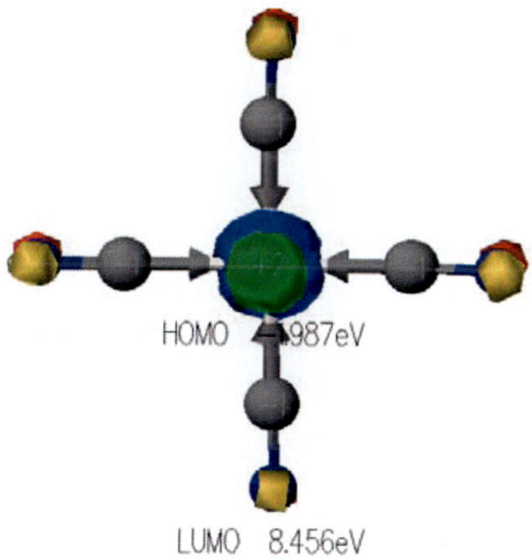


Figure S5. Calculated molecular orbitals associated with HOMO -> LUMO transition.

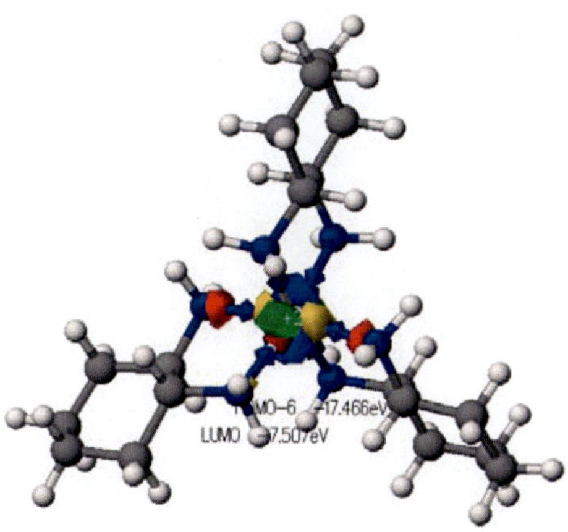

Figure S6. Calculated molecular orbitals associated with HOMO-6 -> LUMO transition.

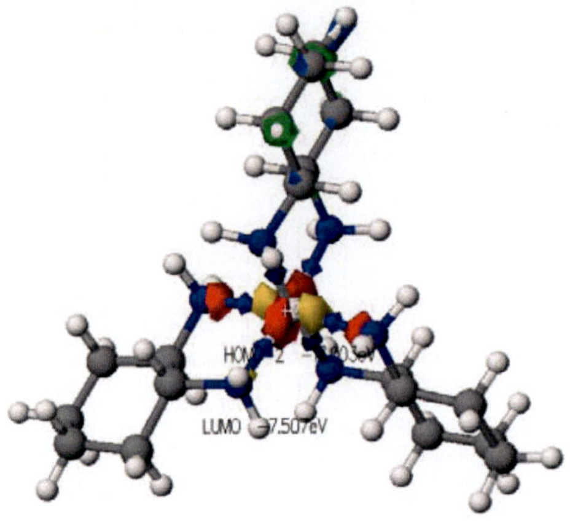

Figure S7. Calculated molecular orbitals associated with HOMO-2 -> LUMO transition.

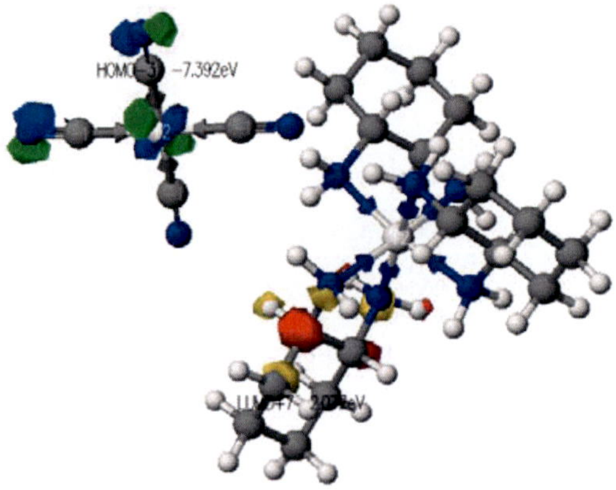

Figure S8. Calculated molecular orbitals associated with HOMO-3 -> LUMO+7 transition.

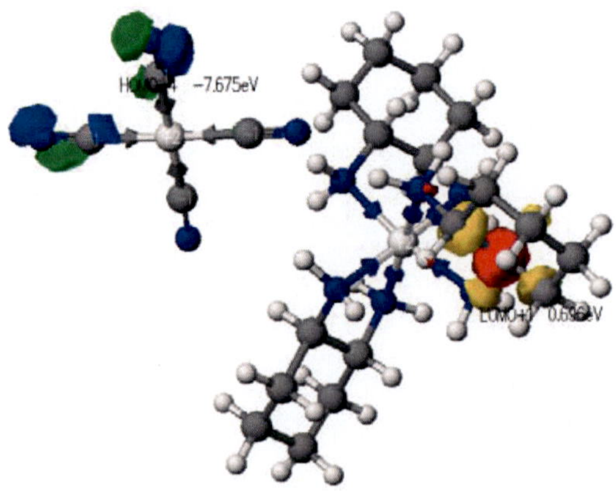

Figure S9. Calculated molecular orbitals associated with HOMO-4 -> LUMO+1 transition.

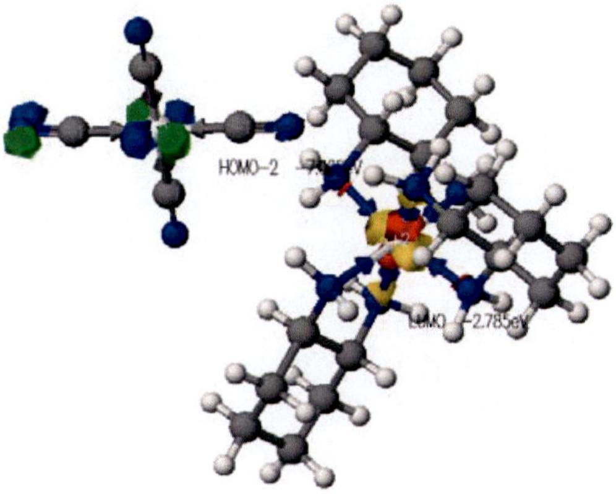


Figure S10. Calculated molecular orbitals associated with HOMO-2 -> LUMO transition.

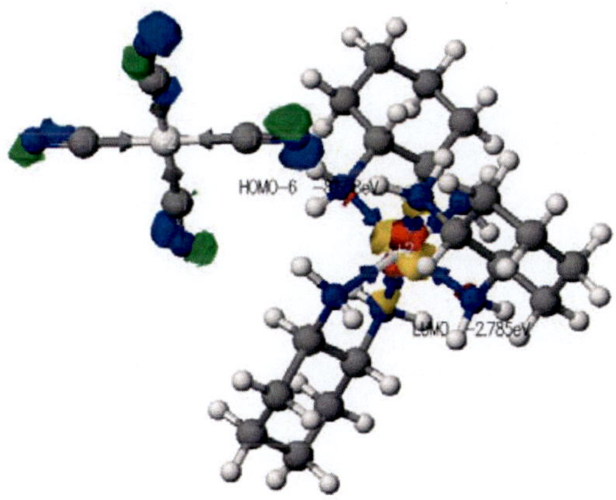


Figure S11. Calculated molecular orbitals associated with HOMO-6 -> LUMO transition.

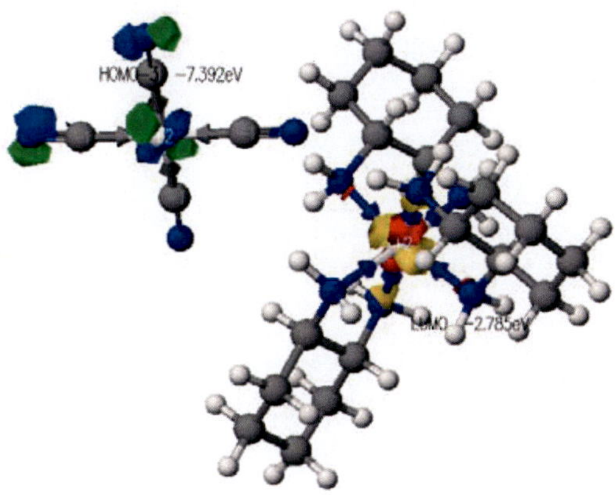


Figure S12. Calculated molecular orbitals associated with HOMO-3 -> LUMO transition.

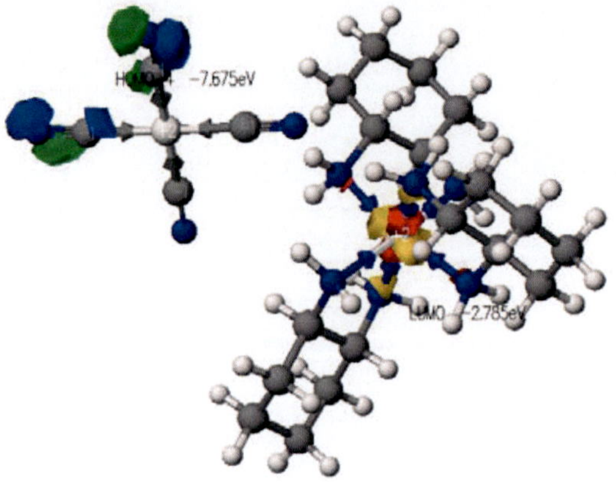

Figure S13. Calculated molecular orbitals associated with HOMO-4 -> LUMO transition.

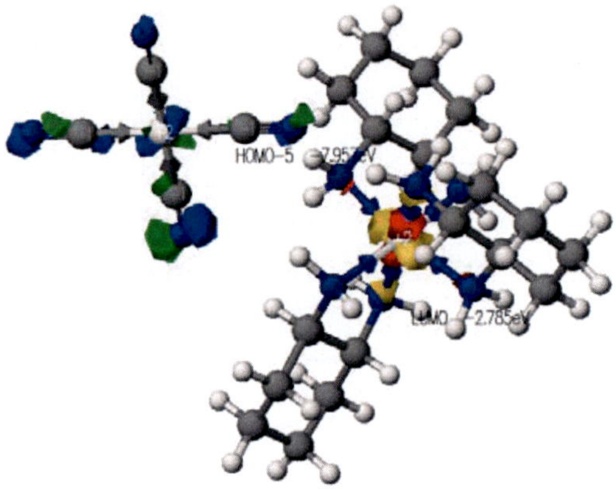

Figure S14. Calculated molecular orbitals associated with HOMO-5 -> LUMO transition.

## ACKNOWLEDGEMENT

This work was supported by Iketani Science and Technology Foundation. The author thanks Professor Hidenari Inoue (keio University) for the use of the CD spectrometer and the Materials Design and Characterization Laboratory, Institute for Solid State Physics, University of Tokyo for the SQUID facilities.

## REFERENCES

[1] Ohba, M.; Okawa, H. *Coord. Chem. Rev.* 2000, 198, 313-328.
[2] Okawa, H.; Ohba, M. *Bull. Chem. Soc. Jpn.* 2002, 75, 1191-1203.
[3] Sato, O. *Acc. Chem. Res.* 2003, 36, 692-700.
[4] Sato O.; Hayami, S.; Einaga, Y.; Gu, Z.-Z. *Bull. Chem. Soc. Jpn.* 2003, 76, 443-470.
[5] Sato, O.; Iyoda, T.; Fujishima, A.; Hashimoto, K. *Science* 1996, 272, 704-705.

[6] Li, G.; Akitsu, T.; Sato, O.; Einaga, Y. *J. Am. Chem. Soc.* 2003, 125, 12396-12397.

[7] Li, G..; Sato, O.; Akitsu, T.; Einaga, Y. *J. Solid State Chem.* 2004, 177, 3835-3838.

[8] Coronado, E.; Gimenez-Saiz, C.; Martinez-Agudo, J. M.; Nuez, A.; Romero, F. M.; Stoeckli-Evans, H. *Polyhedron* 2003, 22, 2435-2440.

[9] Coronado, E.; Gomez-Garcia, C. J.; Nuez, A.; Romero, F. M.; Rusanov, E.; Stoeckli-Evans, H. *Inorg. Chem.* 2002, 41, 4615-4617.

[10] Coronado, E.; Gimenez-Saiz, C.; Nuez, A.; Sanchez, V.; Romero, F. M. Eur. *J. Inorg. Chem.* 2003, 4289-4293.

[11] Kaneko, W.; Kitagawa, S.; Ohba, M. *J. Am. Chem. Soc.* 2007, 129, 248-249.

[12] Inoue, K.; Imai, H.; Ghalasaki, P. S.; Kikuchi, K.; Ohba, M.; Okawa, H.; Yakhmi, J. V. Angew. *Chem. Int. Ed.* 2004, 43, 5618-5621.

[13] Inoue, K.; Kikuchi, K.; Ohba, M.; Okawa, H. Angew. *Chem. Int. Ed.* 2003, 42, 4810-4813.

[14] Akitsu, T.; Einaga, Y. *Inorg. Chem.* 2006, 45, 9826-9833.

[15] Akitsu, T.; Einaga, Y. *Inorg. Chim. Acta* 2007, 360, 497-505.

[16] Falvello, L. R. *J. Chem. Soc., Dalton Trans.* 1997, 4463-4475.

[17] Halcrow, M. A. *Dalton Trans.* 2003, 4375-4384.

[18] Altomare, A.; Cascarano, G..; Giacovazzo, C.; Guagliardi, A.; Brurla, M. C.; Polidori, G..; Camalli, M. *J. Appl. Crystallogr.* 1994, 27, 435-435.

[19] Sheldrick, G. M. (1997) SHELXL97, *Program for the Refinement of Crystal Structures,* University of Goettingen, Germany.

[20] *Molecular Structure Corporation.* (2001) TEXSAN. Version 1.11., MSC, 9009 New Trails Drive, The Woodlands, TX 77381-5209, USA.

[21] Ishida, K.; Tanase, T.; Takahashi, T.; Sato, Y.; Yano, S.; Hidai, M.; Kobayashi, K.; Sakurai, T. *Acta Crystallogr.* C 1998, 44, 83-85.

[22] Morooka, M.; Ohba, S.; Saito, Y.; Miyake, H. *Acta Crystallogr.* B 1991, 47, 910-917.

[23] Mizuta, T.; Tshitani, K.; Miyoshi, K.; Yoneda, H. *Inorg. Chem.* 1990, 29, 3020-3026.

[24] Mizuta, T.; Miyoshi, K. *Bull. Chem. Soc. Jpn.* 1991, 64, 1859-1865.

[25] Paharova, J.; Cernak. J.; Boca, R.; Zak, Z. *Inorg. Chim. Acta* 2003, 346, 25-31.

[26] Smekal, Z.; Cisarova, I.; Mrozinski, *J. Polyhedron* 2001, 20, 3301-3306.

[27] Cernak, J.; Skorsepa, J.; Abboud, K. A.; Meisel, M. W.; Orendac, M.; Orendacova, A.; Feher, A. *Inorg. Chim. Acta* 2001, 326, 3-8.

[28] Zhan, S.-Z.; Guo, D.; Zhang, X.-Y.; Du, C.-X.; Zhu, Y.; Yang, R.-N. *Inorg. Chim. Acta* 2000, 298, 57-62.

[29] Ghoshal, D.; Ghosh, A. K.; Maji, T. K.; Ribas, J.; Mostafa, G..; Zangrando, E.; Chaudhuri, N. R. *Inorg. Chim. Acta* 2006, 359, 593-602.

[30] Muga, I.; Gutierrez-Zorrilla, J. M.; Vitoria, P.; Roman, P.; Lezama, L.; Beitia, J. I. *Eur. J. Inorg. Chem.* 2004, 1886-1893.

[31] Yuge, H.; Mamada, A.; Asai, M.; Nishikiori, S.-i.; Iwamoto, T. *J. Chem. Soc., Dalton Trans.* 1995, 3195-3205.

[32] Cernak, J.; Potocnak, I.; Dunaj-Jurco, M. Z. *Kristallogr.* 1994, 209, 757-758.

[33] Bubanec, J.; Cernak, J.; Potocnak, I.; Drabik, M.; Lipkowski, *J. Chem. Pap.* 2004, 58, 224-231.

[34] Cernak, J.; Chomi, J.; Dunaj-Juro, M.; Kappenstein, C. *Inorg. Chim. Acta* 1984, 85, 219-226.

[35] Cernak, J.; Paharova, J.; Skoresepa, J.; Massa, W. Z. *Anorg. Allg. Chem.* 2002, 628, 344-348.

[36] Rodriguez, V.; Gutierrez-Zorrilla, J. M.; Vitoria, P.; Luque, A.; Roman, P.; Martinez-Ripoll, M. *Inorg. Chim. Acta* 1999, 290, 57-63.

[37] Sakai, K.; Miyabe, Y. *Acta Crystallogr.* E 2004, 60, m69-m70.

[38] Muga, I.; Vitoria, P.; Reinoso, S.; Gutierrez-Zorrilla, J. M. *Acta Crystallogr. E* 2003, 59, m576-m578.

[39] Nakamoto, K. *Infrared and Raman Spectra of Inorganic and Coordination Compounds*, 3rd ed., Wiley, NY, 1978.

[40] Akitsu, T.; Einaga, Y. *Chem. Pap.* 2007, 61, 194-198.

[41] Lever, A. B. P. *Inorganic Electronic Spectroscopy*, 2nd ed.; Elesevier; 1984.

[42] Mantz, Y. A.; Musselman, R. L. *Inorg. Chem.* 2002, 41, 5770-5777.

[43] Cornelius, J. B.; Trapp, R. M.; Delord, T. J.; Fronczek, F. R.; Watkins, S. F.; Orosz, J. J.; Musselman, R. L. *Inorg. Chem.* 2003, 42, 3026-3035.

[44] Zerner, M. C.; Lowe, G. H.; Kirchner, R. F.; Mueller-Westerhoff, U. T. *J. Am. Chem. Soc.* 1980, 102, 589-599.

[45] Head, J. D.; Zerner, M. C. *Chem. Phys. Lett.* 1986, 131, 359-366.

In: Stereochemistry Research Trends
Editor: M. A. Horvat and J. H. Golob

ISBN 978-1-60456-825-7


*Chapter 5*

# (4*R**,5*R**)-4-BENZOYLAMINO-5-PHENYL-3-PYRAZOLIDINONE – A USEFUL BUILDING BLOCK IN THE SYNTHESIS OF FUNCTIONALIZED PYRAZOLES

***Jurij Svete****

Faculty of Chemistry and Chemical Technology, University of Ljubljana, Aškerčeva 5
P.O. Box 537, 1000 Ljubljana, Slovenia

## ABSTRACT

(4*R**,5*R**)-4-Benzoylamino-5-phenyl-3-pyrazolidinone (*1*) is a useful building block, which is available in two steps from *N*-benzoylglycine *via* transformation into the azlactone followed by treatment with excess hydrazine hydrate. Since pyrazolidinone *1* is actually a cyclic $\alpha$-amino acid derivative, it can be successfully employed for the preparation of functionalized pyrazoles containing an $\alpha$-amino acid or a dipeptide structural motif. The most useful derivatives of *1* are stable (1*Z*,4*R**,5*R**)-1-alkylidene-4-benzoylamino-5-phenyl-3-pyrazolidinone-1-azomethine imines *6*, which are obtained in the reactions of *1* with aldehydes and ketones. Azomethine imines *6* derived from aromatic aldehydes are particularly stable and easily available by parallel synthesis. Reduction of azomethine imines *6* with sodium borohydride leads to 1-alkylated 3-pyrazolidinones, which were transformed into (2*R**,3*R**)-3-alkylamino-*N*-benzoyl-3-phenylalaninamides *via* reductive N–N bond cleavage. Similar N–N bond fission was observed also in base-catalyzed reactions of *6* with potassium cyanide. With aliphatic 1,3-diketones in methanol, pyrazolidinone *1* undergoes a 'ring switching' transformation into (2*R**,3*R**)-*N*-benzoyl-3-alkylamino-3-phenyl-3-(1*H*-pyrazol-1-yl)alanine methyl esters. Analogous 3-pyrazolylalanine esters were also obtained in two-steps *via* treatment of azomethine imines *6* with dimethyl acetylenedicarboxylate followed by oxidative opening of the pyrazolidinone ring. So far, [3+2] cycloadditions of (1*Z*,4*R**,5*R**)-1-alkylidene-4-benzoylamino-5-phenyl-3-pyrazolidinone-1-azomethine imines *6* were the most extensively studied reactions witihin the context of (4*R**,5*R**)-4-benzoylamino-5-

* e-mail: jurij.svete@fkkt.uni-lj.si

phenyl-3-pyrazolidinone (*1*) chemistry. Azomethine imines *6* react with typical electron-deficient dipolarophiles, *i.e.* with esters of acrylic, methacrylic, maleic, fumaric, propiolic, and acetylenedicarboxylic acid, as well as with electron-rich $\beta$-keto esters. In addition to broad reactivity (connectivity) spectrum, most of their cycloaddition reactions were highly stereoselective. Stereocontrol depended on (a) stereodirecting phenyl group at position 5, (b) on *ortho*-substituents at the 1'-aryl group in dipole *6*, and (c) on the structure (shape) of the dipolarophile. High selectivity of cycloadditions and simplicity of isolation of the products enable combinatorial solution-phase synthesis of pyrazolo[1,2–*a*]pyrazolone derivatives. In this manner, cycloadditions of dipoles *6* to maleimides and $\beta$-keto esters were carried out to furnish the corresponding libraties of analytically pure cycloadducts.

## 1. INTRODUCTION

3-Pyrazolidinone (Figure 1) and its derivatives are cyclic (internal) hydrazides of 3-hydrazinopropanoic acid, which are commonly available by treatment of $\alpha,\beta$-unsaturated carboxylic acid derivatives with hydrazine hydrate. Despite simplicity of their structure and ease of their preparation, there were almost no reports on 3-pyrazolidinones in the early times of organic chemistry until the second half of 20$^{th}$ century - the first reports on 3-pyrazolidinones date to 1940's [1–6]. Since that time, the importance of 3-pyrazolidinones rose significantly, due to their applicability in industrial processes, and because several pyrazolidin-3-one derivatives exhibit biological activities [7, 8]. An important group of fused pyrazolidinone analogues are 2-acylamino-1-oxo-1*H*,5*H*-pyrazolo[1,2–*a*]pyrazole-7-carboxylates, which are useful scaffolds for the preparation of conformationally constrained peptide mimetics, such as *Eli-Lilly*'s pyrazolo[1,2–*a*]pyrazolone based $\gamma$-lactam antibiotics LY 186826. LY 193239, and LY 255262 [7–12]. Some examples of important 3-pyrazolidinones are depicted on Figure 1.

Most of the early studies in the field of the 3-pyrazolidinone chemistry were performed on achiral and on poorly substituted chiral pyrazolidinones [7–12]. Recent studies, however, established the applicability of chiral polysubstituted 3-pyrazolidinones and their derivatives in the stereoselective synthesis of highly functionalized compounds [13, 14]. Furthermore, 3-pyrazolidinones have also been employed as templates in enantioselective Diels-Alder [15, 16] and Michael [17–20] reactions.

In the last decade, a part of our research interest has also been devoted to the chemistry of 3-pyrazolidinones [21–23]. Most of our work in this field was based on chiral racemic (4*R**,5*R**)-4-benzoylamino-5-phenylpyrazolidin-3-one (*1*) as the model compound, which is easily available from 4-benzylidene-2-phenyl-5(4*H*)-oxazolone (*2*) by heating with excess hydrazine hydrate. The synthesis of 4-benzoylamino-5-phenylpyrazolidin-3-one (*1*) was first reported by Stoddola [3] in 1948 and, later on, by Budovskii and Kochetkov [24] and by Iyengar and Venkataratnam [25]. However, these early reports did not include stereochemical assignment of compound *1*. A decade ago, the latest synthetic variation together with determination of the *trans*-configuration around the C(4)–C(5) bond has been reported [26]. Mechanistically, the reaction most probably starts by opening of the oxazolone ring (1,2-addition) to give the unsaturated hydrazide *3*, which undergoes Michael addition of the second equivalent of hydrazine to afford the saturated hydrazide *4*, which then cyclizes into the pyrazolidinone *1*. This proposed mechanism is supported by the experimental data on reactions of *2* with hydrazine hydrate under various conditions. Thus, treatment of *2* with one equivalent of hydrazine hydrate at room temperature leads to the unsaturated hydrazide *3*, exclusively. On the other hand, the saturated analogue *4* is obtained in the reaction of the oxazolone *2* (or unsaturated hydrazide *3*) with excess hydrazine hydrate. Compound *4* is stable at room temperature and up to ~60 °C, however, it cyclizes into the pyrazolidinone *1* upon heating above 70 °C [3] (Scheme 1).

Stereoselective formation of the (4*R**,5*R**)-isomer *1* is most probably due to thermodynamically controlled formation of the more stable *trans*-isomer *1*. Since pyrazolidinone *1* is a cyclic $\alpha$-amino acid analogue, the proton *H*–C(4) is quite acidic and, consequently, position 4 is epimerizable. Under relatively harsh reaction conditions, *i.e.* heating above 80 °C in the presence of excess hydrazine hydrate, equilibration between the (4*R**,5*R**)-epimer *1* and (4*S**,5*R**)-epimer *1'* *via* the (5*R**)-tautomer *1''* seems reasonable an favors formation of less strained and thermodynamically more stable *trans*-isomer *1* (Scheme 2).

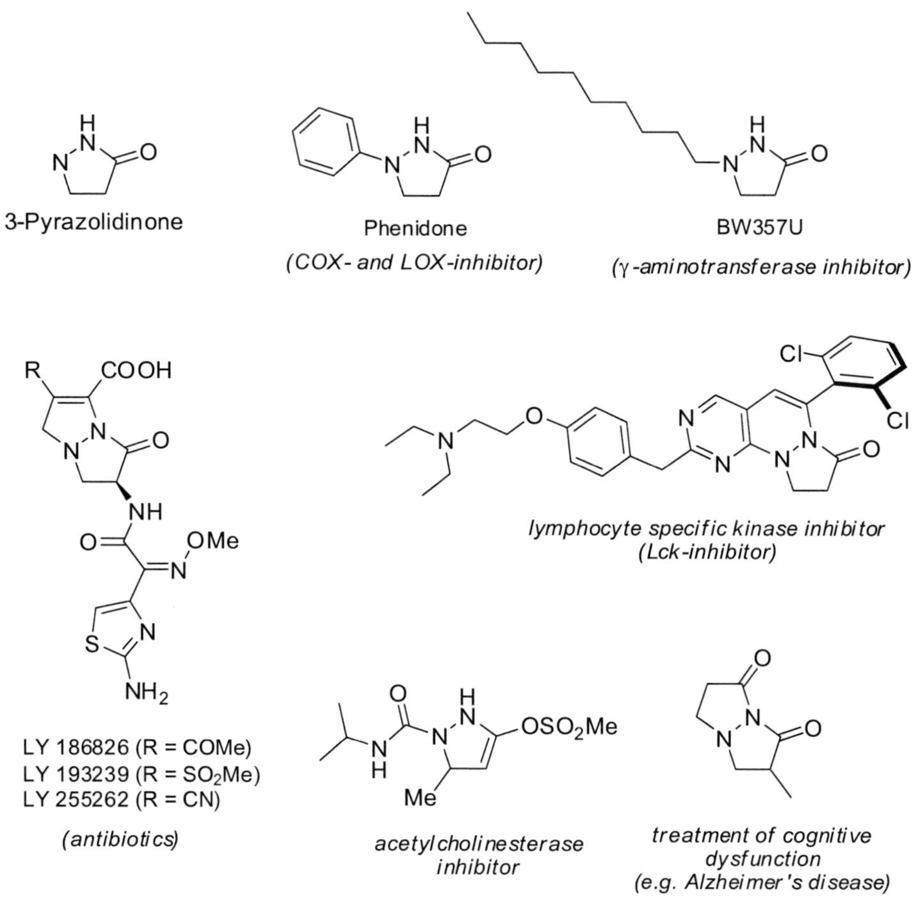


Figure 1.

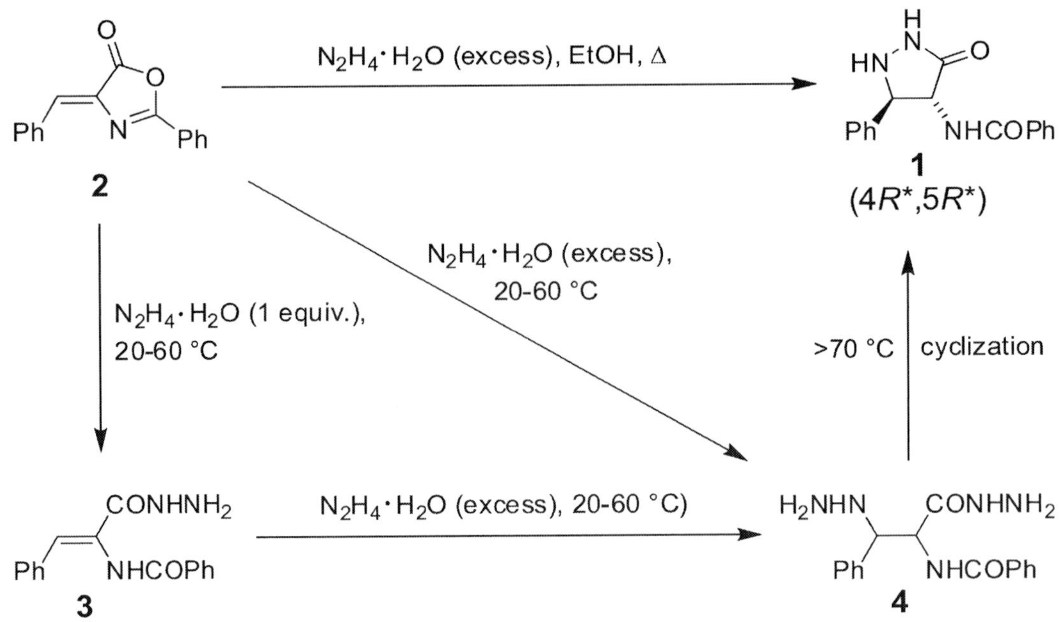


Scheme 1.

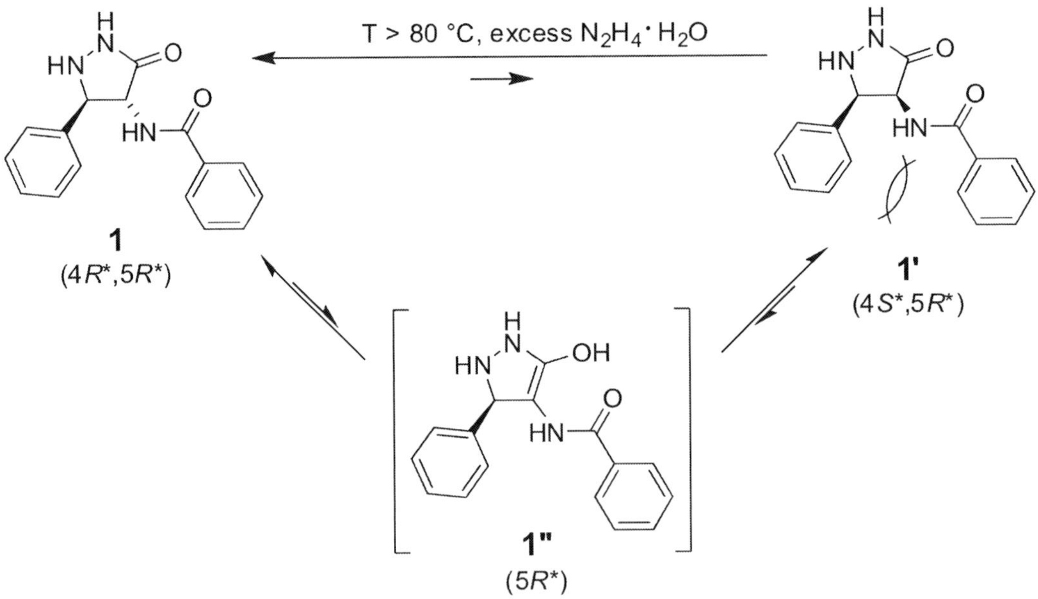


Scheme 2.

Initially, we became interested in (4*R**,5*R**)-4-benzoylamino-5-phenyl-3-pyrazolidinone (*2*) as an easily available and useful model intermediate for the synthesis of unnatural amino acid derivatives [21–23]. Within this context, transformations of *1* with aromatic aldehydes *5* into azomethine imines *6* followed by treatment of dipoles *6* with dimethyl acetylenedicarboxylate (DMAD) and some other dipolarophiles revealed high degree of regioselectivity and stereoselectivity of these 1,3-dipolar cycloadditions [26]. Consequently, this prompted us to continue with a more detailed study on cycloadditions of (4*R**,5*R**)-4-benzoylamino-5-phenyl-3-pyrazolidinone-1-azomethine imines *6* to various olefinic and acetylenic dipolarophiles. From a practical point of view, these reactions enable an easy access to polysubstituted 2-amino-1-oxopyrazolo[1,2–*a*]pyrazolidine-7-carboxylates as heterocyclic dipeptide analogues, where the amino acid sequence can be controlled by the choice of the dipole and the dipolarophile [23] (Figure 2).

The aim of this ongoing study is to explore the reactivity and selectivity pattern in 1,3-dipolar cycloaddition reactions within combinations of various 4-acylamino-3-pyrazolidinon-1-azomethine imines and dipolarophiles. In extension, the results can be further utilized in combinatorial synthesis of libraries of pyrazolo[1,2–*a*]pyrazole peptide mimetics with variable amino acid sequence and variable, yet predictable configuration. Herein, a review and recent results on this ongoing stereochemical study is given.

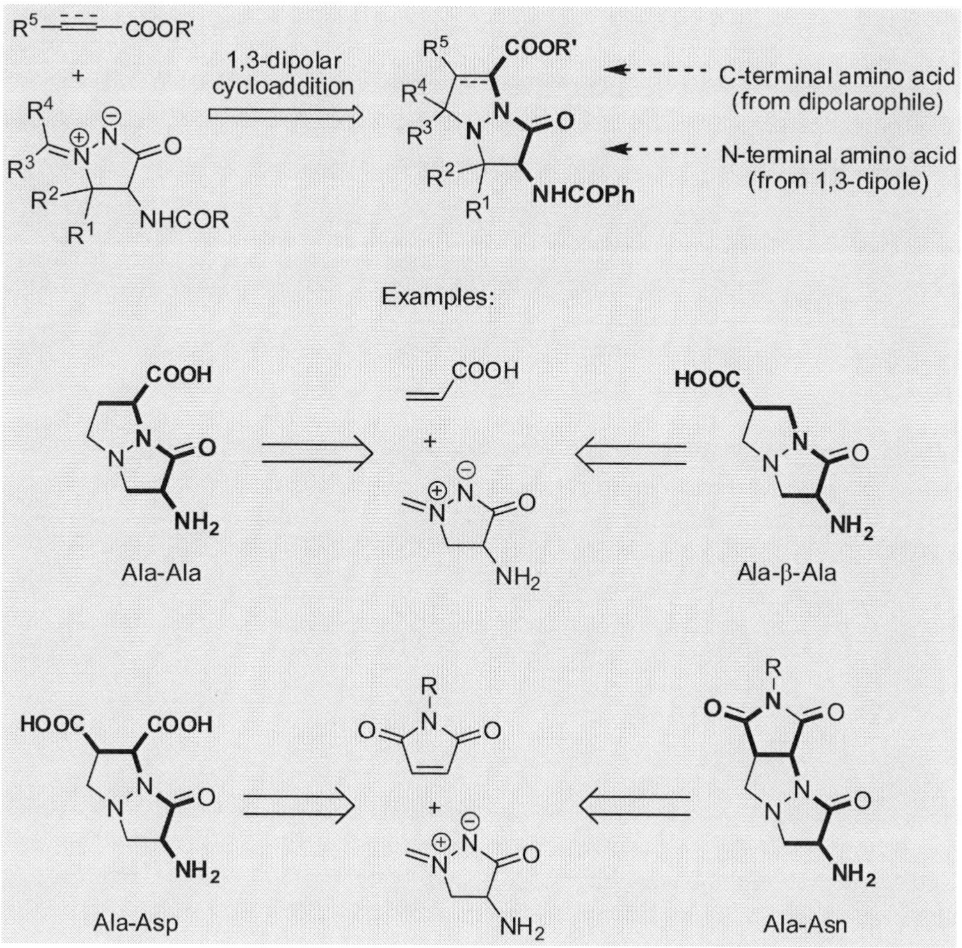


Figure 2.

## 2. Synthesis of (4*R**,5*R**)-1-Alkylidene-4-benzoylamino-5-phenyl-3-pyrazolidinone-1-azomethine Imines

Like other 3-pyrazolidinones [7, 8], (4*R**,5*R**)-4-benzoylamino-5-phenyl-3-pyrazolidinone (*1*) also readily reacts with aromatic aldehydes *5a–p* to give the corresponding azomethine imines *6a–p*. Typically, reactions are carried out in refluxing ethanol in the presence of catalytic amounts of trifluoroacetic acid to afford dipoles *6* as stable crystalline solids in good yields [26–31]. This procedure is completely general and also allows parallel solution-phase preparation of (1*Z*,4*R**,5*R**)-1-arylmethylidene-4-benzoylamino-5-phenyl-3-pyrazolidinon-1-azomethine imines *6* [29, 30] (Scheme 3).

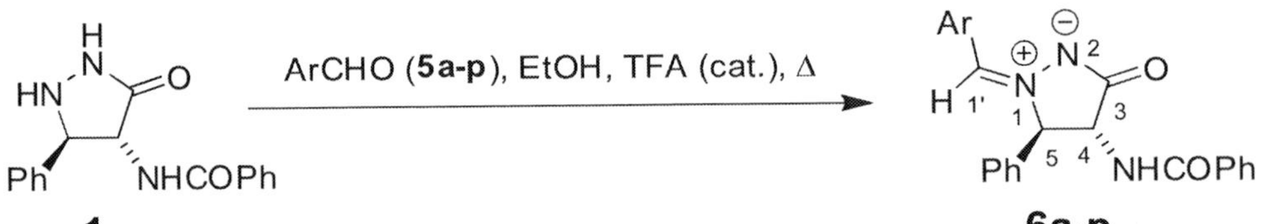


| Compound | Ar | Yield (%) | Ref. |
|---|---|---|---|
| **6a** | phenyl | 89 | 26 |
| **6b** | 4-methylphenyl | 81 | 27 |
| **6c** | 4-methoxyphenyl | 78 | 26 |
| **6d** | 3,4,5-trimethoxyphenyl | 92 | 28 |
| **6e** | 4-dimethylaminophenyl | 93 | 26 |
| **6f** | 3-nitrophenyl | 90 | 26 |
| **6g** | 4-nitrophenyl | 76 | 26 |
| **6h** | 3-fluorophenyl | 67 | 31 |
| **6i** | 2-furyl | 88 | 27 |
| **6j** | 3-methyl-2-thienyl | 75 | 31 |
| **6k** | 2-methoxyphenyl | 81 | 31 |
| **6l** | 2,4-dichlorophenyl | 80 | 30 |
| **6m** | 2,6-dichlorophenyl | 88 | 26 |
| **6n** | 2,4,6-trimethylphenyl | 93 | 28 |
| **6o** | 2,4,6-trimethoxyphenyl | 79 | 31 |
| **6p** | styryl | 95 | 27 |

Scheme 3.

Azomethine imines *6a–p* derived from aromatic aldehydes exist in DMSO-$d_6$ solution as pure (Z)-isomers *6a–p*. So far, existence of the (*E*)-isomer *6'* has not been observed yet. The (*Z*)-configuration around the exocyclic C=N double bond in dipoles *6a* [26] and *6m* [28] was established by NOESY spectroscopy on the basis of n.O.e. between *H*–C(1') and *H*–C(5). Furthermore, n.O.e. between *H*–C(1') and *H*–C(5) in *6m* was measured in DMSO-$d_6$ solution at 29, 62, 100, and 150 °C. Intensity of n.O.e. between *H*–C(1') and *H*–C(5) decreased with increasing temperature: from 3.7% (29 and 62 °C) to 2.2% (100 °C) and 0% (150 °C). These measurements were in agreement with stable (*Z*)-configuration around the exocyclic C=N double bond up to ~100 °C. This was also supported by the fact, that all four $^1$H NMR spectra, even at 150 °C, exhibited single sets of signals. On the other hand, disappearance of n.O.e. between *H*–C(1') and *H*–C(5) at 150 °C indicated, that the *Z*/*E*-isomerization around the exocyclic C=N double bond cannot be excluded and might be possible at elevated temperatures. The (Z)-configuration of other azomethine imines *6* was established on the basis of correlation of $^1$H NMR data [26–30] (Scheme 4).

Similarly, reactions of *1* with aliphatic aldehydes and ketones also led to formation of the corresponding azomethine imines *6*. However, efficient isolation of pure dipoles was usually quite difficult and, in most cases, accompanied with substantial loss of material. In practice, these dipoles were, either prepared and reacted *in situ*, or isolated by evaporative workup to afford the crude azomethine imines, which were used for further transformations without purification. Exceptions were azomethine imines *6q* and *6r*, which were obtained in good yields upon acid-catalyzed treatment of the pyrazolidinone *1* with acetone (*5q*) and butyraldehyde (*5r*), respectively. In contrast to their aromatic analogues *6a–p*, the non-symmetrical 1-propylidene azomethine imine *6r* exists in DMSO-d$_6$ solution as a 1:1 mixture of the (*E*)- and the (*Z*)-isomer [32]. Similarly, treatment of *1* with 2-indanone (*7*) gave the enamine *8* in 65% yield [21, 33], while reaction of *1* with enantiopure enaminone *9* furnished a mixture of two diastereomeric enamino pyrazolidinones, *10* and *10'* [34] (Scheme 5).

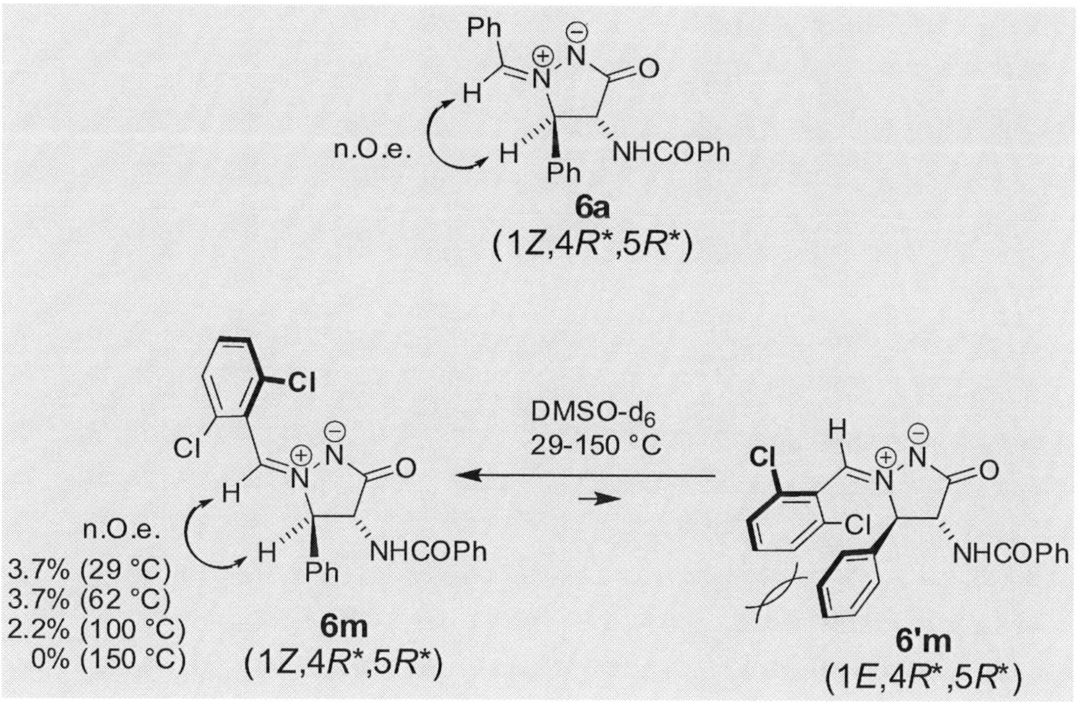


Scheme 4.

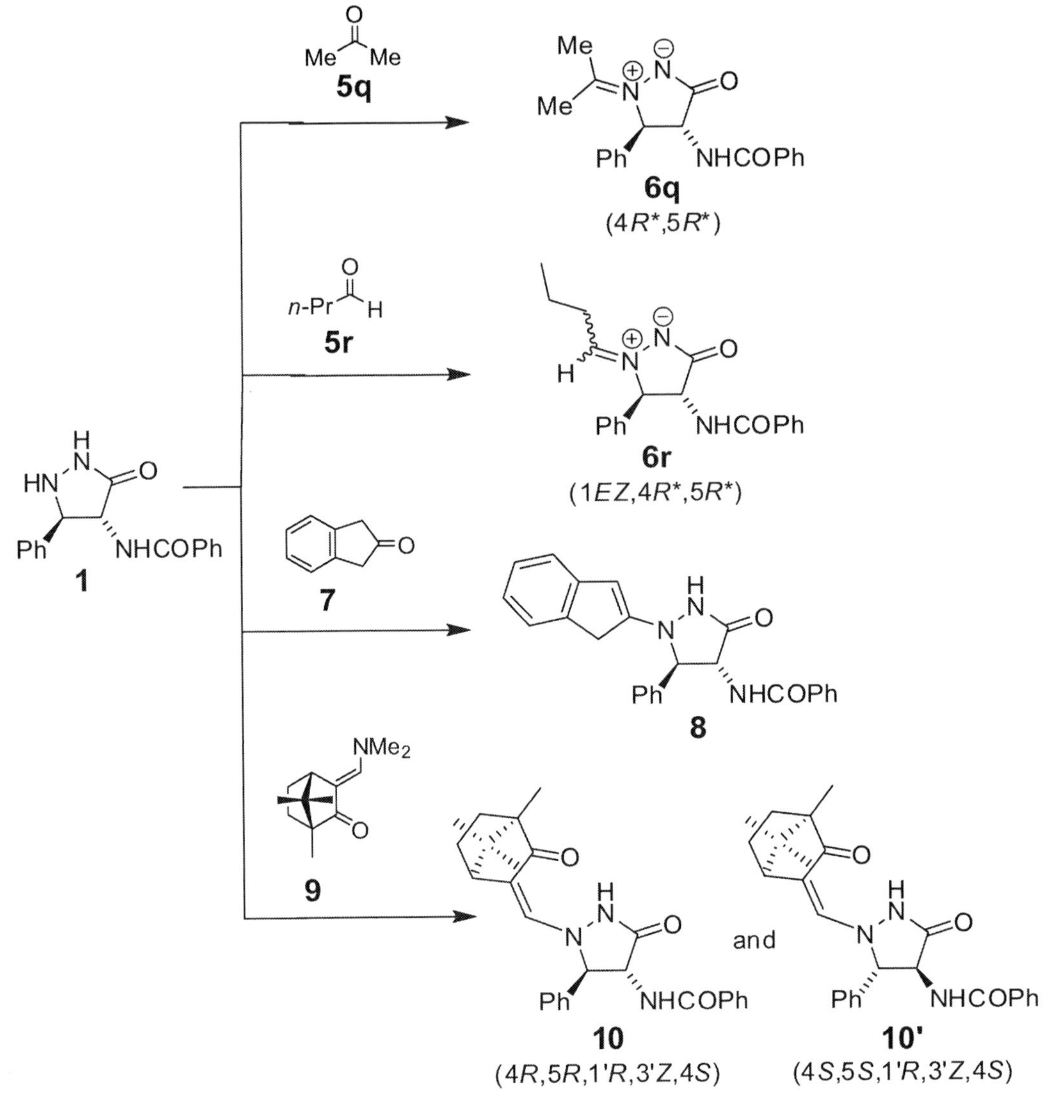

Scheme 5.

## 3. Synthesis of (2*R**,3*R**)-3-Alkylamino-3-phenylalaninamides *via* Reductive Alkylation of (4*R**,5*R**)-4-Benzoylamino-5-phenyl-3-pyrazolidinone Followed by Reductive N–N Bond Cleavage

Treatment of azomethine imines *6a,b,i,p,q* with sodium borohydride in methanol at room temperature resulted in smooth reduction of the exocyclic C=N double bond to give the corresponding 1-alkylated derivatives *11a,b,i,p,q*. Similarly, dipoles *6s–y* were formed *in situ* from *1* and propanal (*5s*), cyclopentanone (*5t*), and aldoses *5u–y*, respectively. Subsequently treatment of *6s–y* with sodium borohydride afforded the corresponding 1-alkyl-3-pyrazolidinones *11s–y*. In the case of dipole *6p*, no reduction of the C=C double bond was observed [27, 31] (Scheme 6, Table 1).

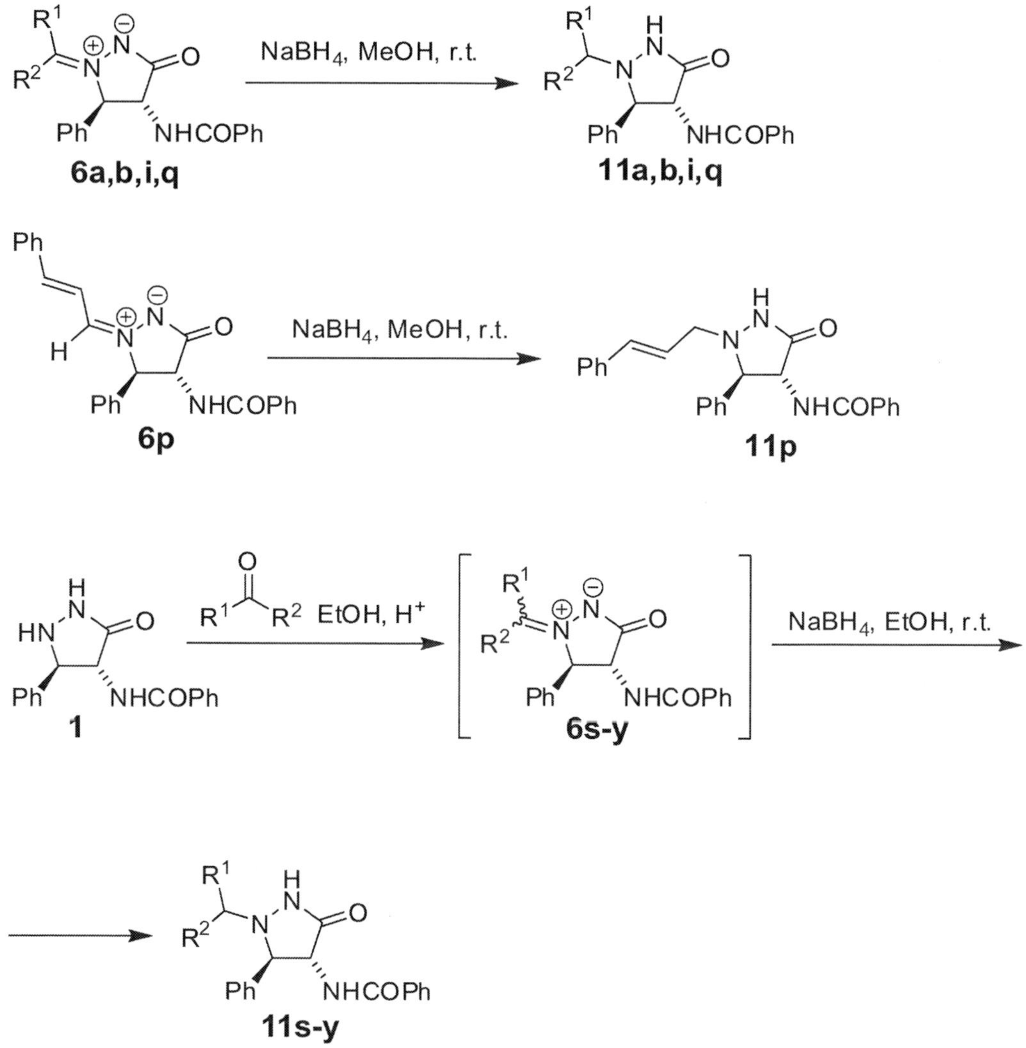


Scheme 6.

**Table 1. 1-Alkylated 3-pyrazolidinones 11**

| Compound | $R^1$ | $R^2$ | Yield (%) | Ref. |
|---|---|---|---|---|
| **11a** | Ph | H | 87 | 27 |
| **11b** | 4-methylphenyl | H | 95 | 27 |
| **11i** | 2-furyl | H | 82 | 27 |
| **11p** | styryl | H | 85 | 27 |
| **11q** | Me | Me | 81 | 27 |

| | | | | |
|---|---|---|---|---|
| **11s** | Et | H | 50 | 27 |
| **11t** | $-(CH_2)_4-$ | | 87 | 27 |
| **11u** | 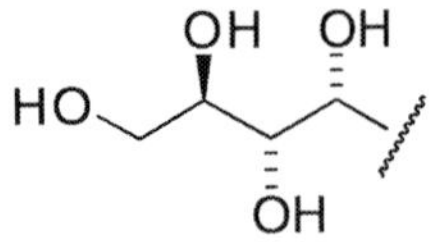 (D-*arabino*) | H | 76 | 26 |
| **11v** | 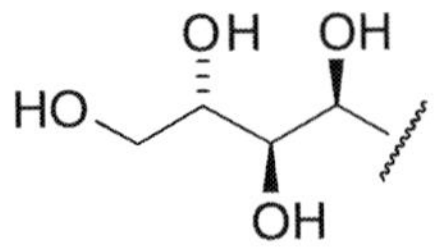 (L-*arabino*) | H | 76 | 26 |
| **11w** | 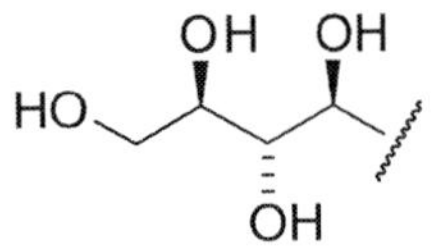 (D-*ribo*) | H | 25 | 26 |
| **11x** | 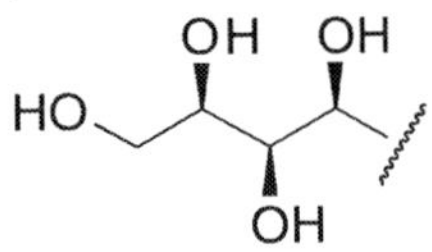 (D-*xylo*) | H | 46 | 26 |
| **11y** | 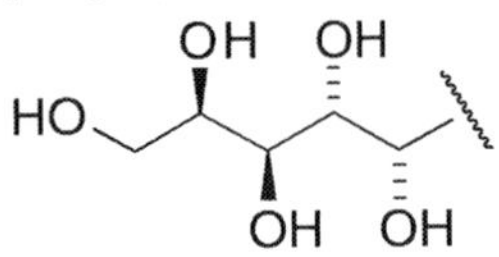 (D-*galacto*) | H | 39 | 26 |

Treatment of 1-alkylpyrazolidinones *11a,b,p,q,s* with Raney-nickel alloy in a solution of potassium hydroxide in aqueous methanol resulted in fission of the N–N bond to furnish the corresponding (2*R**,3*R**)-3-alkylamino-*N*-benzoyl-3-phenylalaninamides *12a,b,p,q,s*. Under these reaction conditions, also reduction of the C=C double bond in *11p* took place to give the 3-(3-phenylpropyl)amino substituted compound *12p* [27] (Scheme 7).

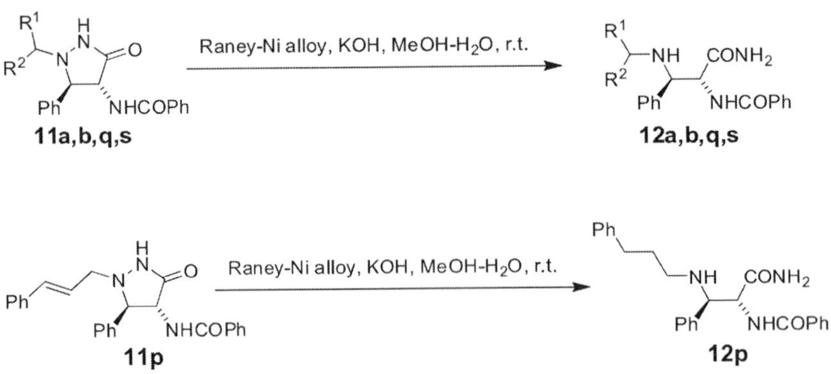


Scheme 7 (Continued on next page.)

| Compound | $R^1$ | $R^2$ | Yield (%) |
|---|---|---|---|
| **12a** | Ph | H | 56 |
| **12b** | 4-methylphenyl | H | 67 |
| **12p** | 3-phenylpropyl | H | 56 |
| **12q** | Me | Me | 71 |
| **12s** | Et | H | 40 |

Scheme 7

Somehow surprising result was obtained, when azomethine imines *6a,c,d,m,n* were reacted with 1.5 equivalent of potassium cyanide in methanol in the presence of acetic acid. All products precipitated from the reaction mixtures and were isolated by filtration and washing. According to spectral and analytical data of the products and due to almost neglibile C≡N absorption at ~2200 $cm^{-1}$ we initially beleived, that the corresponding cycloadducts, *i.e.* pyrazolo[1,2–*a*][1,2,3]triazoles were formed. However, the X-ray diffraction analysis unambigously showed, that *N*-[(1*R**,2*R**)-3-amino-2-benzoylamino-3-oxo-1-phenylpropyl]-3,4,5-trimethoxybenzimidoyl cyanide (*13d*) was obtained in the reaction of *6d* with HCN. The same structure was then assigend to compounds *13a,c,m* on the basis of correlation of $^1$H NMR data. This quite unexpected transformation could be explained by initial addition of the cyanide anion to the exocyclic C=N double bond to give the intermediate *12* followed by cleavage of the N–N single bond *via* elimination of the carboxamido group. This proposed mechanism is supported by the isolation of the intermediate *12n*, which which did not undergo ring opening. Higher stability of the mesityl derivative *12n* could be explained by steric hindrance as a result of two *ortho*-methyl groups, which prevent deprotonation and eliminative ring-opening. Alternative explanation could be, that *β*-elimination step requires an antiperiplanar conformation around the C–N bond, which is not feasible in the case of a bulky mesityl group [35] (Scheme 8).

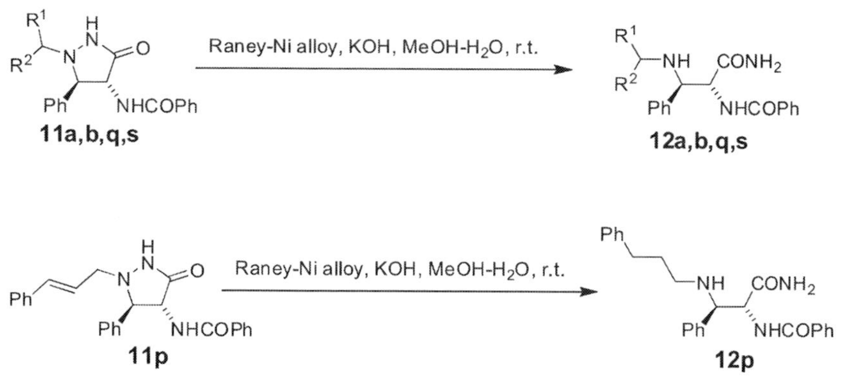

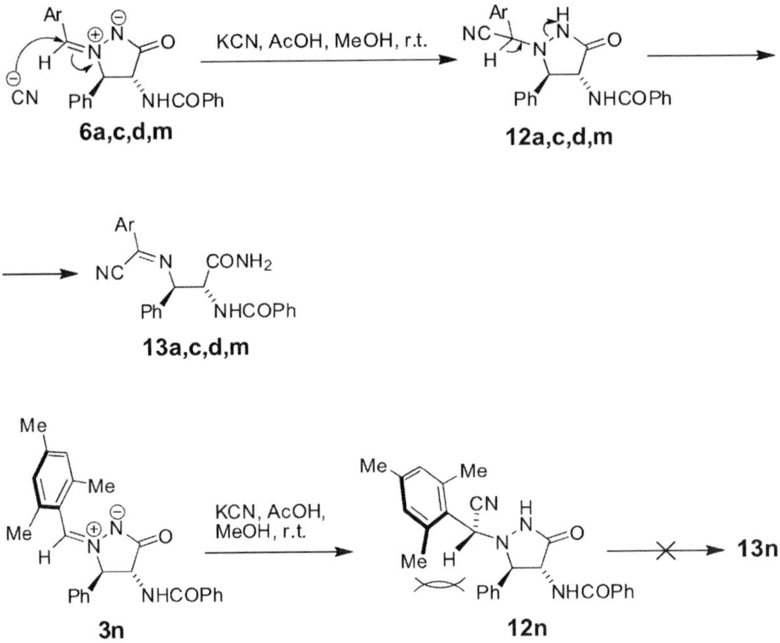


| Compound | Ar | Yield (%) |
|---|---|---|
| **11a** | phenyl | 85 |
| **11c** | 4-methoxyphenyl | 35 |
| **11d** | 3,4,5-trimethoxyphenyl | 64 |
| **11m** | 2,6-dichlorophenyl | 28 |
| **10n** | 2,4,6-trimethylphenyl | 76 |

Scheme 8.

## 4. Synthesis of (2*R**,3*R**)-*N*-benzoyl-3-phenyl-3-(1*H*-pyrazol-1-yl)alanine Esters

(4*R**,5*R**)-4-Benzoylamino-5-phenyl-3-pyrazolidinone (*1*) was also used as starting compound in the 'ring switching' synthesis of (2*R**,3*R**)-3-phenyl-3-(1*H*-pyrazol-1-yl)alanine esters *14a–d*. Acid-catalyzed treatment of *1* with 1,1,3,3-tetraethoxypropane, 3-oxobutyraldehyde dimethylacetal, and acetylacetone in methanol at room temperature afforded the corresponding (2*R**,3*R**)-*N*-benzoyl-3-phenyl-3-(3,5-disubstituted 1*H*-pyrazol-1-yl)alanine methyl esters *14a–c*, respectively, in 62–88% yields. Similarly, treatment of *1* with 2,4-pentanedione in ethanol afforded (2*R**,3*R**)-*N*-benzoyl-3-phenyl-3-(3,5-dimethyl-1*H*-pyrazol-1-yl)alanine ethyl ester (*14d*) in 48% yield. The reaction mechanism could be explained by initial formation of azomethine imine *15*, which is in equilibrium with the non-dipolar bicyclic intermediate *16*. Nucleophilic attack of methanol to the lactam carbonyl group results in opening of the pyrazolidinone ring and elimination of water to afford the final amino acid ester *14* [36] (Scheme 9).

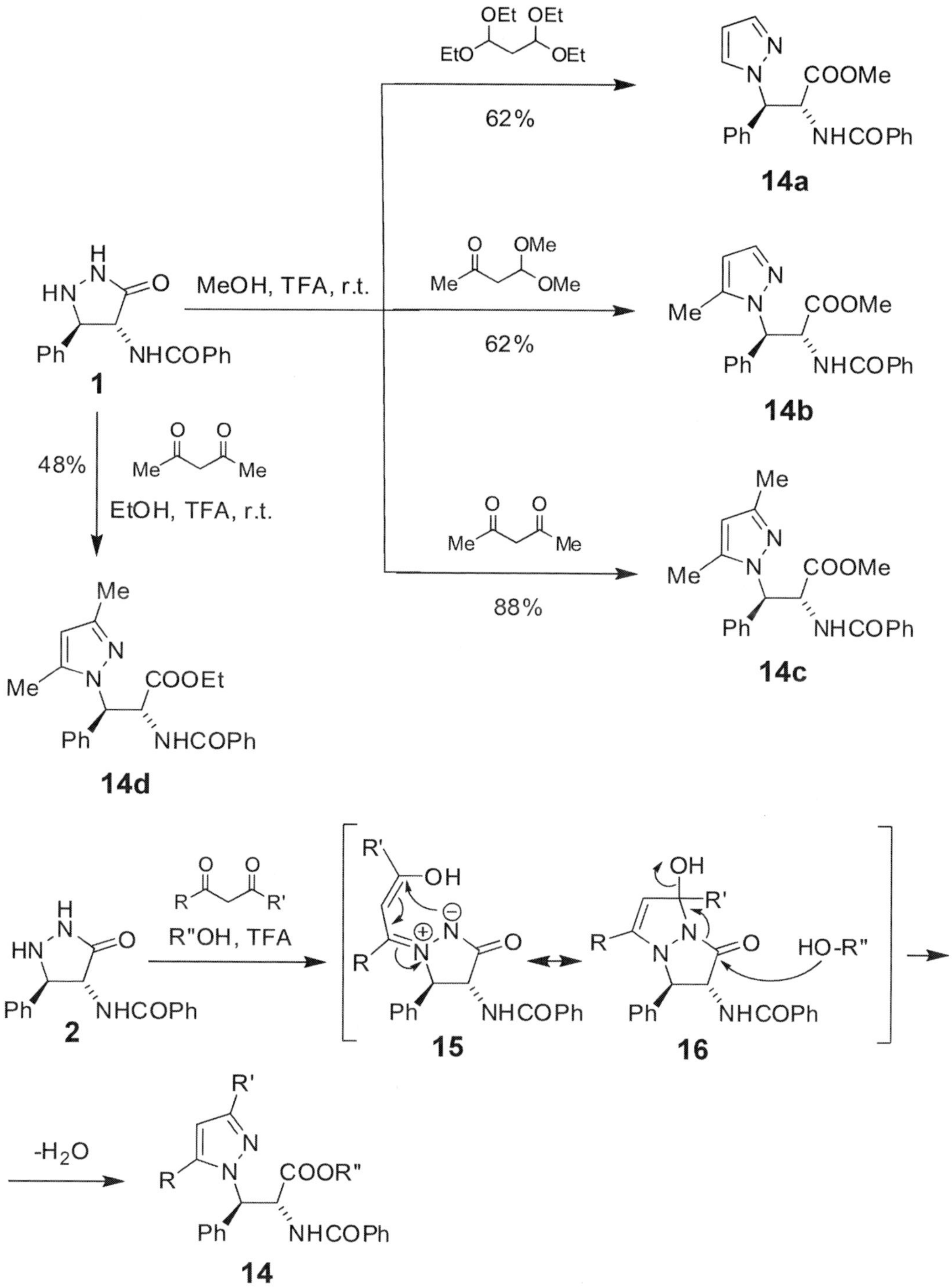
MeOH, TFA, r.t.
62%
62%
88%
48%
EtOH, TFA, r.t.
R"OH, TFA
-H2O
1
2
14a
14b
14c
14d
14
15
16

Scheme 9.

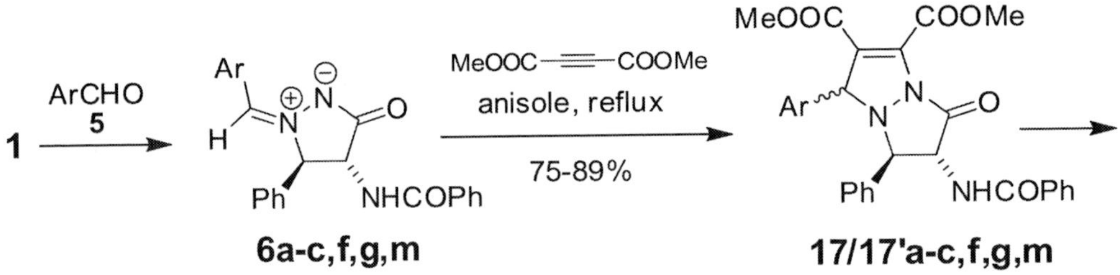

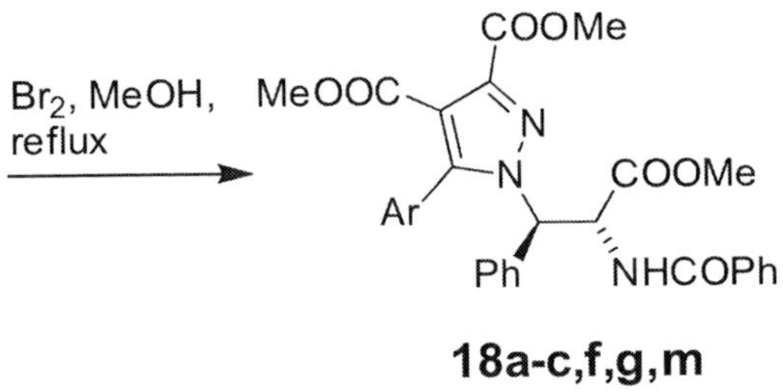

|  | Ar | Yield (%) |
|---|---|---|
| **18a** | phenyl | 87 |
| **18b** | 4-methylphenyl | 94 |
| **18c** | 4-methoxyphenyl | 93 |
| **18f** | 3-nitrophenyl | 83 |
| **18g** | 4-nitrophenyl | 86 |
| **18m** | 2,6-dichlorophenyl | 89 |

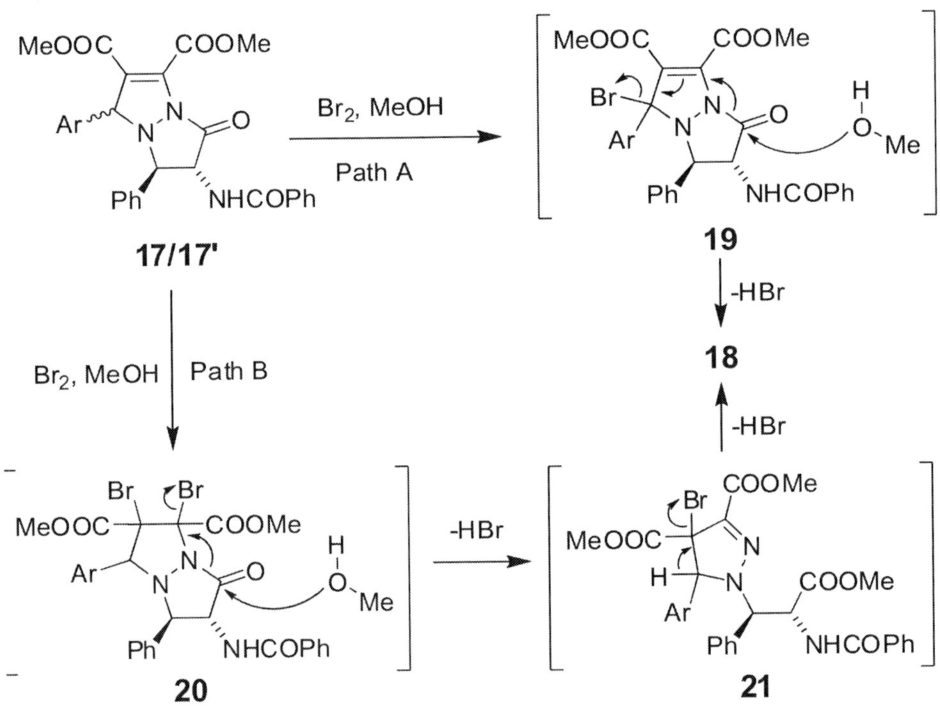

Scheme 10.

Another application of the pyrazolidinone *1* in the amino acid chemistry is a three-step synthesis of pyrazolylalanines *18a–c,f,g,m*. First, starting compound *1* was transformed with aromatic aldehydes *5* into the corresponding azomethine imines *6a–c,f,g,m*. Next, treatment of dipoles *6a–c,f,g,m* with dimethyl acetylenedicarboxylate (DMAD) in refluxing anisole give the bicyclic pyrazolidinones *17a–c,f,g* and *17'm*.[1] Finally, oxidative ring-opening of cycloadducts *17/17'* with bromine in methanol under reflux affords the final (4*R**,5*R**)-*N*-benzoyl-3-phenyl-3-[5-aryl-3,4-bis(methoxycarbonyl)-1*H*-pyrazol-1-yl]alanine methyl esters *18a–c,f,g,m*. Transformation of pyrazolo[1,2–*a*]pyrazolones *17/17'* into the final compounds *18* could be explained by the following proposed mechanism. Bromination of *17/17'* can take place , either at the allylic position (Path A) to give the intermediate *19*, or at the double bond (Path B) to give the intermediate *20*. Nucleophilic attack of methanol to the intermediates *19* and/or *20* followed by elimination of hydrogen bromide then leads to opening of the pyrazolidinone ring and formation of the pyrazolylalanine ester *18* [37] (Scheme 10).

## 5. [3+2] Cycloadditions of (4*R**,5*R**)-1-Alkylidene-4-benzoylamino-5-phenyl-3-pyrazolidinone-1-azomethine Imines: Synthesis of Functionalized Pyrazolo[1,2–*a*]pyrazolones

1,3-Dipolar cycloadditions are useful for the preparation of five-membered heterocycles, because they enable access to polyfunctional compounds with multiple asymmetric centers, usually with excellent stereocontrol. Despite widely reported asymmetric cycloadditions in various chiral nitrone, nitrile oxide, and azomethine ylide series, few examples have been reported in the chiral azomethine imine series [38–41]. Within this context, stable chiral azomethine imines have been prepared from pyridazine-diazoalkane cycloadducts [42–46], pyrazolidin-3-ones [7, 8, 11, 13, 14, 21–23, 29, 30, 40, 41], and 1,3,4-oxadiazin-2-ones [47–50]. Generally, 1,3-dipolar cycloadditions of cyclic chiral azomethine imines to various dipolarophiles have been accompanied by high facial and *endo*/*exo*-selectivity and afforded the corresponding functionalized fused pyrazolones with a bridgehead N–N structural element [7–14, 21–23, 26–50].

1,3-Dipolar cycloadditions of (4*R**,5*R**)-1-alkylidene-4-benzoylamino-5-phenyl-3-pyrazolidinon-1-azomethine imines *6* are important reactions, since they enable an easy access to 2-acylamino-1-oxo-1*H*,5*H*-pyrazolo[1,2–*a*]pyrazole-7-carboxylates as useful scaffolds for the preparation of conformationally constrained peptidomimetics [8, 12, 13, 51–53]. In most cases, reactions of (4*R**,5*R**)-1-alkylidene-4-benzoylamino-5-phenyl-3-pyrazolidinon-1-azomethine imines *6* with various dipolarophiles were highly stereoselective and regioselective as well. In the case of azomethine imines *6a–o* derived from aromatic aldehydes, stereocontrol did not depend only on the stereodirecting phenyl group at position 5, but also on the *ortho*-substituents at the aromatic ring attached at position 1'. Generally, cycloadditions of benzaldehyde derived dipoles were quite selective and stereocontrol was often predictable. On the other hand, cycloadditions of dipoles derived from aliphatic aldehydes and ketones are less elaborated and shall require more attention in future.

---

[1] The configuration at the newly formed stereocented in compounds *17* and *17'* was not relevant in this case, since this stereocenter was lost in the next, oxidation step. For details on stereochemistry and configuration of cycloadducts *17* and *17'* see Section 5.1.

## 5.1. Cycloadditions to Acetylenic Dipolarophiles

Dimethyl acetylenedicarboxylate (DMAD) is probably one of the most common dipolarophiles used in 1,3-dipolar cycloaddition chemistry. Therefore, it is not surprising, that (1*Z*,4*R**,5*R**)-1-arylmethylidene-4-benzoylamino-5-phenyl-3-pyrazolidinon-1-azomethine imines *6a,c,e–g,m* reacted readily with DMAD in refluxing anisole to give the corresponding cycloadducts *17a,c,e–g* and *17'm* as single isomers [26, 28, 37]. However, the stereochemistry of cycloadducts was dependent on the *ortho*-substituents at the aromatic ring attached at position 1'. Thus, azomethine imines *6a,c,e–g* without *ortho*-substituents at the 1'-aryl group gave the (3*S**,5*R**,6*R**)-isomers *17a,c,e–g*, where the aryl groups at positions 3 and 5 were *syn*-oriented. On the other hand, dipole *6m* with two *ortho*-substituents at the 1'-aryl group gave the (3*S**,5*R**,6*R**)-isomer *17'm*[2] with *anti*-oriented aryl groups at positions 3 and 5 [28] (Scheme 11).

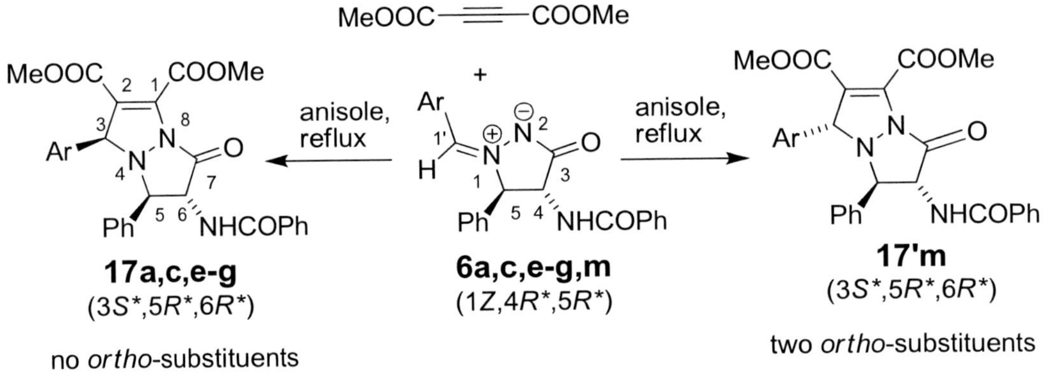


| | Ar | Yield (%) |
|---|---|---|
| **17a** | phenyl | 86 |
| **17c** | 4-methoxyphenyl | 75 |
| **17e** | 4-dimethylaminophenyl | 81 |
| **17f** | 3-nitrophenyl | 89 |
| **17g** | 4-nitrophenyl | 80 |
| **17'm** | 2,6-dichlorophenyl | 77 |

Scheme 11.

In addition to the above mentioned cycloadditions of 1-arylmethylidene substituted dipoles *6a,c,e–g,m*, cycloadditions of azomethine imines *6q* and *6r* derived from acetone and butyraldehyde, respectively, were also investigated. Heating of dipole *6q* with 1 equiv. of DMAD followed by chromatographic separation gave compounds *17q* and 22 in 33% and 16% yield, respectively. Similarly, treatment of *3r* with DMAD afforded a mixture of compound *22* and epimeric cycloadducts *17r* and *17'r*. Further chromatographic separation

[2] Formally, the configurations at C(3) in compounds *17a,c,e–g* and *17'm* are the same. This is due to chlorine atoms at the *ortho*-positions, which change the order of priority in compound *17'm*. Essentially, the configuration at C(3) in all compounds *17* and *17'* is the opposite.

then furnished compound 22 in 34% yield and a mixture of *17r* and *17'r* in a ratio of 29:71, respectively, in 8% yield [32] (Scheme 12).

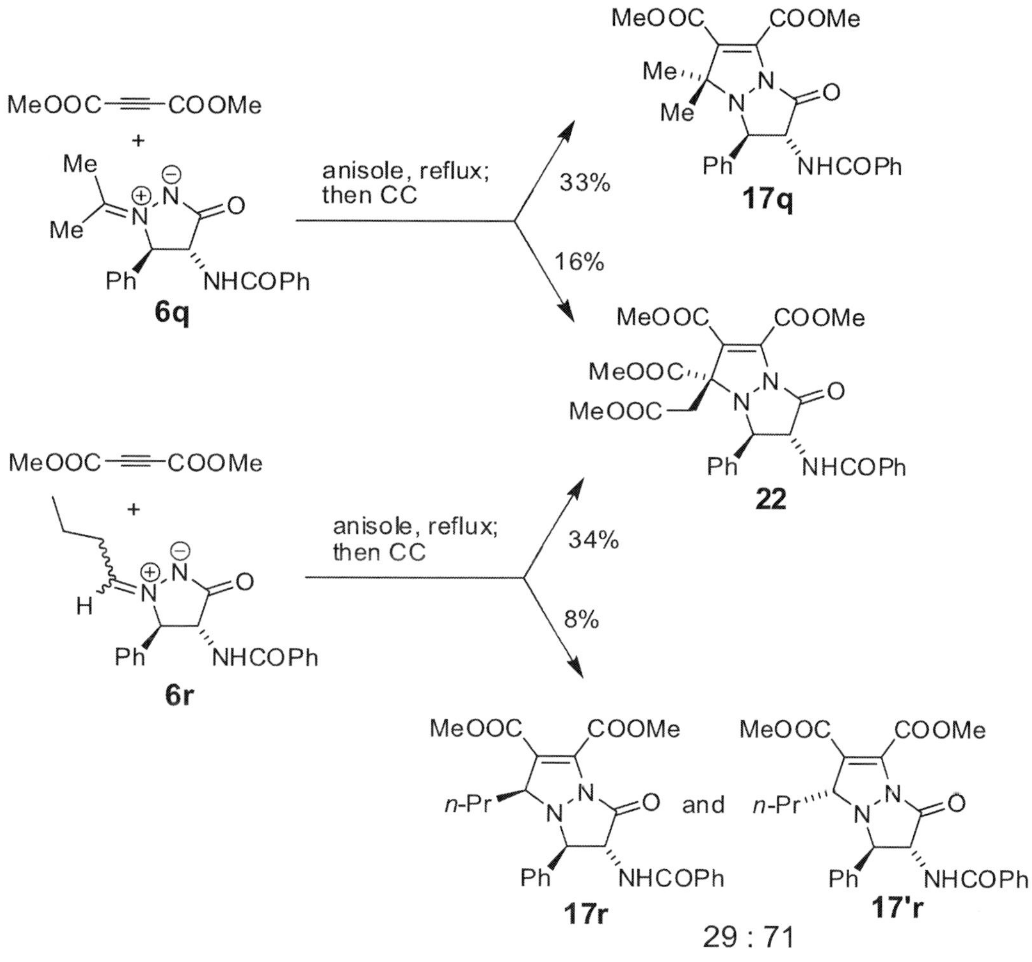

Scheme 12.

Formation of *22* could be explained by hydrolysis of azomethine imine *6* into the starting pyrazolidinone *1*, which undergoes 1,4-addition to DMAD to give the Michael-adduct *23*. Isomerization of *23* gives the dipole *24*, which reacts with the second equivalent of DMAD to furnish compound *22*. The proposed mechanism was supported by another experiment, where *1* was treated with two equivalents of DMAD in refluxing anisole to afford *22* as the only product in 71% yield [32] (Scheme 13).

Similarly, reaction of the enamine *8* with DMAD in refluxing anisole furnished the corresponding spiro cycloadduct *26*, most probably *via* the dipolar intermediate *25* [21, 33] (Scheme 14).

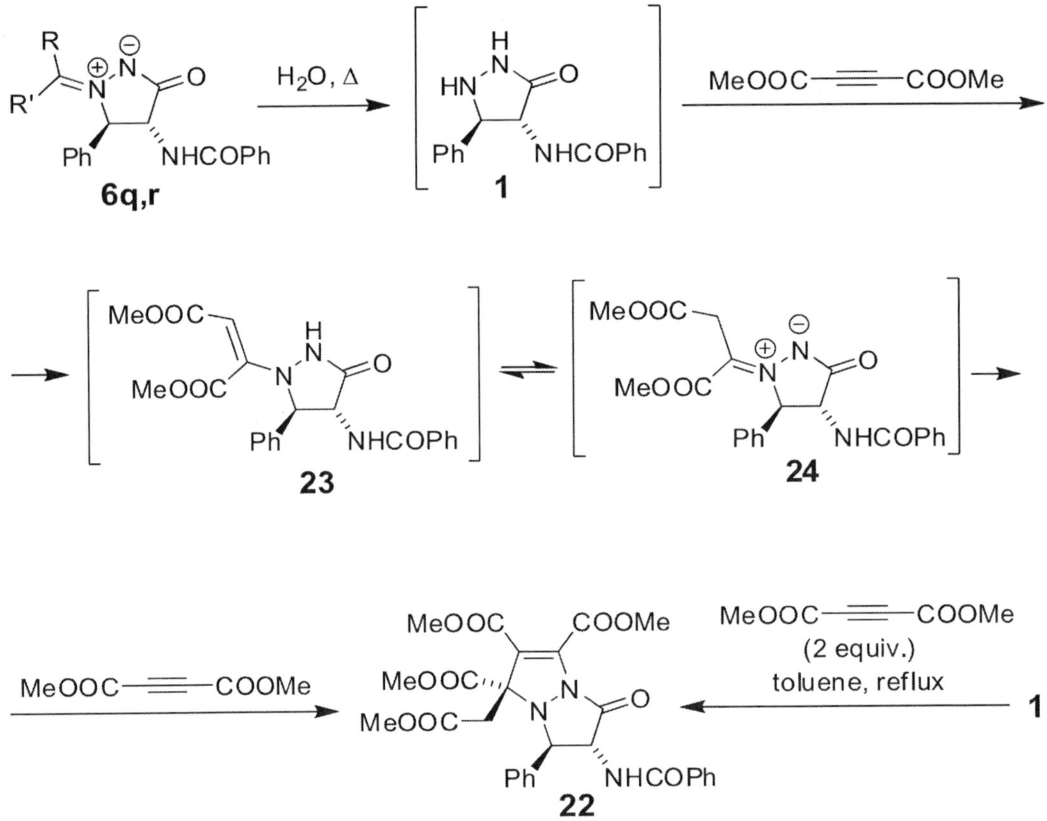


Scheme 13.

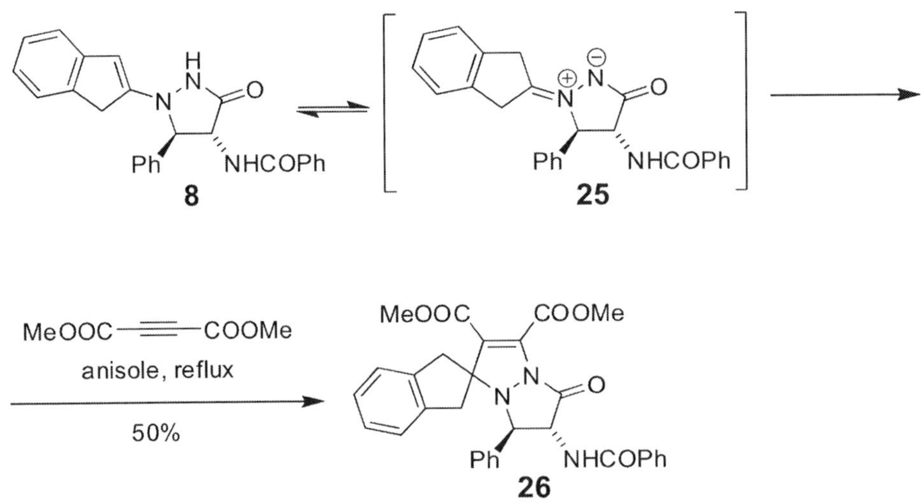


Scheme 14.

In contrast to cycloaddition to DMAD, reactions of dipoles *6a,c,g,m,n* with ethyl propiolate were much less selective. Preliminary experiment was carried out with dipole 6a in refluxing anisole. Under these reaction conditions, the conversion of azomethine imine *6a* was relatively slow and furnished dark reaction mixture containing several by-products. On the other hand, the conversion was much faster and cleaner under microwave irradiation. Therefore, cycloadditions dipoles *6a,c,g,m,n* to ethyl propiolate were carried out under microwave irradiation in anisole at 150 °C to furnish the corresponding mixtures of the isomeric cycloadducts *27/27'/28/28'a,c,g,m,n*. Primary purification of the crude reaction mixtures by flash chromatography (FC) furnished pure compounds *27g* and *27m* in 28% and 8% yield, respectively, and purified mixtures of isomeric cycloadducts *27/27'/28/28'a,c,n*. Additional purification of these isomeric mixtures by medium pressure liquid chromatography (MPLC) then afforded pure major isomers *27a,c,n* in 15-32% yields. All cycloadditions gave the major 2-COOEt regioisomers *27/27'* and the major (1*S**6*R**,7*R**)-diastereomers *27* with *syn*-oriented aryl groups at positions 3 and 5 [54] (Scheme 15).

Copper(I)-catalysed 1,3-dipolar cycloadditions azides [55, 56] to terminal alkynes are probably the most outstanding examples within the emerging field of 'click' chemistry. In contrast to the non-catalysed reactions, copper(I)-catalysed cycloadditions proceed regioselectively and under mild reaction conditions. Presumably, these reactions are not the concerted reactions and proceed by a stepwise Huisgen cycloaddition process [55, 57–59]. Recently, Fu and coworkers reported regioselective and enantioselective copper(I)-catalysed [3+2] coupling of achiral 3-pyrazolidinon-1-azomethine imines with terminal acetylenes to generate optically active pyrazolo[1,2–*a*]pyrazolone derivatives in 74–98% e.e. [60]. To improve the low selectivity of microwave-assisted cycloadditions of dipoles *6a,c,g,m,n* to ethyl propiolate, reactions were carried out in refluxing dichloromethane in the presence of 20% of copper(I) iodide, following the procedure of Fu and co-workers [60]. Under these reaction conditions, conversions were mostly fast and clean; however, partial decomposition of dipoles *6* also took place as the competitive reaction. Dipoles *6g* and *6n* did not react at all and eventually decomposed upon prolonged heating. On the other hand, dipoles *6a,c,m* reacted regio- and stereoselectively to form the corresponding cycloadducts *27a*, *27c*, and *27'm* as single isomers, which were then isolated in 51%, 60% and 8% yield, respectively [31] (Scheme 16).

## 5.2. Cycloadditions to Olefinic Dipolarophiles

Reactions of (1*Z*,4*R**,5*R**)-1-arylmethylidene-4-benzoylamino-5-phenyl-3-pyrazolidinon-1-azomethine imines *6a,c,d,g,m,n* with dimethyl maleate in refluxing anisole proceeded stereoselectively giving the corresponding cycloadducts *29a,c,d,g* and *30m,n* as single isomers. Stereochemistry of cycloadditions was dependent on the *ortho*-substituents at the aromatic ring. In the *ortho*-unsubstituted compounds *29a,c,d,g*, the aryl groups at positions 3 and 5 were *syn*-oriented, whilst *anti*-orientation of these two aryl groups was established in *ortho*-disubstituted compounds *30m,n* [28]. Similarly, reactions of 1-alkylidene azomethine imines *6q* and *6r* afforded cycloadducts *29q* and *29r* in moderate yields [32] (Scheme 17).

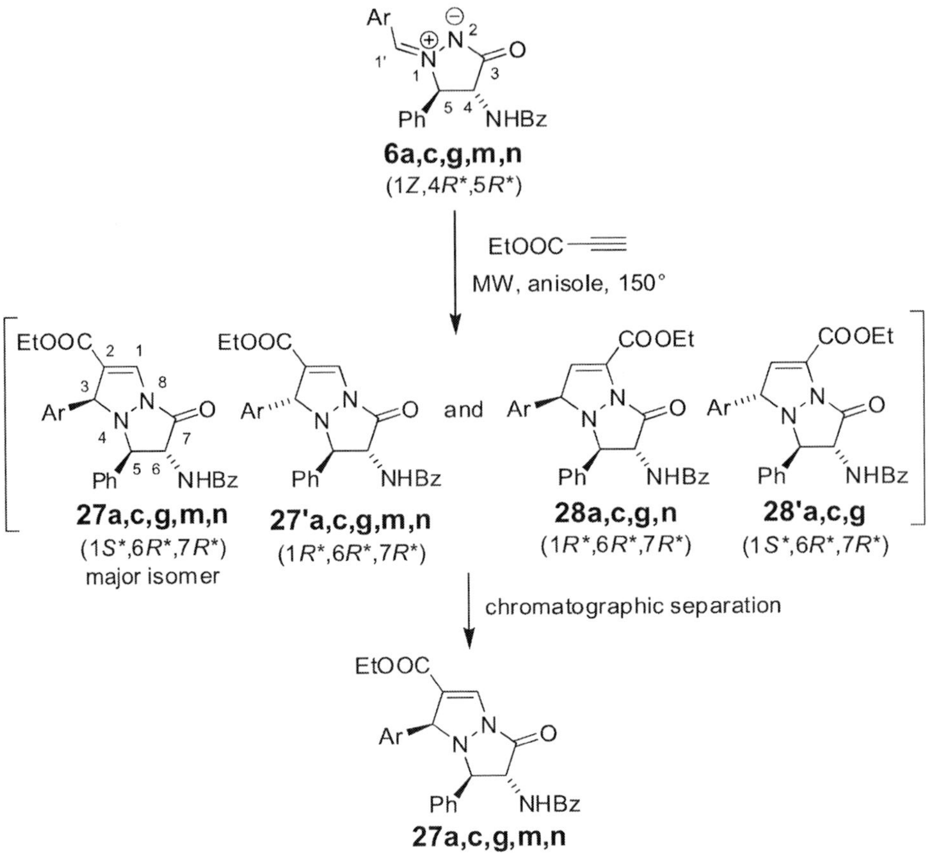

| Compound | Ar | **27:27':28:28'** | Yield (%) |
|---|---|---|---|
| **27/27'/28/28'a** | phenyl | 36:10:29:25 | 15 |
| **27/27'/28/28'c** | 4-methoyphenyl | 48:21:31:0 | 23 |
| **27/27'/28/28'g** | 4-nitrophenyl | 55:7:34:4 | 28 |
| **27/27'/28/28'm** | 2,6-dichlorophenyl | 85:15:0:0 | 8 |
| **27/27'/28/28'n** | 2,4,6-trimethylphenyl | 62:15:23:0 | 32 |

Scheme 15.

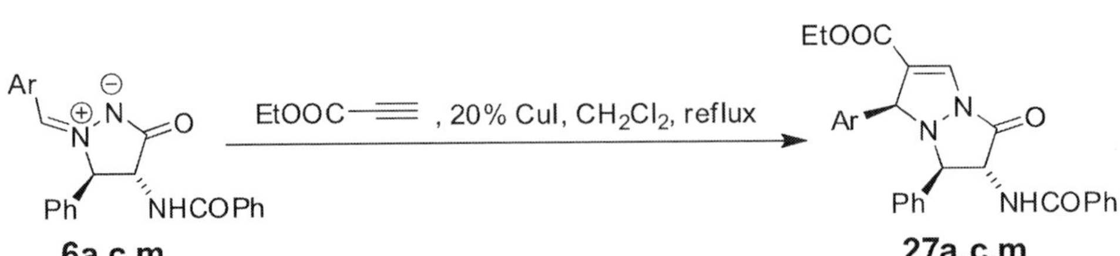

| | Ar | Yield (%) |
|---|---|---|
| **27a** | phenyl | 51 |
| **27c** | 4-methoxyphenyl | 60 |
| **27m** | 2,6-dichlorophenyl | 8 |

Scheme 16.

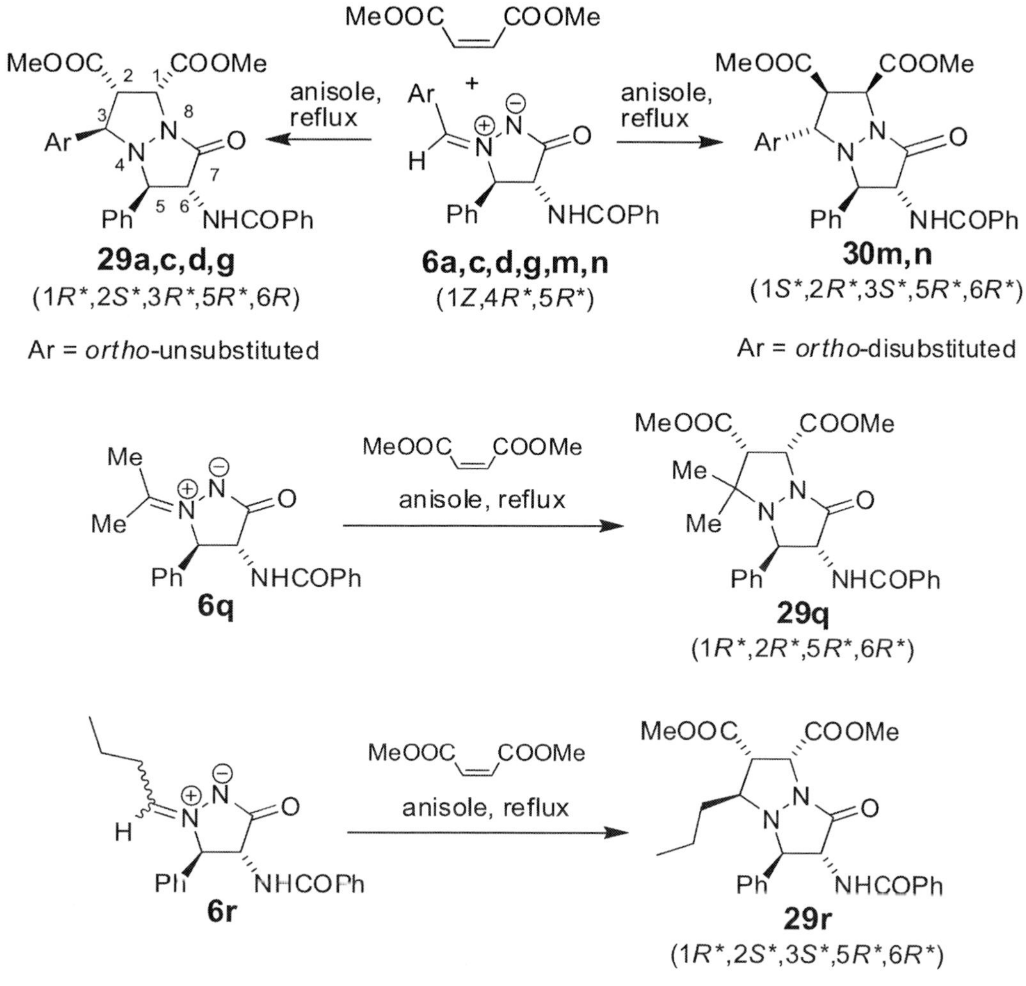

| | Ar | Yield (%) |
|---|---|---|
| **29a** | phenyl | 49 |
| **29c** | 4-methoxyphenyl | 49 |
| **29d** | 3,4,5-trimethoxyphenyl | 68 |
| **29g** | 4-nitrophenyl | 55 |
| **29q** | - | 18 |
| **29r** | - | 37 |
| **30m** | 2,6-dichlorophenyl | 81 |
| **30n** | 2,4,6-trimethylphenyl | 73 |

Scheme 17.

In order to investigate preparation of enantiopure cycloadducts *via* treatment of racemic dipoles with enantiopure dipolarophile followed by resolution of diastereomeric cycloadducts, cycloadditions of azomethine imines *6* to di-(–)-menthyl maleate were also studied. Quite expectedly, reaction of 1-benzylidene substituted dipole *6a* with di-(–)-menthyl maleate afforded a 1:1 mixture of two diastereomeric cycloadducts, *31a* and *31'a*. Further crystallization of this isomeric mixture from methanol furnished diastereomerically pure cycloadduct in 29% yield. Upon this promising result, a series of cycloadditions to di-(–)-menthyl maleate were carried out with *ortho*-unsubstituted azomethine imines *6b–e,g–i,k* and *ortho*-monosubstituted azomethine imine *6l*. Quite expectedly, reactions of dipoles *6g–i* and *6l* gave two diastereomeric cycloadducts, *31/31'g,i* and *32/32'h,l* in a ratio of ~1:1. To our surprise, dipoles *6b–e,k* with electron-releasing substituents at the 1'–Ar group furnished mixtures of four diastereomers, *31*, *31'*, *32*, and *32'*, as a consequence of diminished *exo/endo*-selectivity. Further separation of these mixtures of diastereomers was achieved, either by crystallization, or/and by medium pressure liquid chromatography (MPLC). In most cases, such separation of diastereomers was successful and afforded at least one pure diastereomer. while compounds *31/31'/32/32'd*, *32/32'h*, and *31/31'h* could not be separated [31] (Scheme 18, Table 2).

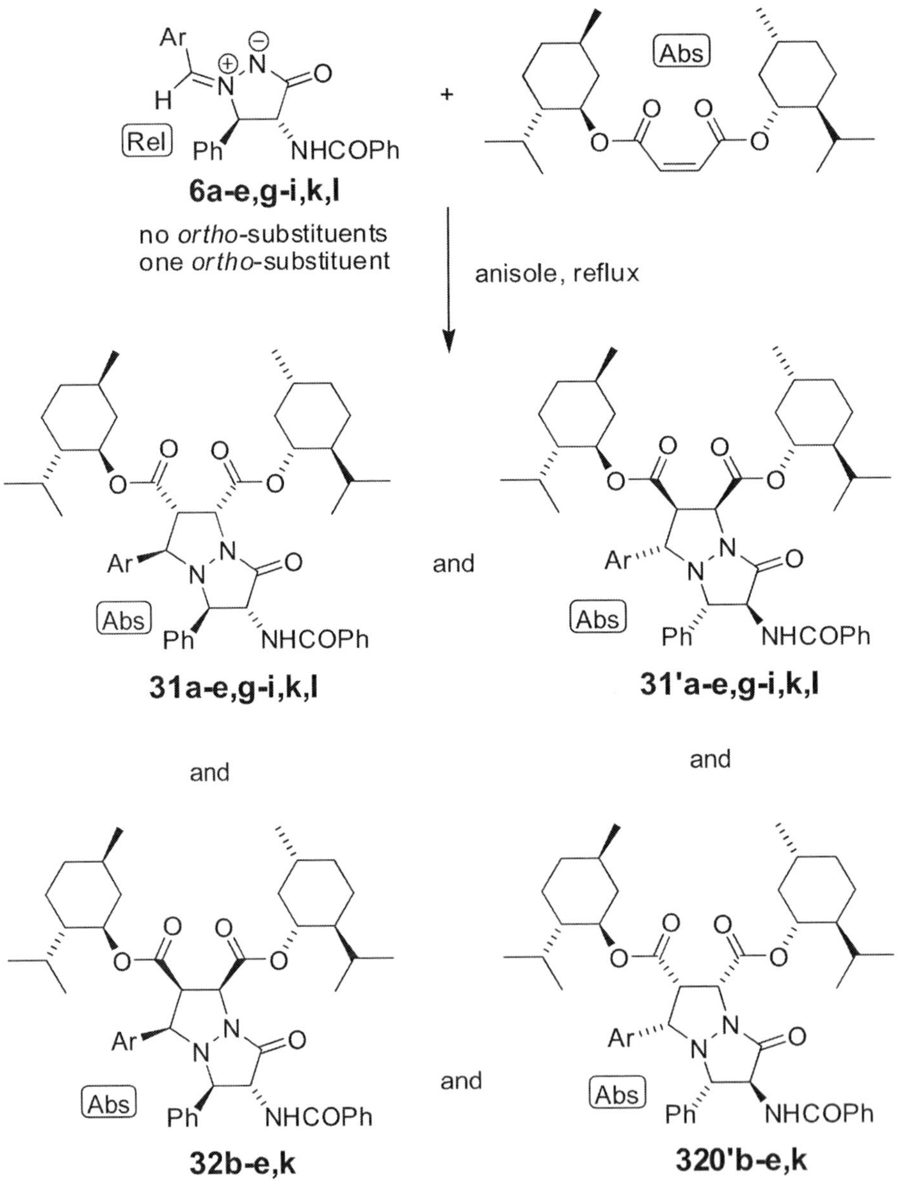


Scheme 18.

**Table 2. Selected experimental data for compounds 31, 31', 32, 32', 33, and 33'**

| Compound | Ar | Ratio of Isomers | Separation | Yield (%)[a] |
|---|---|---|---|---|
| **31/31'a** | phenyl | 56:44 | cryst. | 29 |
| **31/31'/32/32'b** | 4-methylphenyl | 39:37:13:11 | cryst., MPLC | 1, 2 |
| **31/31'/32/32'c** | 4-methoxyphenyl | 31:29:20:20 | MPLC | 7, 3, 5, 8 |
| **31/31'/32/32'd** | 3,4,5-trimethoxyphenyl | 39:26:18:17 | [b] | - |
| **31/31'/32/32'e** | 4-dimethylaminophenyl | 43:27:16:14 | cryst. | 5 |
| **31/31'g** | 4-nitrophenyl | 54:46 | cryst. | 32, 42 |
| **32/32'h** | 3-fluorophenyl | 51:49 | [b] | - |
| **31/31'i** | 2-furyl | 50:50 | cryst. | 6 |
| **31/31'/32/32'k** | 2-methoxyphenyl | 28:25:25:22 | MPLC | 6 |
| **32/32'l** | 2,4-dichlorophenyl | 50:50 | cryst. | 21 |
| **33/33'm** | 2,6-dichlorophenyl | 51:49 | [b] | - |
| **33/33'n** | 2,4,6-trimethylphenyl | 56:44 | [b] | - |

[a] Yield of diastereomerically pure compounds isolated upon separation.
[b] Isomers could not be separated.

On the other hand, both *ortho*-disubstituted dipoles, *6m* and *6n*, reacted selectively and furnished mixtures of two diastereomeric cycloadducts, *33/33'm* and *33/33'n*, respectively. Attempts to separate diastereomeric cycloadducts *33/33'm* and *33/33'n* were not successful [31] (Scheme 19, Table 2).

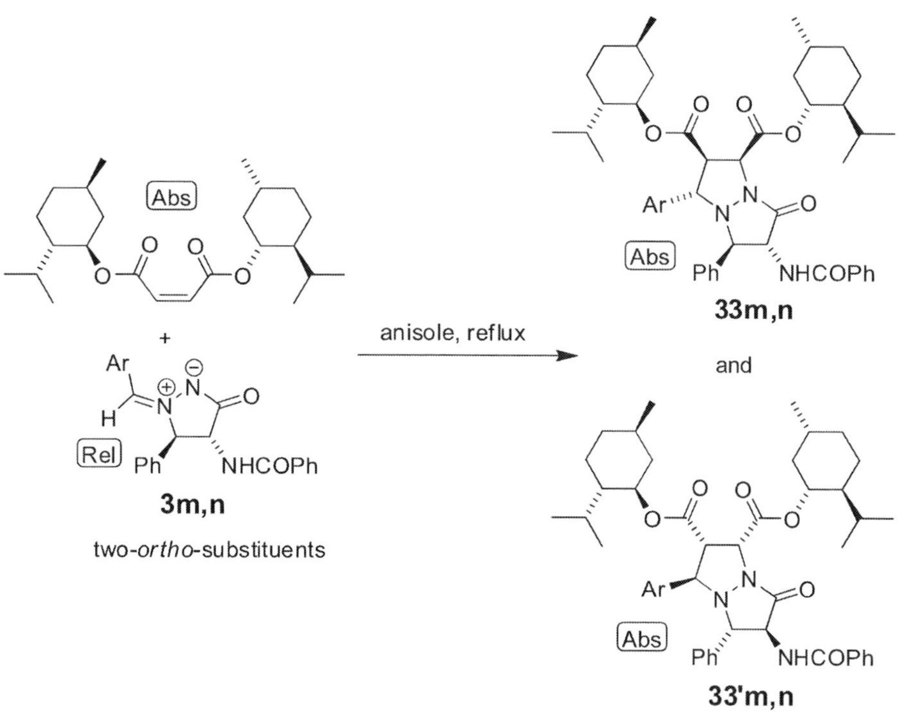


Scheme 19.

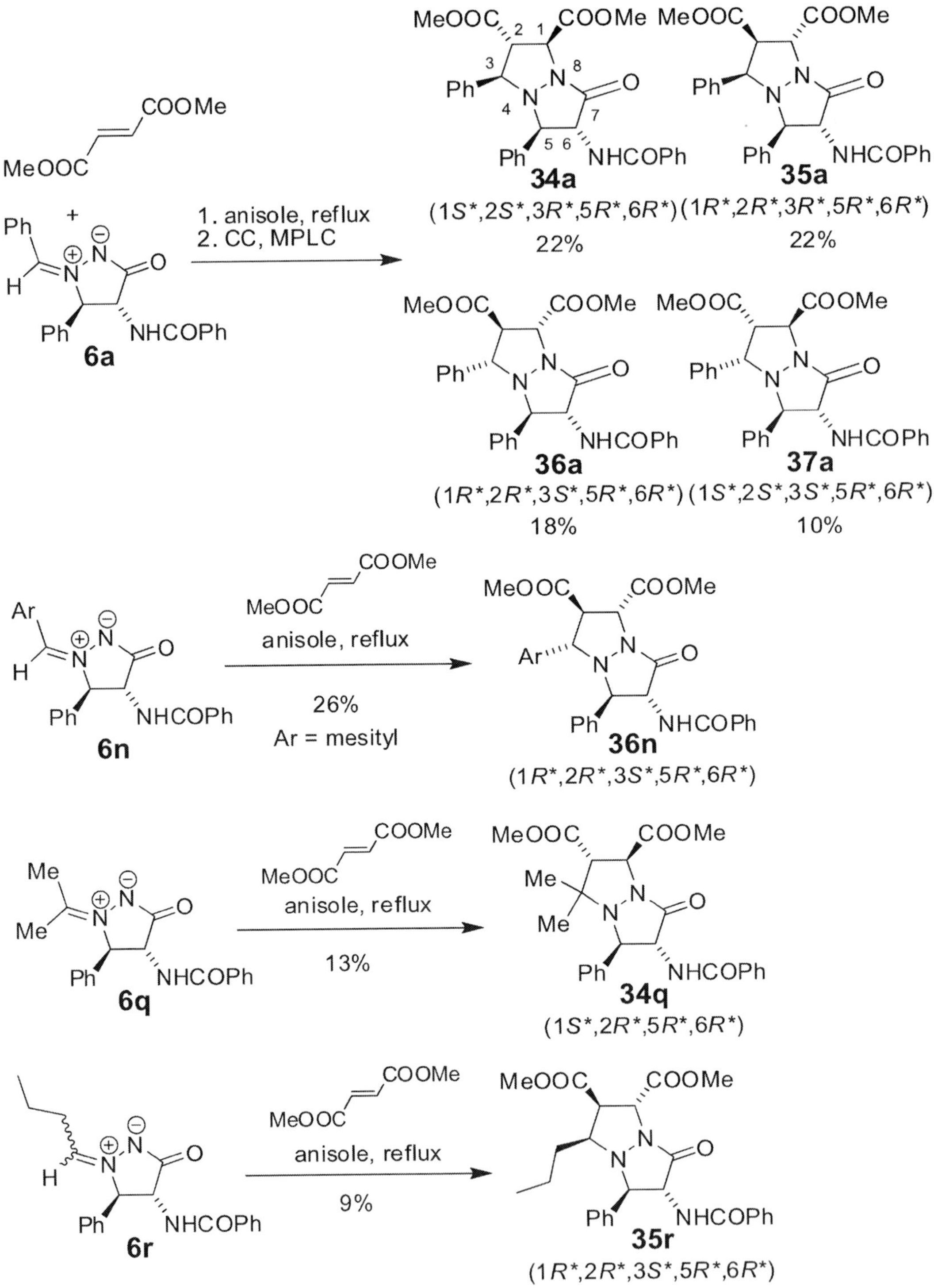

COOMe
MeOOC
Ph
H
6a
1. anisole, reflux
2. CC, MPLC
34a
(1S*,2S*,3R*,5R*,6R*)
22%
35a
(1R*,2R*,3R*,5R*,6R*)
22%
36a
(1R*,2R*,3S*,5R*,6R*)
18%
37a
(1S*,2S*,3S*,5R*,6R*)
10%
NHCOPh
6n
anisole, reflux
26%
Ar = mesityl
36n
(1R*,2R*,3S*,5R*,6R*)
6q
13%
34q
(1S*,2R*,5R*,6R*)
6r
9%
35r
(1R*,2R*,3S*,5R*,6R*)


Scheme 20.

In summary, cycloadditions of racemic dipoles *6* to di-(–)-menthyl maleate as enantiomerically pure dipolarophile indicated, that preparation of enantiopure cycloadducts *via* resolution of diastereomeric cycloadducts is feasible, although not very efficient. So far, we were not able to determine the absolute configuration of optically active cycloadducts by X-ray diffraction, since all attempts to obtain suitable single crystals of any of the isomerically pure cycloadducts failed. Nevertheless, the results of this study also showed, that steric effects of the dipolarophile and electronic effects of substituents at the 1'–Ar residue in dipole *6* may influence selectivity of cycloadditions. In our case, increased bulkiness of the dipolarophile (di-(–)-menthyl maleate *vs.* dimethyl maleate) significantly lowered the *exo/endo*-selectivity in cycloadditions of dipoles with electron-releasing groups at the 1'–Ar residue (Schemes 18 and 19, Table 2).

Next, cycloadditions to dimethyl fumarate were investigated. Treatment of *ortho*-unsubstituted dipole *6a* with dimethyl fumarate in refluxing anisole gave a mixture of four isomeric cycloadducts *34a–37a* in a ratio of 33:30:22:15, which were then separated by chromatography (CC followed by MPLC) to furnish compounds *34a*, *35a*, *36a*, and *37a* in 22%, 22%, 18%, and 10% yield, respectively. On the other hand, cycloaddition of dimethyl fumarate to 1'-mesityl substituted dipole *6n* was selective and gave compound *36n* as the only product in 26% yield [28]. Similarly, both aliphatic dipoles, *6q* and *6r*, reacted selectively and gave the corresponding cycloadducts *34q* and *35r* in 13% and 9% yield, respectively [32] (Scheme 20).

Reactions of *ortho*-unsubstituted dipoles *6a,d,g* with methyl acrylate were regioselective and stereoselective as well and gave the corresponding methyl (1*R**,3*R**,5*R**,6*R**)-6-benzamido-7-oxo-3,5-diphenylhexahydropyrazolo[1,2–*a*]pyrazole-1-carboxylates *38a,d,g* in 68–88% yield. On the other hand, cycloaddition of *ortho*-disubstituted dipole *6n* was non-selective and gave a mixture of four isomeric cycloadducts *39n–42n*, which were then isolated upon thorough chromatographic separation [28]. Reactions of acetone and butyraldehyde derived dipoles *6q* and *6r* with methyl acrylate afforded cycloadducts *38q* and *38r* as single isomers in 37% and 17% yield, respectively [32] (Scheme 21).

In contrast to the above described cycloadditions to methyl acrylate, reactions of azomethine imines *6a,c,g,m,n* with methyl methacrylate were regioselective and stereoselective, regardless the *ortho*-substituents. Reactions of *ortho*-unsubstituted dipoles *6a*, *6c*, and *6g* gave methyl (1*S**,3*S**,5*R**,6*R**)-3-aryl-6-benzamido-1-methyl-7-oxo-5-phenylhexahydropyrazolo-[1,2–*a*]pyrazole-1-carboxylates *43a*, *43c*, and *43g*, whilst cycloadditions of their *ortho*-disubstituted analogues *6m* and *6n* furnished their (1*R**,3*S**,5*R**,6*R**)-epimers *43'm* and *43'n*. Such stereochemical outcome was quite surprising: in contrast to stereocontrol in reactions with other olefinic dipolarophiles [28], the aryl groups at positions 3 and 5 were *anti*-oriented in both types of cycloadducts, *43* and *43'*. On the other hand, the *ortho* substituents apparently had a major impact on *endo/exo*-selectivity: cycloadditions of *ortho*-unsubstituted dipoles *6a*, *6c*, and *6g* were *exo*-selective, whilst reactions of dipoles *6m* and *6n* with two *ortho*-substituents were *endo*-selective [31] (Scheme 22).

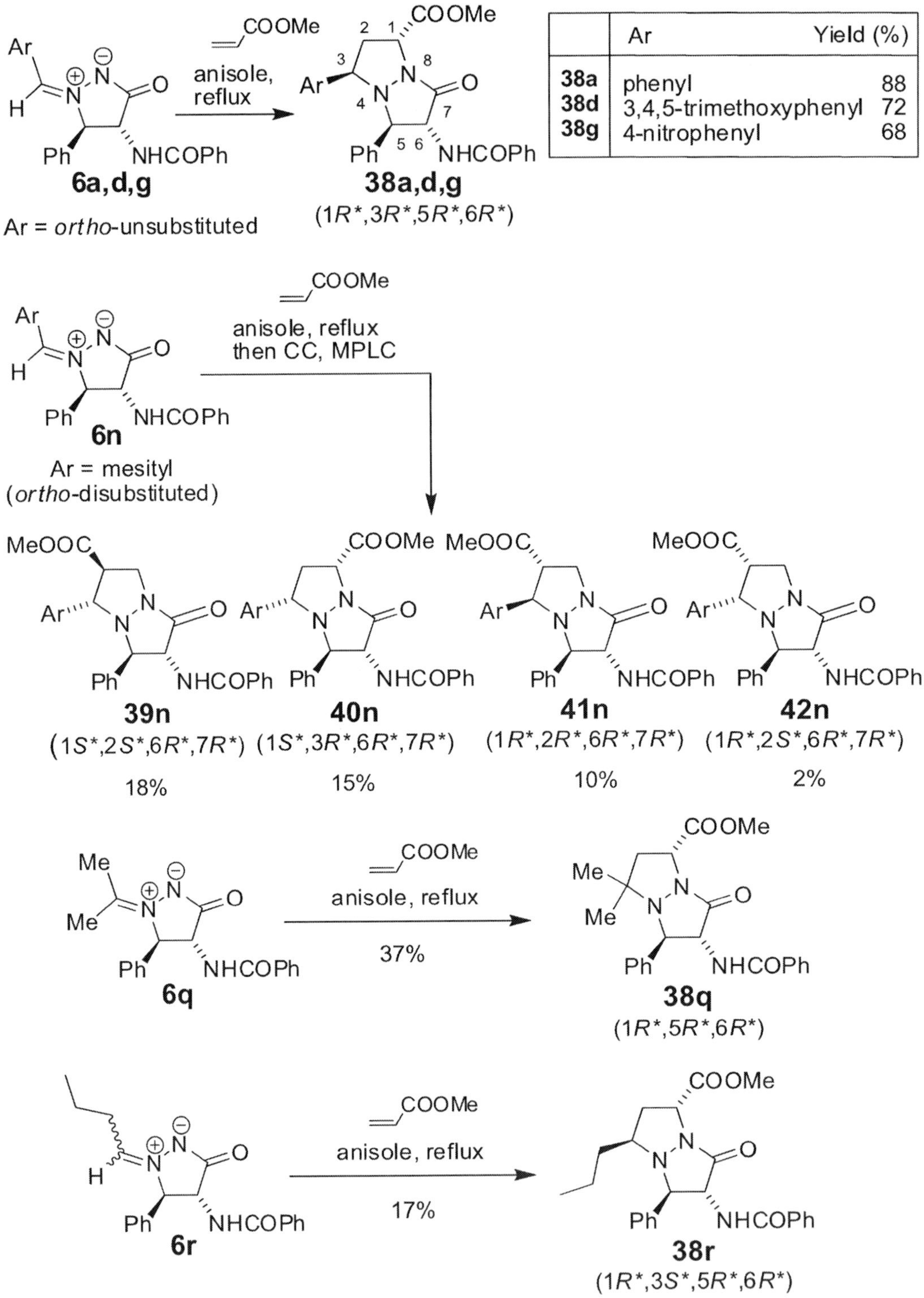


| | Ar | Yield (%) |
|---|---|---|
| **38a** | phenyl | 88 |
| **38d** | 3,4,5-trimethoxyphenyl | 72 |
| **38g** | 4-nitrophenyl | 68 |

Scheme 21.

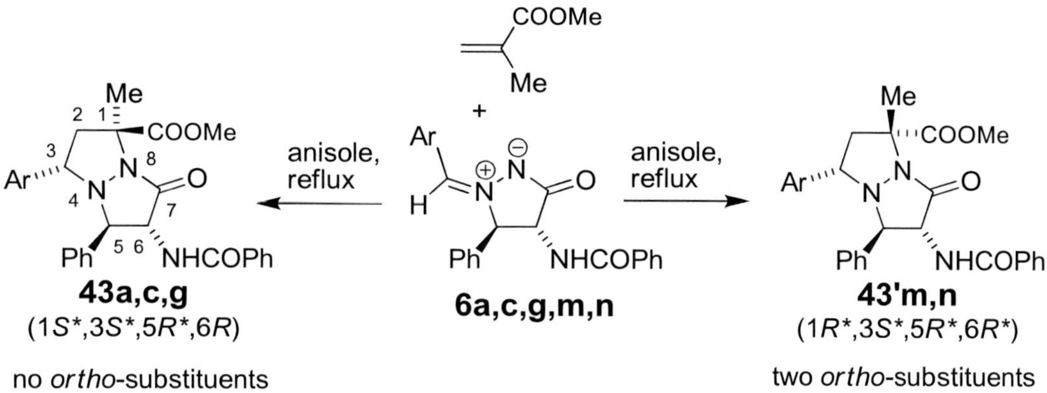


| | Ar | Yield (%) |
|---|---|---|
| **43a** | phenyl | 35 |
| **43c** | 4-methoxyphenyl | 84 |
| **43g** | 4-nitrophenyl | 63 |
| **43'm** | 2,6-dichlorophenyl | 38 |
| **43'n** | 2,4,6-trimethylphenyl | 4 |

Scheme 22.

### 5.3. Combinatorial Cycloadditions to Maleimides and β-keto Esters

Cycloadditions of azomethine imines *6a–p* derived from aromatic aldehydes to various olefinic and acetylenic dipolarophiles exhibited several common points: (a) they were carried out under identical reaction condition, *i.e.* in anisole under reflux, (b) regioselectivity and stereoselectivity were dependent on the *ortho*-substituents at the 1'-Ar group, and (c) in the case of selective reactions, the products were usually isolated simply by filtration. These observations led us to conclusion, that cycloadditions of 1-arylmethylidene substituted dipoles *6a–p* could also be studied combinatorially. Since reactions with DMAD and dimethyl maleate were selective regardless the *ortho*-substituents of the dipole *6*, we considered, that maleimides as symmetrical, cisoid dipolarophiles should also react selectively. Thus, a combinatorial solution-phase synthesis of 15 tetrahydro-5*H*-pyrazolo[1,2–*a*]pyrrolo[3,4–*c*]pyrazole-1,3,5(2*H*,3a*H*)-triones, *44*{1–3; 1–3} and *44'*{4,5; 1–3}, was carried out by reacting five azomethine imines *6a,d,g,m,n* (dipoles 1–5) with three maleimides (dipolarophiles 1–3). Azomethine imines *6a,d,g,m,n* were prepared in a parallel manner following the general literature procedures for conventional, single-vessel synthesis [26, 28] and then treated combinatorially with maleimides in refluxing anisole. Upon a simple workup, *i.e.* parallel evaporation and trituration of the residues with ether, the precipitated products were collected by filtration to give a library of 15 cycloadducts. Nine (3a*S**,6*R**,7*R**,9*R**,9a*R**)-isomers *44*{1–3; 1–3} and six (3a*S**,6*R**,7*R**,9*S**,9a*R**)-epimers *44'*{4,5 1–3} were obtained in 18–89% yields and in analytical purity [29]. In contrast to the *endo*-selective cycloadditions to dimethyl maleate [28], reactions of *ortho*-unsubstituted dipoles *6a,d,g* with maleimides 1–3 were *exo*-selective [29] (Scheme 23, Table 3).

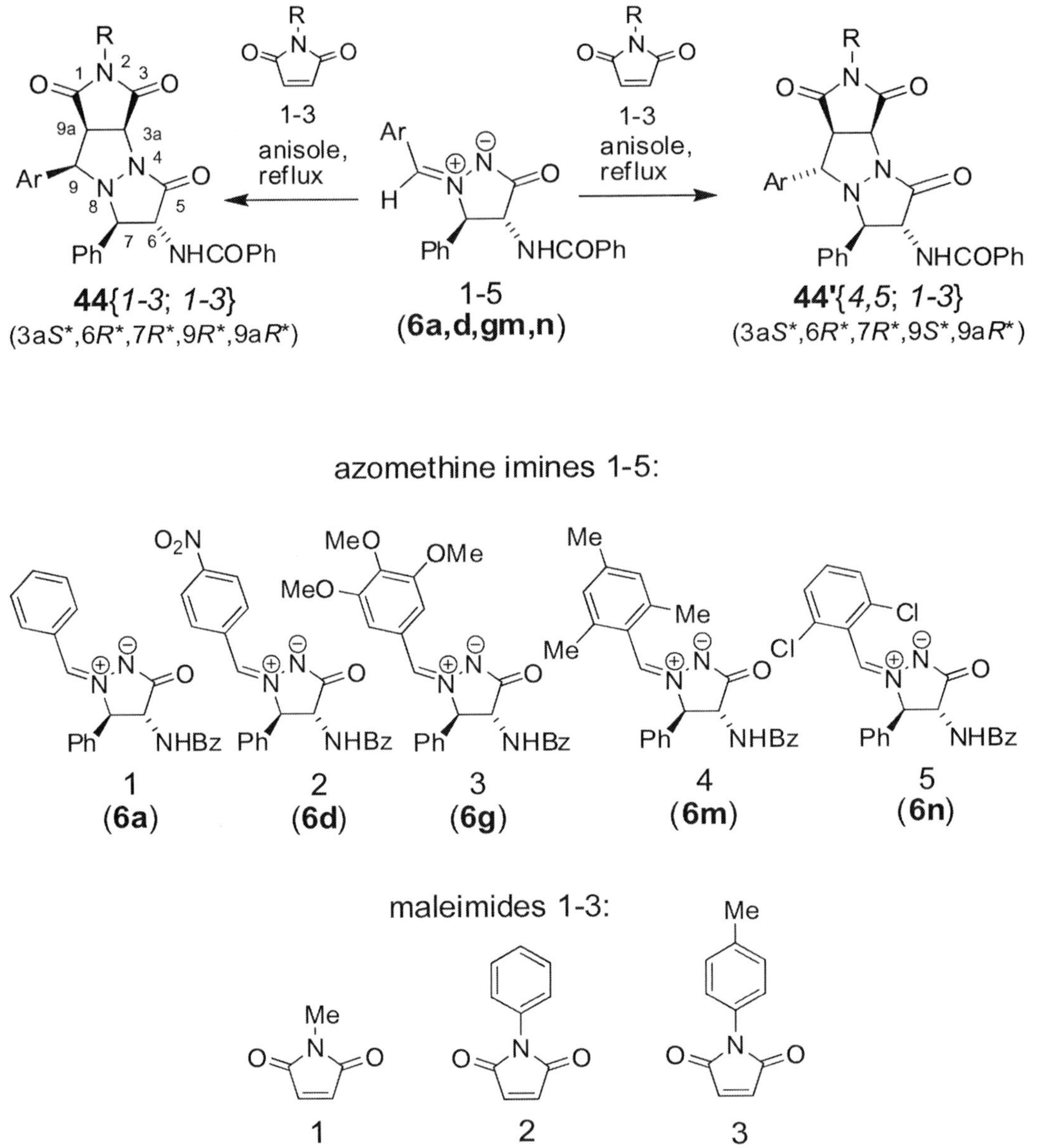


Scheme 23.

Reactions of 1-alkylidene substituted azomethine imines *6q* and *6r* with *N*-phenylmaleimide were also carried out in a conventional way. Thus, treatment of *6q* and *6r* with *N*-phenylmaleimide in refluxing anisole followed by chromatographic workup furnished the corresponding tetrahydro-5*H*-pyrazolo[1,2–*a*]pyrrolo[3,4–*c*]pyrazole-1,3,5(2*H*,3a*H*)-triones *44q* and *44r*, stereoselectively [32] (Scheme 24).

**Table 3. Combinatorial library of (3a*S**,6*R**,7*R**,9*R**,9a*R**)-tetrahydro-5*H*-pyrazolo[1,2–*a*]pyrrolo[3,4–*c*]pyrazole-1,3,5(2*H*,3a*H*)-triones 44{*1–3*; *1–3*} and their (9*S**)-epimers 44'{*4,5*; *1–3*}**

| Yield (%) | 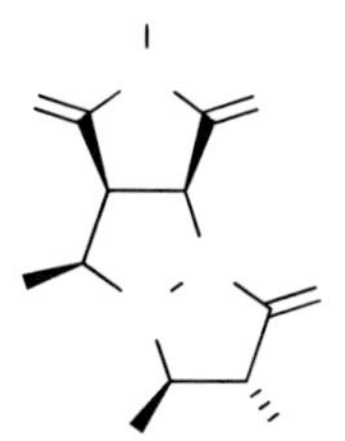 **44**{*1–3*; *1–3*} | | | 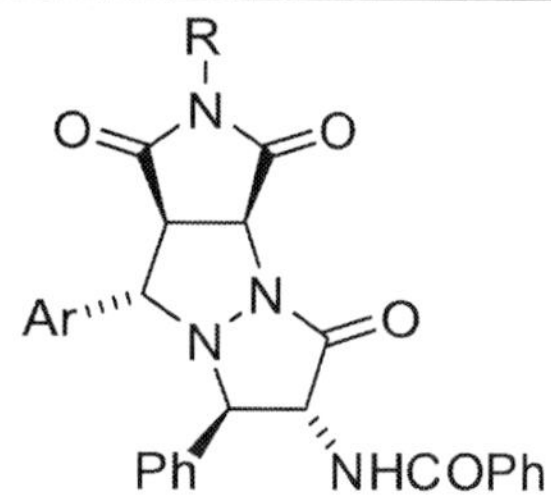 **44'**{*4,5*; *1–3*} | |
|---|---|---|---|---|---|
| | ← Ar → | | | | |
| | 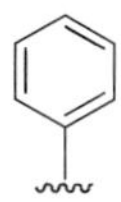 1 | 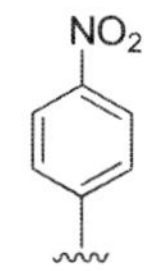 2 | 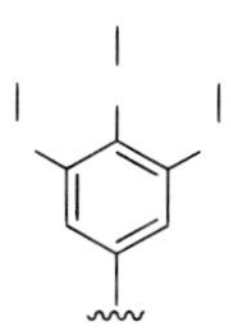 3 | 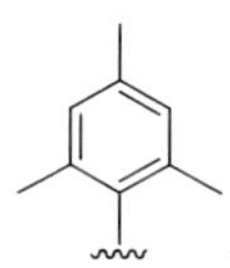 4 | 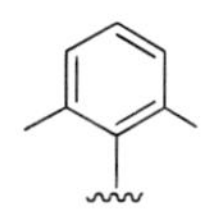 5 |
| ↑ R ↓ Me— 1 | 55 | 73 | 85 | 18 | 84 |
| 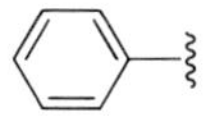 2 | 74 | 84 | 75 | 63 | 82 |
| 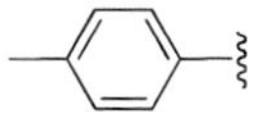 3 | 48 | 57 | 76 | 55 | 89 |

Finally, combinatorial cycloadditions of *ortho*-unsubstituted azomethine imines *6a,c,g* (dipoles 1–3) and *ortho*-monosubstituted azomethine imine *6l* (dipole 4) to *β*-keto esters 1–9 were investigated. *ortho*-Disubstituted dipoles were not included in this combinatorial study, since the preliminary experiments unambiguously showed, that *ortho*-disubstituted dipoles did not react. Also in this case, azomethine imines 1–4 were first prepared by a parallel solution-phase synthesis form the pyrazolidinone *1* and the corresponding benzaldehydes. In summary, a matrix of 36 combinatorial solution-phase cycloadditions was carried out. Ten reactions failed, since no conversion was detected by TLC, however, the majority of reactions preceded smoothly and afforded a library of 26 cycloadducts *45/45'* upon simple filtration workup. In a few cases, the products contained minor amounts of the unreacted azomethine imines, which were then removed by FC. All products were isolated in >90% purity according to $^1$H NMR, 25 of them were analytically pure. Stereocontrol of cycloadditions was the same as that established previously in cycloadditions of *ortho*-unsubstituted dipoles to acyclic olefinic dipolarophiles [28]: all reactions were *endo*-selective and proceeded predominantly from the less hindered face of dipole. Anomeric center at C(1) allows epimerization of *45/45'* in solution and, consequently, most of the products were obtained as mixtures of the major (1*S**)-epimers *45* and the minor (1*R**)-epimers *45'* [30] (Scheme 25, Table 4).

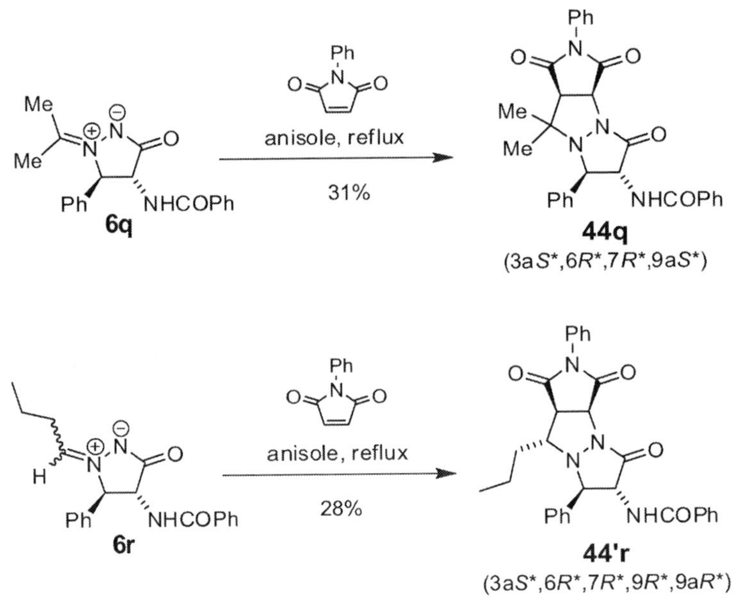


Scheme 24.

**Table 4. Yields of compounds 45/45'{*1–4*; *1–9*} (ratio of isomers, 45:45')**

| R¹ R² ↓ / ← Ar → | 1 (Ph) | 2 (4-OMe) | 3 (4-NO$_2$) | 4 (Cl, Cl) |
|---|---|---|---|---|
| Yield (%) Ratio of Isomers (45:45') | 1 | 2 | 3 | 4 |
| OEt 1 | 76 (74:26) | 0[a] | 75 (81:19) | 0[a]. |
| OBn 2 | 60 (57:43) | 34 (58:42) | 89 (80:20) | 0[a]. |
| OBu$^t$ 3 | 78 (73:27) | 80 (84:16) | 86 (93:7) | 66 (88:12) |
| MeOOC ... OMe 4 | 76 (100:0) | 0[a] | 86 (100:0) | 0[a]. |
| OMe 5 | 60 (86:14) | 29 (86:14) | 85 (93:7) | 44 (88:12) |

**Table 4. (Continued)**

| | | ← Ar → | | | |
|---|---|---|---|---|---|
| | 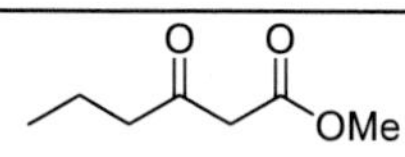 6 | 19 (91:9) | 22 (87:13) | 85 (90:10) | 54 (84:16) |
| | | ← Ar → | | | |
| | 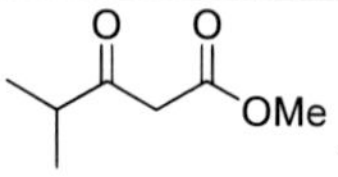 7 | 32 (100:0) | 43 (100:0) | 49 (100:0) | 23 (100:0) |
| | 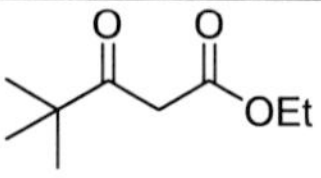 8 | 0[a] | 0[a] | 41 (100:0) | 0[a] |
| | 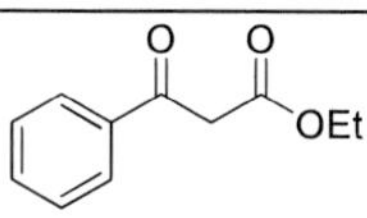 9 | 0[a] | 0[a] | 14 (100:0) | 6 (100:0) |

[a] No conversion was detected by TLC.

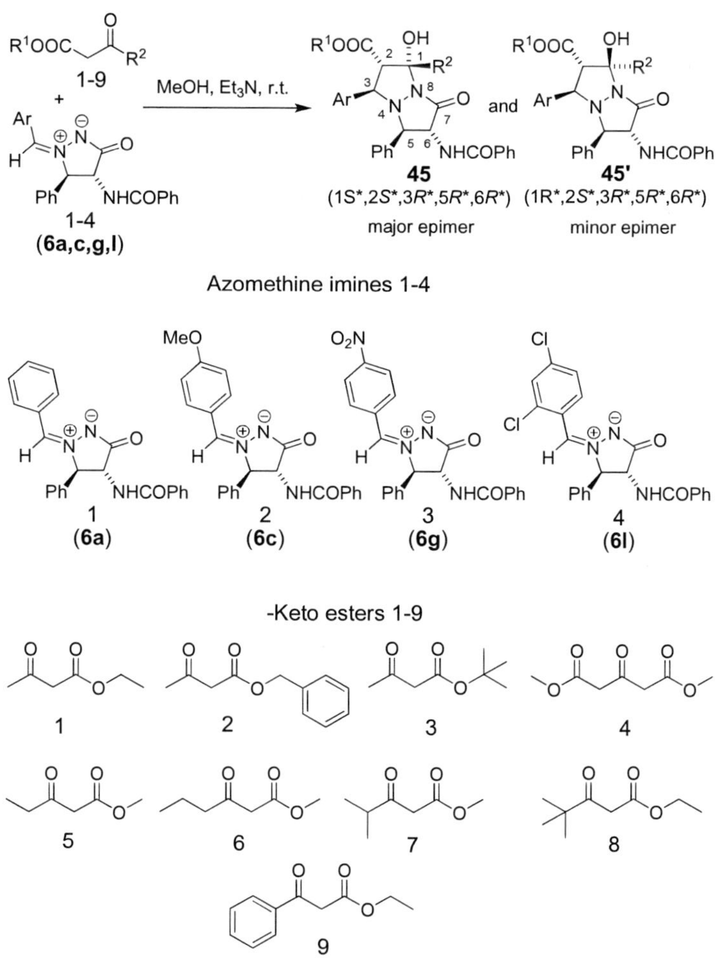


Scheme 25.

## 5.4. Reactivity, Selectivity, Stereocontrol, and Mechanistic Considerations

Most of cycloaddition reactions of (4*R**,5*R**)-4-benzoylamino-5-phenyl-3-pyrazolidinone (1) derived azomethine imines *6* were performed with 1-arylmethylidene-3-pyrazolidinone-1-azomethine imines *6a–o*, derived from 1 and aromatic aldehydes *5*. In this series, steric and electronic effects of substituents did influence the reactivity (connectivity) to some extent, nevertheless, azomethine imines *6a–o* were reactive enough to undergo cycloadditions to the most typical dipolarophiles: dimethyl acetylenedicarboxylate, ethyl propiolate, dimethyl maleate, methyl acrylate, methyl methacrylate, maleimides, and *β*-keto esters. In most cases, the conversions were fast and the yields of the isolated cycloadducts were good. The only general exception was found within base-promoted cycloadditions to *β*-keto esters, which did not take place, if both *ortho*-positions in the dipole *6* were occupied (see Section 5.3).

In contrast to reactivity, selectivity was strongly dependent on the structure of dipole and dipolarophile. Within the context of this section, reactions yielding a single cycloadduct are regarded as selective, *i.e.* stereoselective and regioselective. On the contrary, reactions affording at least two isomeric cycloadducts are considered as non-selective. Stereoselectivity and regioselectivity in reactions of 1-arylmethylidene-3-pyrazolidinone-1-azomethine imines *6a–o* can be summarized in the following way:

a. Reactions with symmetrical cisoid dipolarophiles (dimethyl acetylenedicarboxylate, dimethyl maleate, and maleimides) and reactions with methyl methacrylate and β-keto esters were all selective regardless the structure of a dipole *6*.
b. Reactions with dimethyl fumarate were selective only in the case of two *ortho*-substituents.
c. Reactions with methyl acrylate were selective only in the case of no *ortho*-substituent.
d. Reactions with ethyl propiolate were non-selective. However, regioselectivity and stereoselectivity were significantly improved, when the reactions were catalyzed by copper(I) iodide.

Selectivity pattern found in cycloadditions to 1-arylmethylidene-3-pyrazolidinone-1-azomethine imines *6a–o* to various dipolarophiles is depicted in Table 5.

**Table 5. Selectivity of Cycloadditions to Azomethine Imines 6a–o[a]**

| Dipole→<br>Dipolarophile↓ | 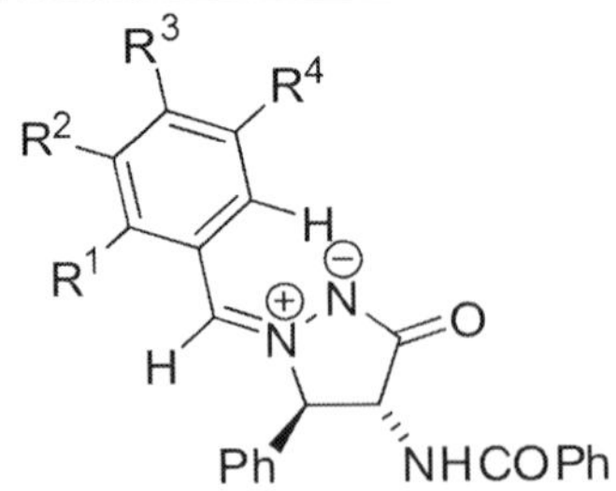<br>0 or 1 *ortho*-substituent | 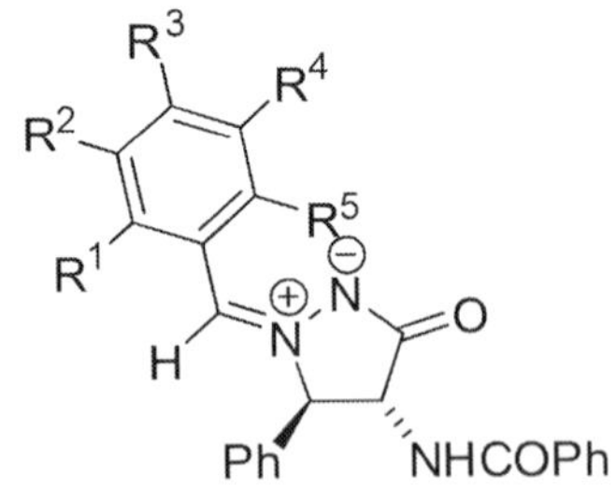<br>2 *ortho*-substituents |
|---|---|---|
| MeOOC—≡—COOMe | + | + |

**Table 5. (Continued)**

| | | |
|---|---|---|
| 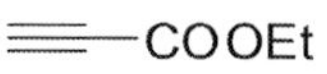 | –/+ [b] | –/+ [b] |
| 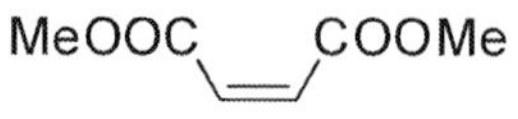 | + | + |
| 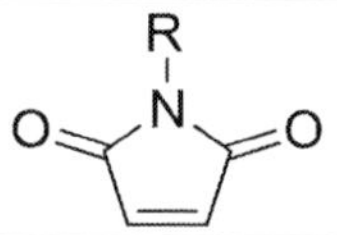 | + | + |
| 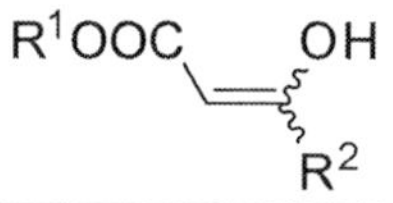 | + | 0[c] |
| 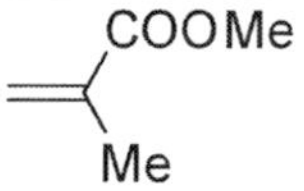 | + | + |
| 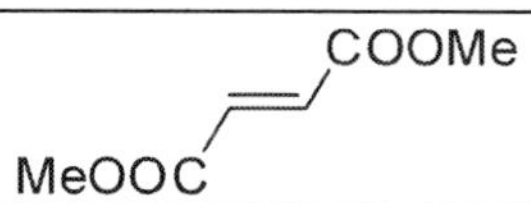 | – | + |
| 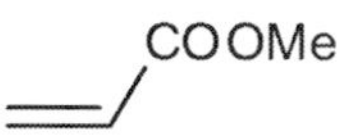 | + | – |

[a] Selective reaction: +; non-selective reaction: –.
[b] Cycloaddition was selective only in the presence of copper(I) iodide.
[c] No conversion.

Stereochemistry and regiochemistry of cycloadditions of 1-arylmethylidene-3-pyrazolidinone-1-azomethine imines *3a–o* was controlled (a) by the stereodirecting phenyl group at position 5, (b) by the *ortho*-substituents at the 1'–Ar group, and (c) by the structure (shape) of the dipolarophile. Generally, dipoles without *ortho*-substituents or with only one *ortho*-substituent favored formation of the major isomers with *syn*-oriented aryl groups at positions 3 and 5. On the contrary, dipoles with two *ortho*-substituents gave the major isomers with *anti*-oriented aryl groups Scheme 26).

So far, only two exceptions to this general stereocontrol were observed. First, stereochemistry of non-selective cycloadditions to ethyl propiolate was not controlled by the *ortho*-substituents - these reactions always led to the major 3,5-*syn*-cycloadducts. Similarly, selective cycloadditions to methyl methacrylate produced the 3,5-*anti*-cycloadducts, exclusively regardless the *ortho*-substituents at the 1'-aryl group in dipole *3* (Scheme 27).

Another important issue in cycloadditions to olefinic dipolarophiles is configuration around the newly formed C–C bond, which depends on *exo*/*endo*-selectivity. Generally, reactions of both types of dipoles to olefinic dipolarophiles afforded the major isomers with *trans*-configuration around the C(2)–C(3) bond. The first exceptions were cycloadditions of *ortho*-unsubstituted azomethine imines to maleimides, which furnished tricyclic compounds *47* with *cis*-configuration around the newly formed C–C bond (*c.f.* Scheme 26). The other exceptions were cycloadditions to methyl methacrylate, where the *exo*/*endo*-selectivity depended on *ortho*-substituents at the 1'-aryl group (*c.f.* Scheme 27).

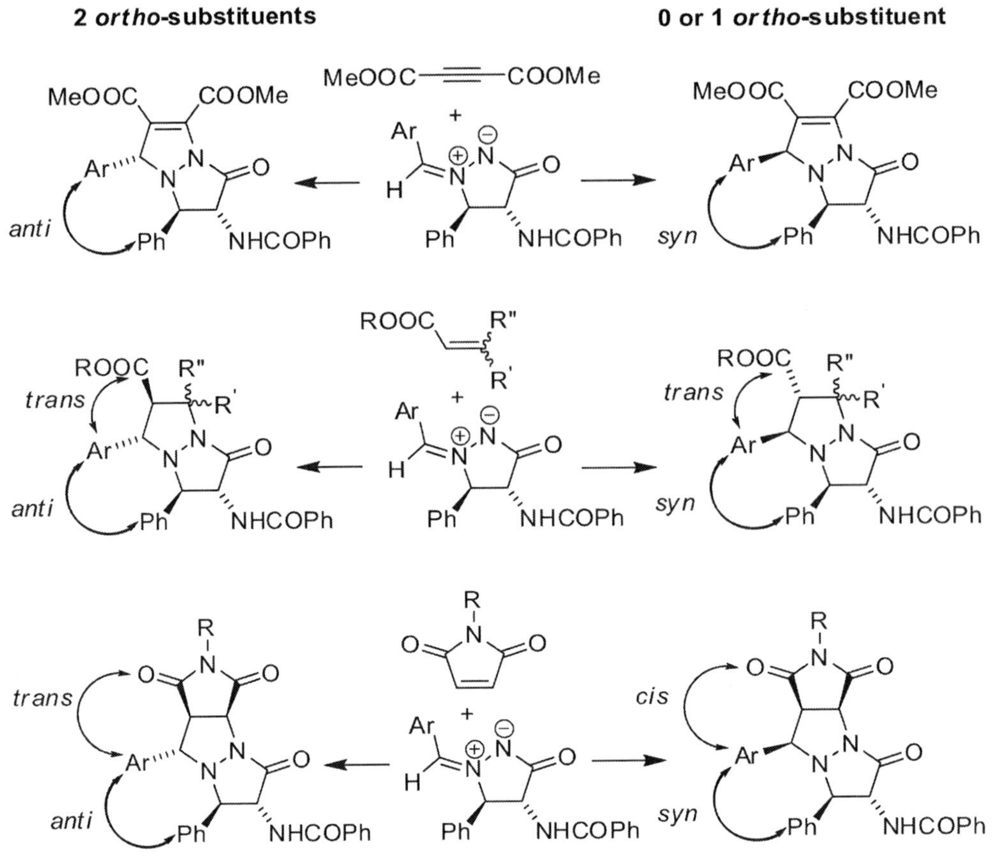


Scheme 26.

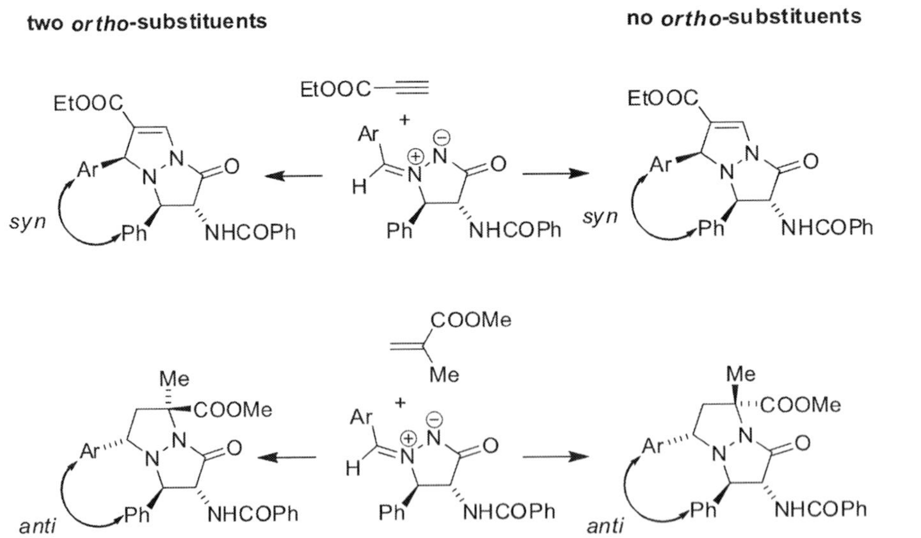


Scheme 27.

So far, we do not have a firm mechanistic explanation for the stereoselectivity and regioselectivity observed in cycloadditions of 1-arylmethylidene-3-pyrazolidinone-1-azomethine imines *6a–o*. Possible mechanisms could be exemplified at best by stereoselective cycloadditions to dimethyl maleate, which were highly selective regardless the structure of a dipole. Dipoles *6a–l* with at least one free *ortho*-position can adopt a planar conformation *6** allowing the transition state for the concerted 1,3-dipolar cycloaddition. Consequently, formation of the major isomers *29* with *syn*-oriented aryl groups at positions 3 and 5 could be explained by preferential *endo*-approach of the dipolarophile from the sterically less hindered (1'*Re**)-face of the (1*Z*,4*R**,5*R**)-dipole *6 via* the proposed transition state *46*. Thus, facial selectivity is controlled by the phenyl ring at position 5 as the stereodirecting group [28] (Scheme 28).

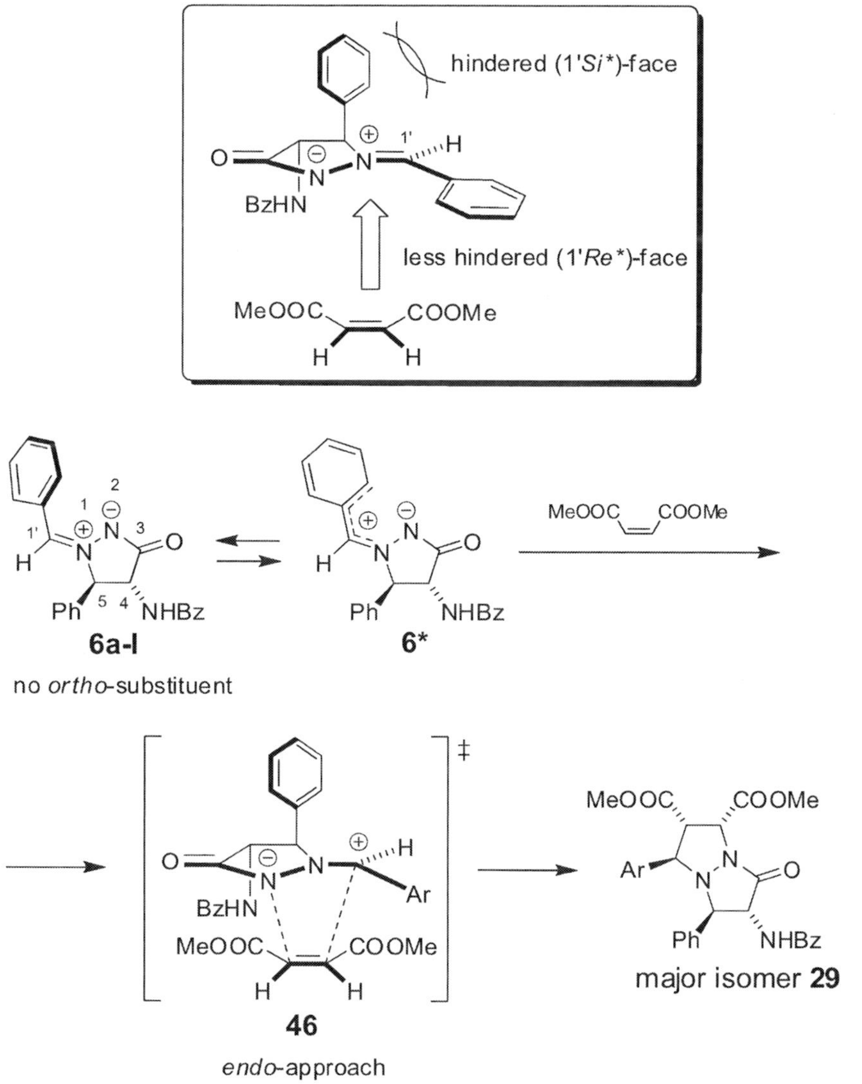


Scheme 28.

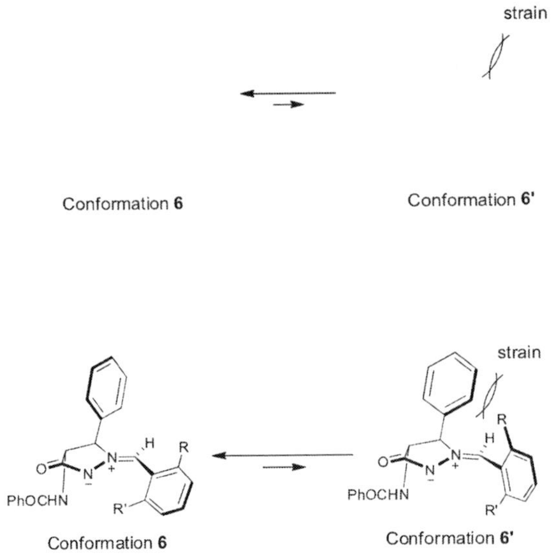


Figure 3.

According to the concerted mechanism, formation of the *ortho*-disubstituted major isomer *30* with *anti*-oriented aryl groups at positions 3 and 5 could take place in two ways:

(a) by the *exo*-approach of the dipolarophile from the (1'*Si*)-face of the (*E*)-dipole *6' via* transition state *47* (Scheme 29, Path A) or
(b) by the *endo*-approach of the dipolarophile from the (1'*Si**)-face of the (*Z*)-dipole *6 via* the proposed transition state *48* (Scheme 29 Path B).

Most probably, Path A can be ruled out, because the *E/Z*-isomerization around the exocyclic C=N double bond was not confirmed by NMR in DMSO-$d_6$ solution at 150 °C [28] (*c.f.* Scheme 4) and because at least two diastereomeric cycloadducts would be obtained if the *Z/E*-isomerization of dipole *6* would have taken place. However, Path B seems to be a possible explanation. Since *ortho*-disubstituted (*Z*)-dipole *6* is unable to adopt a planar conformation *6**, it exists in a preferred 'twisted' conformation *6*, where the (1'*Re**)-face is strongly hindered by the *ortho*-substituent R' close to the N(2)-position. The other 'twisted' conformation *6'*, where the *ortho*-substituent R' shields the (1'*Si*)-face of dipole *6* is much less stable, due to steric repulsion (or strain) between the R' and the phenyl ring at position 5 (Figure 3).

On the other hand, the (1'*Si**)-face of dipole is less hindered by the phenyl group. Consequently, stereocontrol is reversed and (1'*Si**)-*endo*-selective cycloaddition to dimethyl maleate yields the major isomer *30* (Scheme 29).

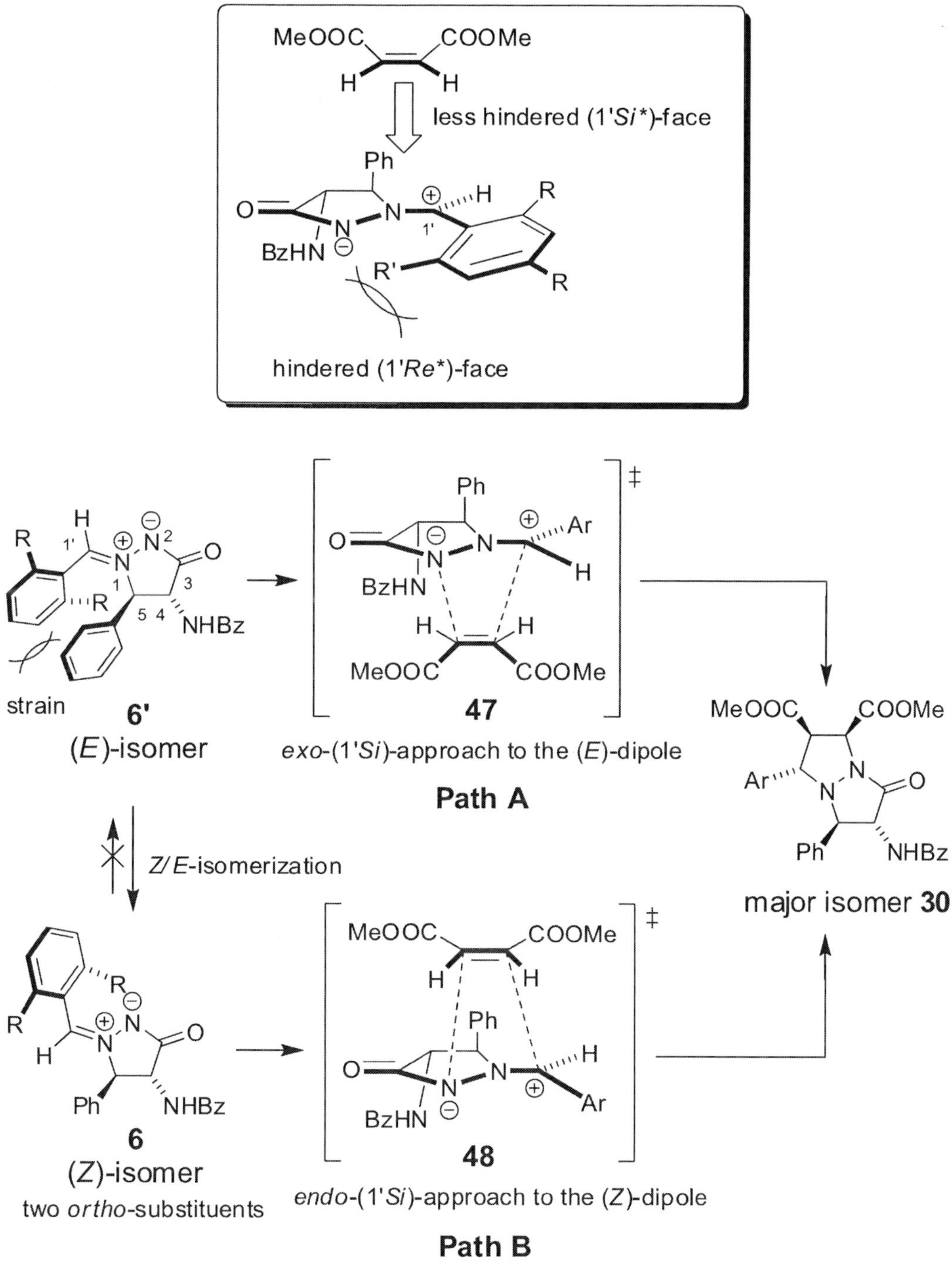


Scheme 29.

Alternatively, formation of the major isomers *30* in cycloadditions of *ortho*-disubstituted dipoles *6m* and *6n* to dimethyl maleate could also be explained by a two-step 1,4-addition-cyclization mechanism. In the mesomeric structure *6'*, rotation around the C–N single bond gives the rotamer *6''* with a bulky aryl group twisted away from the phenyl ring at position 3.

Conformer *6''* then undergoes Michael-type *anti*-addition to give the zwitterion (or a biradical [61, 62]) *49*, which then cyclizes into the final cycloadduct *30* (Scheme 30).

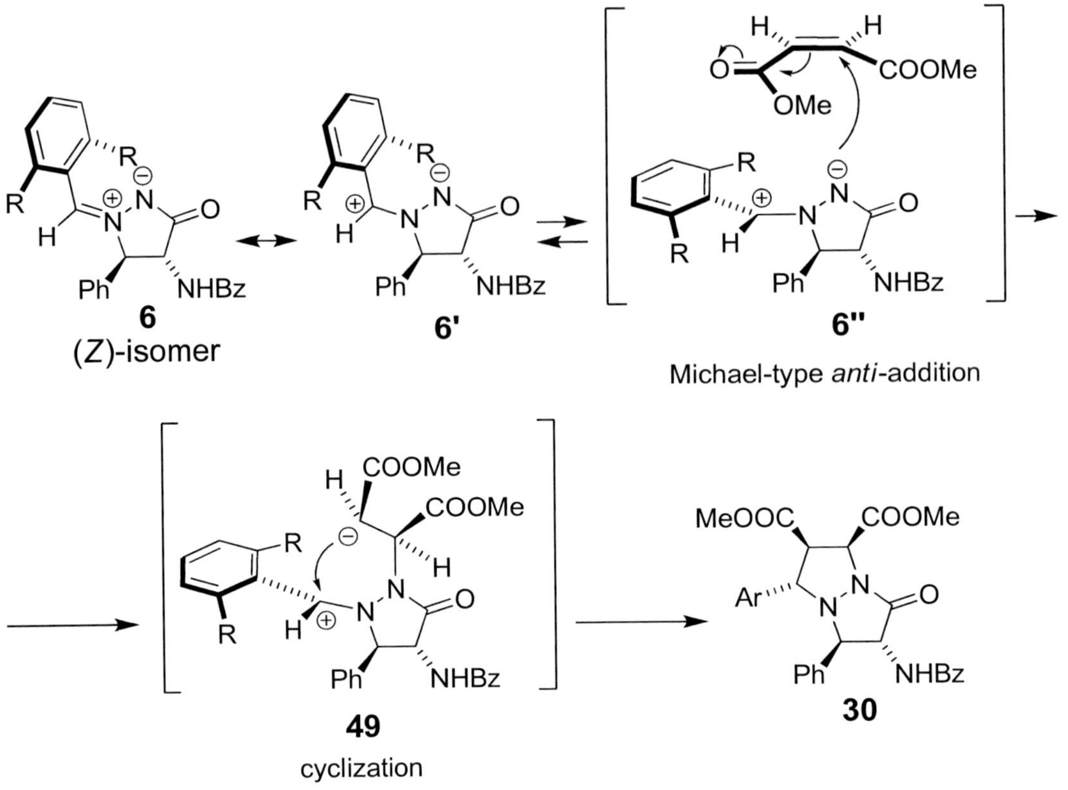


Scheme 30.

In analogous way, all other selective cycloadditions can be explained: (a) cycloadditions to dimethyl acetylenedicarboxylate and maleimides, (b) cycloadditions of *ortho*-unsubstituted dipoles to methyl acrylate and *β*-keto esters, and (c) cycloadditions of *ortho*-disubstituted dipole *6n* to dimethyl fumarate. Generally, cycloadditions of *ortho*-unsubstituted dipoles *3a–j* and *ortho*-monosubstituted dipoles *6k,l* to olefinic dipolarophiles were *endo*-selective. Exceptions were *exo*-selective cycloadditions to maleimides (*c.f.* Scheme 26). Most probably, preferential *exo*-approach of a maleimide is due to its cyclic shape, since rigidity of a maleimide makes the *endo*-approach sterically quite demanding, due to steric repulsion between the benzoylamino group in dipole *6* and the substituent R at position 1 in maleimide. In contrast, there is no such steric hindrance in the case of the *exo*-approach (Scheme 31).

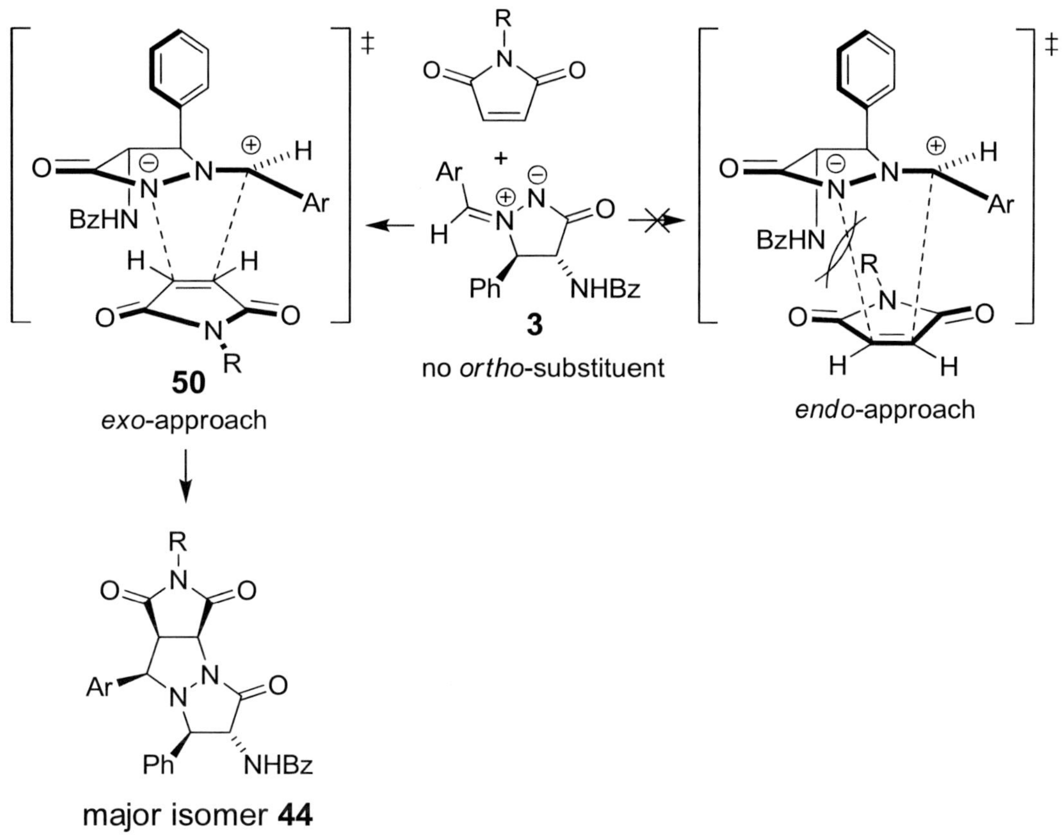


Scheme 31.

Microwave-assisted cycloadditions to ethyl propiolate which gave the major isomers with *syn*-oriented 3-H and 5-H regardless the *ortho*-substituents [54] (see Section 5.1, Scheme 13, Table 7). A possible explanation for this might be, that under microwave irradiation also the *ortho*-disubstituted dipoles *6m* and *6n* adopt a planar (or almost planar) conformation *6**. Accordingly, cycloadditions of *ortho*-disubstituted dipoles exhibit the same regiochemistry and stereochemistry as cycloadditions of their *ortho*-unsubstituted analogues. This hypothesis was then checked by performing reactions of dipoles *6a,c,g,m,n* with ethyl propiolate under conventional heating in anisole. The ratios of *ortho*-unsubstituted isomers *27/27'a,c,g* did not change much, while the ratios of isomers changed significantly in the case of two *ortho*-substituents *27/27'/m,n*: from *27:27'* > 5:1 in the presence of microwaves to *27:27'* ~ 1:1 under conventional heating. Decreased facial selectivity under conventional heating was in agreement with the microwave-induced planarization of *ortho*-disubstituted dipoles *6m* and *6n* proposed above. Namely, in the absence of microwaves a planar conformation *6** is not accessible for dipoles *6m,n* with two *ortho*-substituents and the (1'*Re**)-face becomes hindered by the *ortho*-substituent close to the N(2)-position. Consequently, facial selectivity is lost and almost equal amounts diastereomers *27* and *27'* are formed. On the other hand, two other experiments were not in agreement with the influence of microwaves on selectivity. When *ortho*-disubstituted dipole *6m* was reacted with dimethyl acetylenedicarboxylate and dimethyl maleate under microwave irradiation in anisole at 150 °C, both reactions were highly selective and furnished the same cycloadducts as reported previously under conventional heating as single isomers [26, 28] (see also Schemes 11 and 17). In summary, the concerted 1,3-dipolar mechanism seems to be a suitable mechanistic explanation for cycloadditions to ethyl propiolate, since it allows formation of both regioisomers, *27/27'* and *28/28'*. Nevertheless, the regioselectivity of cycloadditions of *ortho*-disubstituted dipoles *6m,n* under conventional heating is also in agreement with a two-step mechanism as proposed previously for related cycloadditions of 1-arylmethylidene-5,5-dimethyl-3-pyrazolidinone-1-azomethine imines to methyl propiolate [54, 63] (Scheme 32, Table 6).

**Table 6. Ratio of isomers 27, 27', 28, and 28' obtained under different reaction conditions**

| Compound | Ar | Ratio of Isomers **27**:**27'**:**28**:**28'** | |
|---|---|---|---|
| | | Microwave irradiation | Conventional heating |
| **27/27'/28/28'a** | phenyl | 36:10:29:25 | 50:16:25:9 |
| **27/27'/28/28'c** | 4-methoxyphenyl | 48:21:31:0 | 57:20:17:6 |
| **27/27'/28/28'g** | 4-nitrophenyl | 55:7:34:4 | 61:9:20:10 |
| **27/27'/28/28'm** | 2,6-dichlorophenyl | 85:15:0:0 | 53:47:0:0 |
| **27/27'/28/28'n** | 2,4,6-trimethylphenyl | 62:15:23:0 | 53:43:4:0 |

Another exception with respect to the usual stereocontrol was observed in cycloadditions of azomethine imines *6a,c,g,m,n* to methyl methacrylate. Regardless the *ortho*-substituents, the 1-COOMe regioisomers *43/43'* with *anti*-oriented aryl groups at positions 3 and 5 were obtained in all cycloadditions. Exclusive formation of the 1-COOMe regioisomers *43/43'* rules out a two-step mechanism, since the electronic distribution in azomethine imines *3* is represented at best by the mesomeric forms, where the negative charge is located at the N(2) atom [7, 8, 38–41]. Therefore, cycloadditions of dipoles *6a,c,g,m,n* to methyl methacrylate most probably proceed by the concerted 1,3-dipolar cycloaddition mechanism *via* preferential approach of the dipolarophile from the less hindered (1'*Si*)-face of the (*Z*)-dipole *3* and the proposed transition state *53*. The 'reversed' stereocontrol can be explained by the structure (shape) of the dipolarophile. Since methyl methacrylate is a 1,1-disubstituted ethene, predominant steric interactions with the dipole *6* occur with the methyl group and/or the ester group. Accordingly, it is the benzoylamino group at position 4 (and not the phenyl group at position 5), which acts as the stereodirecting group in the reaction with azomethine imine *6*. On the other hand, the correlation between the *endo/exo*-selectivity and the *ortho*-substituents, which was observed in these cycloadditions, remains unexplained [31] (Scheme 33).

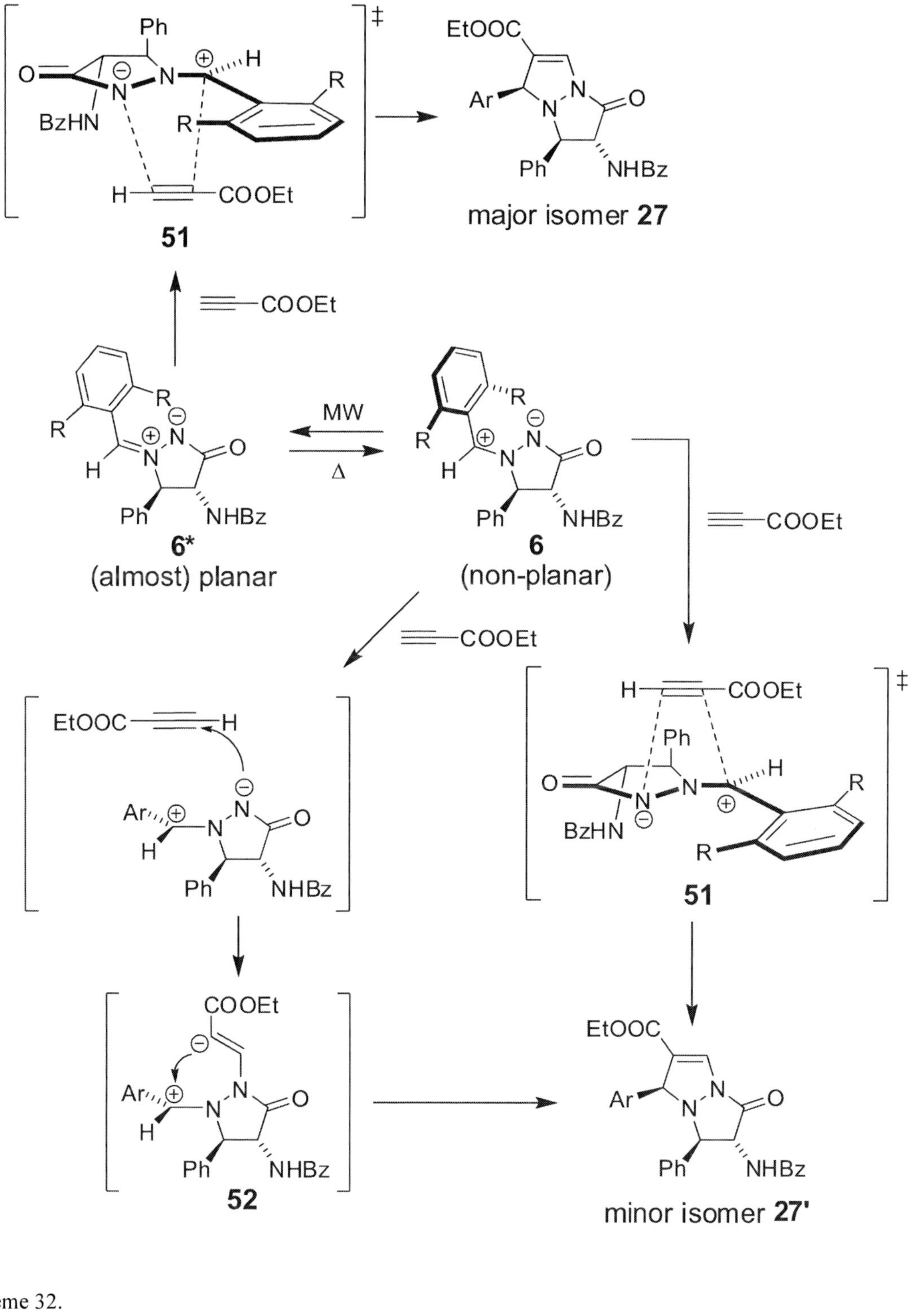
51
major isomer 27
COOEt
MW
Δ
6*
(almost) planar
6
(non-planar)
COOEt
COOEt
51
52
minor isomer 27'

Scheme 32.

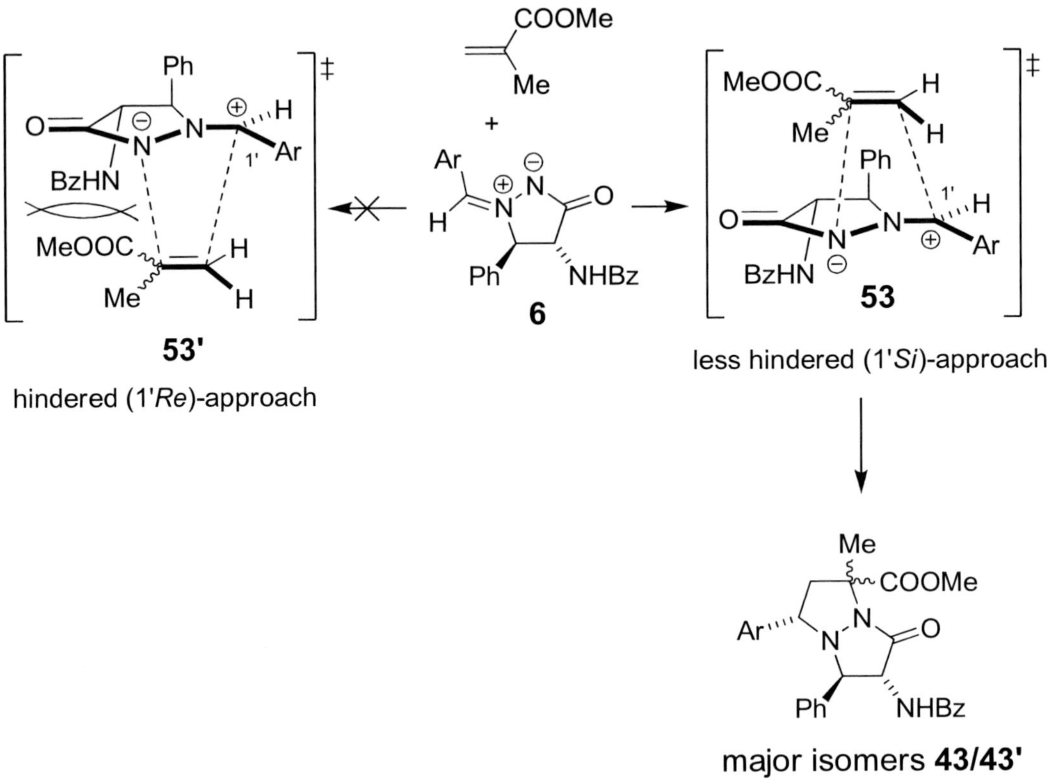


Scheme 33.

In contrast to dipoles derived from aromatic aldehydes, general reactivity, selectivity, and stereocontrol are not so well explored in 3-pyrazolidinon-1-azomethine imines derived from aliphatic aldehydes and ketones. So far, cycloadditions were studied with only two dipoles, *3q* and *3r*, obtained from *2* and acetone (*5q*) and butyraldehyde (*5r*), respectively. The results are summarized in Table 7.

At first glance, it seems that cycloadditions of dipoles *6q* and *6r* proceeded selectively; with exception of *17/17'r* (*c.f.* Scheme 11), all other cycloadducts were isolated as single diastereomers. However, isolation of single diastereomers itself does not prove stereoselectivity of cycloadditions, since the isolated yields of cycloadducts were only moderate. Unfortunately, unambiguous determination of stereoselectivity by $^1$H NMR was not possible, because the $^1$H NMR spectra of the crude reaction mixtures were very complex due to presence various by-products and impurities. Nevertheless, stereoselective cycloadditions might seem reasonable for reactions of dipole *6q* lacking a prochiral center at C(1'). For cycloadditions of the (1'*EZ*)-dipole *6r*, however, formation of at least two diastereomeric cycloadducts would be expected. Therefore, isolation of single diastereomers is most probably due to the loss of the minor isomer(s) during workup. The structures of the isolated cycloadducts *17q*, *17/17'r*, *29q,r*, *34q*, *35r*, *38q,r*, *44q*, and *44'r* indicated, that stereocontrol in cycloadditions of 1-alkylidene substituted dipoles *6q* and *6r* were mostly analogous to that, observed previously in reactions to the *ortho*-unsubstituted 1-arylmethylidene analogues [28–30]. Thus, cycloadditions of *6q* and *6r* to methyl acrylate and dimethyl maleate were *endo*-selective, whilst cycloaddition to *N*-phenylmaleimide was *exo*-

selective. Reactions of both dipoles, *6q* and (*EZ*)-*6r*, can be explained by the concerted 1,3-dipolar cycloaddition mechanism *via* preferential approach of dipolarophiles from the less hindered face of dipoles *6q* and *6r* [32] (Scheme 34).

**Table 7. Selectivity of Cycloadditions of Azomethine Imines 6q and 6r[a]**

| Dipole→<br>Dipolarophile↓ | 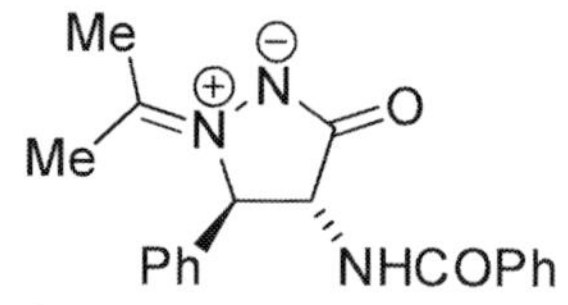<br>**6q** | 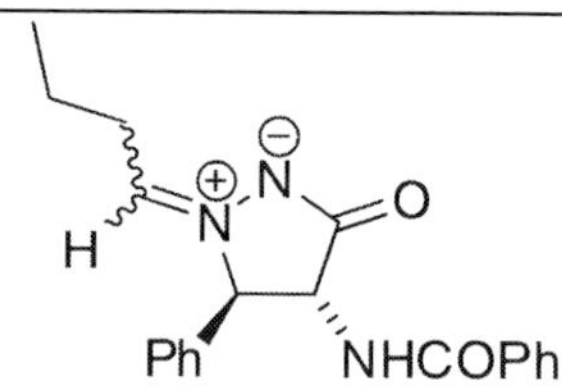<br>**6r** |
|---|---|---|
| MeOOC—≡—COOMe | + | +/– [b] |
| ≡—COOEt | + | + |
| 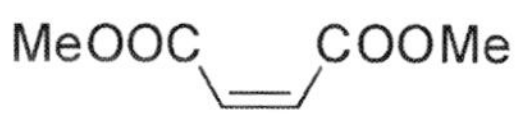 | + | + |
| 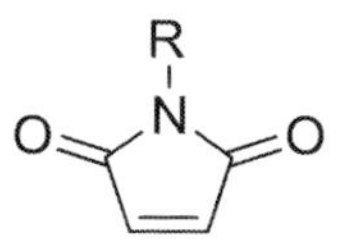 | + | + |
| 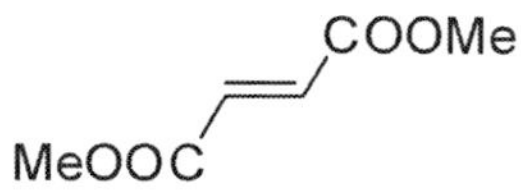 | + | + |
| 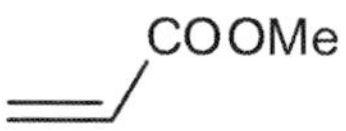 | + | + |

[a] Selectivity, i.e. ratio of isomers, was established upon chromatographic workup.
[b] The major diastereomer was obtained in 42% d.e.

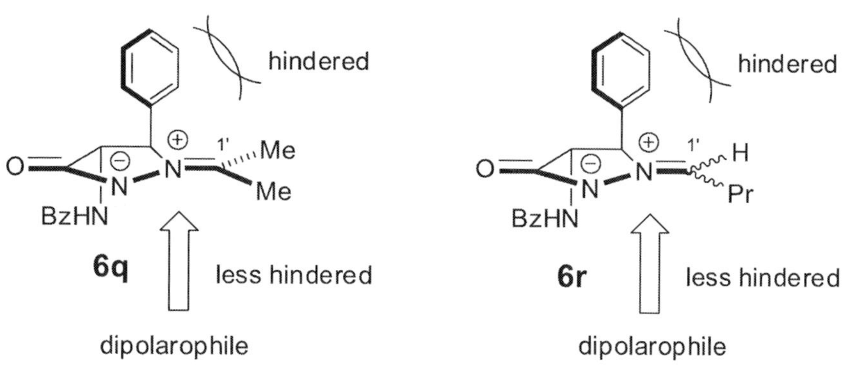


Scheme 34.

## 5.5.Structural Features of Cycloadducts

The structures of cycloadducts *29a, 30m, 35a, 36a* [28], *38q* [32], *44a* [31], and *45{1; 4}* [30] were determined by X-ray diffraction (Scheme 35).

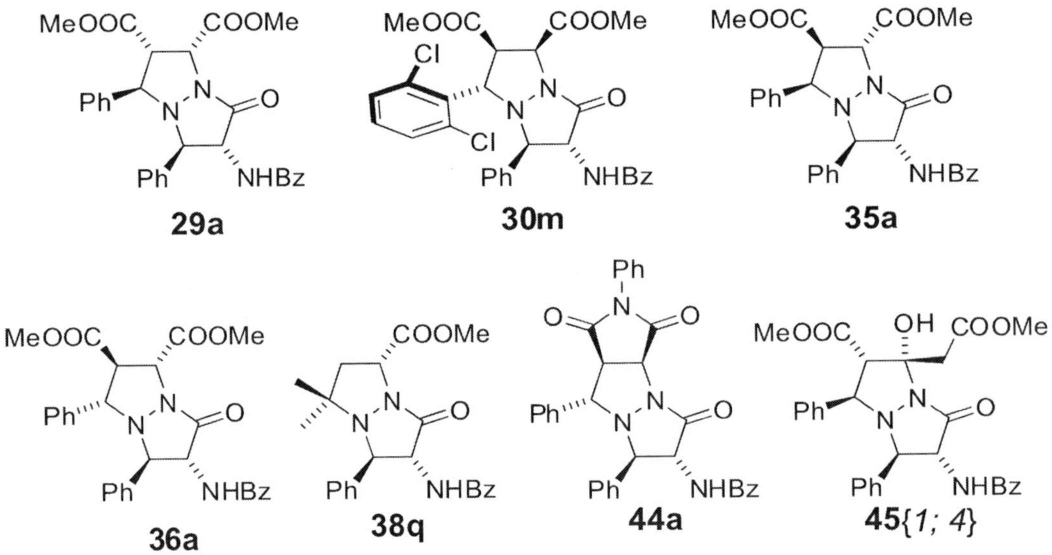

Scheme 35.

Structures of cycloadducts were established by $^1$H NMR and by NOESY spectroscopy. The configuration at the newly formed stereocenters at positions 1–3 were usually essential for the structure determination. Based on presumption, that the stereocenter at position 4 in azomethine imines is epimerizable, whilst stereocenter at position 5 is not, the (4*R**)-configuration at the benzylic position in dipole *6* was set as the stereochemistry reference point for structure determination by NMR. NOESY spectroscopy was used for determination of relative configuration at the newly formed stereocenter across the bridgehead nitrogen atom, N(4). In the case of the *syn*-oriented protons, n.O.e. between 3–H and 5–H was observed, while absence of n.O.e. indicated the *anti*-orientation between these two nuclei. In this manner, the *syn/anti*-orientation between the protons 3-H and 5-H and, consequently, the *syn/anti*-configuration at position 3 in all representative cycloadducts has been established. Furthermore, NOESY spectroscopy has also been used for determination of configuration around the C(1)–C(2) and C(2)–C(3) single bonds, because the magnitudes of vicinal coupling constants, $^3J_{H\text{-}H}$, alone could not be used for an unambiguous determination of configuration at the newly formed chiral centers at positions 1–3 (Table 8).

The configuration at positions 1–3, 5, and 6 were also determined by $^1$H NMR on the basis of vicinal coupling constants, $J_{H1\text{-}H2}$, $J_{H2\text{-}H3}$, and $J_{H5\text{-}H6}$. In most of the cycloadducts, as well as in the starting pyrazolidinone *1*, a large vicinal coupling constant, $^3J_{H\text{-}H}$ = 10–12 Hz, was observed between the *trans*-oriented protons at positions 4 and 5 [26–30]. In bicyclic cycloadducts, coupling constants showed the following characteristic values: (a) $J_{H1\text{-}H2}$ = 8–11 Hz (*cis*) > $J_{H1\text{-}H2}$ = 1–7.5 Hz (*trans*), (b) $J_{H2\text{-}H3}$ = 5–8 Hz (*cis*) < $J_{H2\text{-}H3}$ = 8–12 Hz (*trans*), and (c) $J_{H5-H6}$ = 8–12 Hz (*trans*) [26–30]. In tricyclic cycloadducts, on the other hand, typical magnitudes of coupling constants between protons at the newly formed C–C bond were different: $J_{H9-H9a}$ = 8–9 Hz (*cis*) > $J_{H9-H9a}$ = 2–3 Hz (*trans*) [29] (Table8, Figure 4).

**Table 8. Structure determination by NOESY spectroscopy**

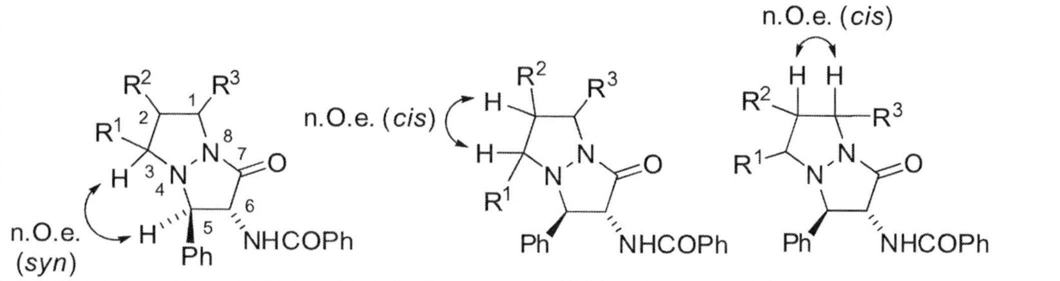


| Compound | Ref. | Compound | Ref. |
|---|---|---|---|
| 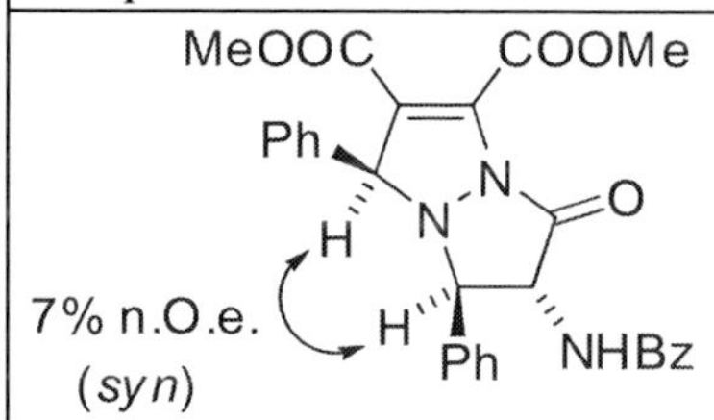 7% n.O.e. (syn) **17a** | 26 | 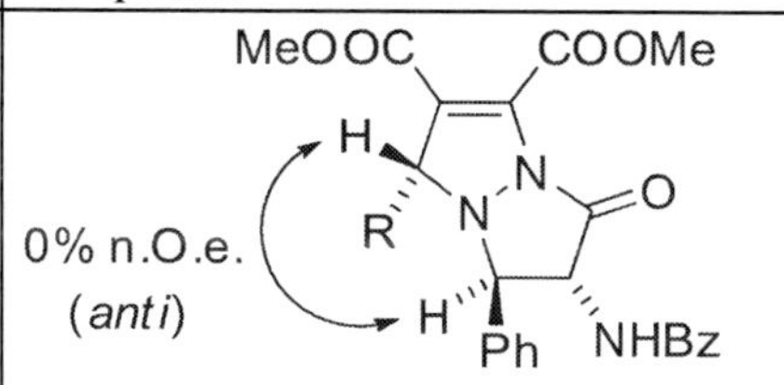 0% n.O.e. (anti) **17'm,r** | 28 |
| 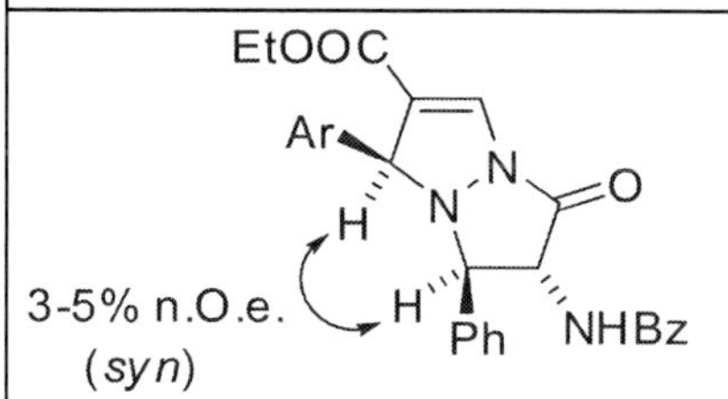 3-5% n.O.e. (syn) **27a,c,g,m,n** | 54 | 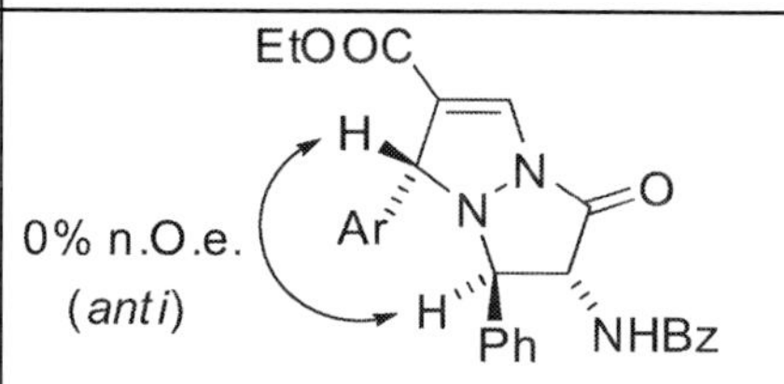 0% n.O.e. (anti) **27'a,c,n** | 54 |
| 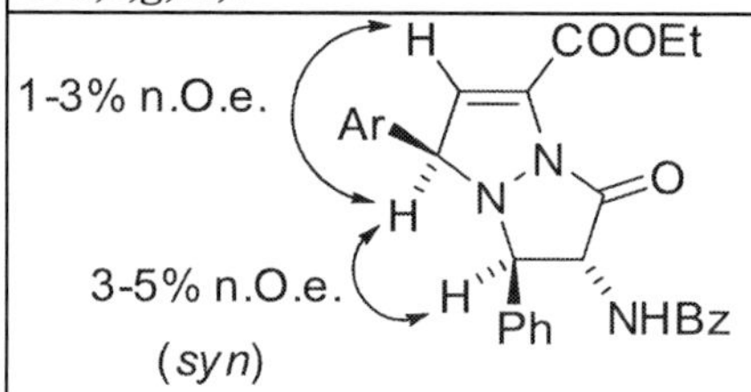 1-3% n.O.e.; 3-5% n.O.e. (syn) **28a,c,n** | 54 | 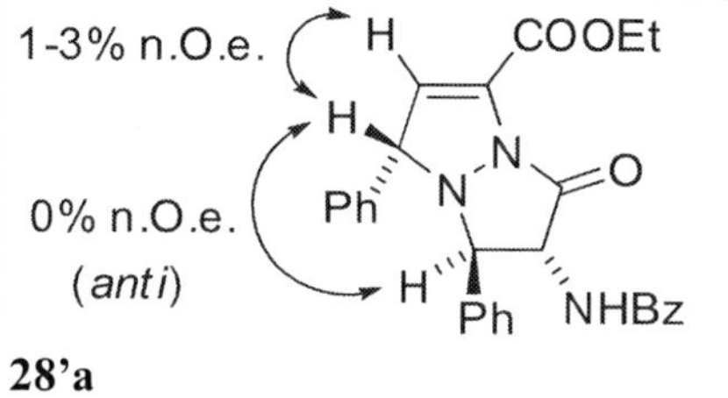 1-3% n.O.e.; 0% n.O.e. (anti) **28'a** | 54 |
| 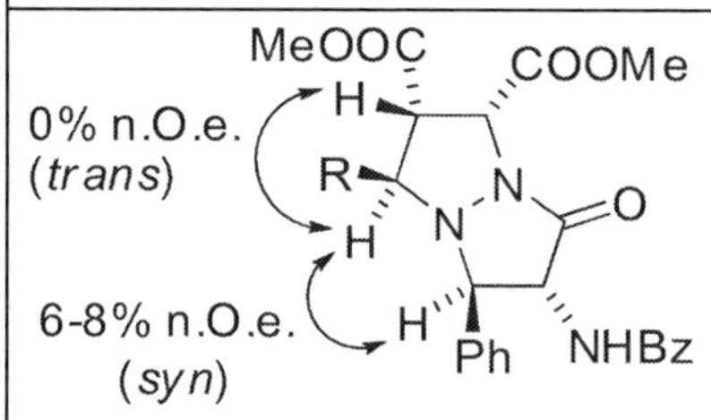 0% n.O.e. (trans); 6-8% n.O.e. (syn) **28g,r** | 28, 32 | 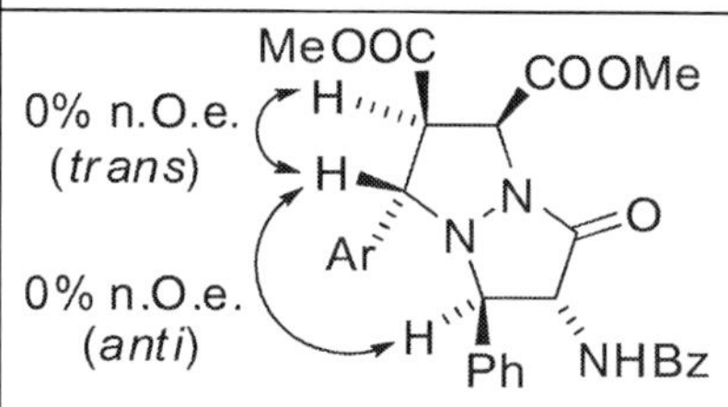 0% n.O.e. (trans); 0% n.O.e. (anti) **30m,n** | 28 |

**Table 8. Continued**

<table>
<tr>
<td>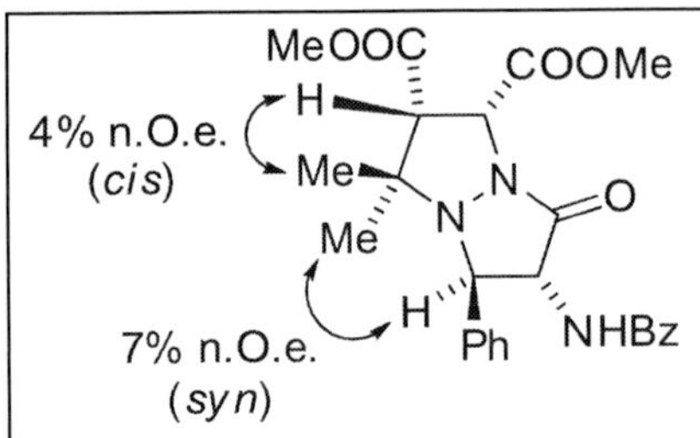

29q</td>
<td>32</td>
<td>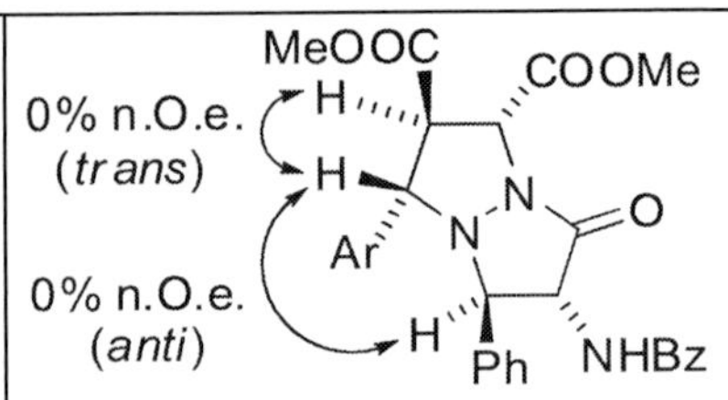

36a,n</td>
<td>28</td>
</tr>
<tr>
<td>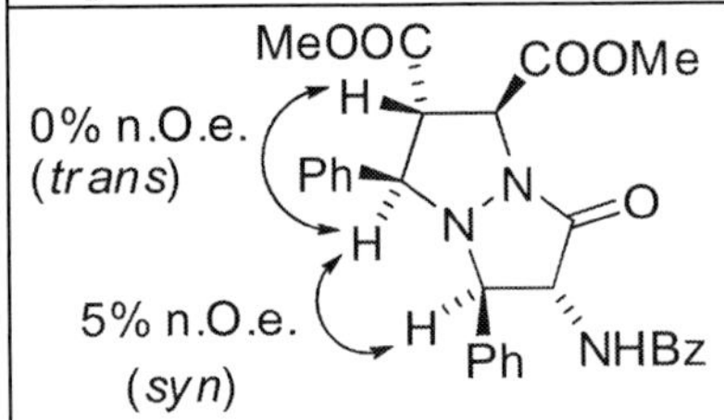

34a</td>
<td>28</td>
<td>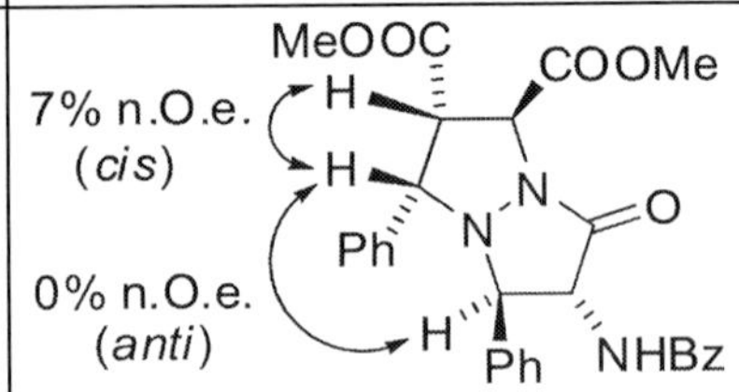

37a</td>
<td>28</td>
</tr>
<tr>
<td>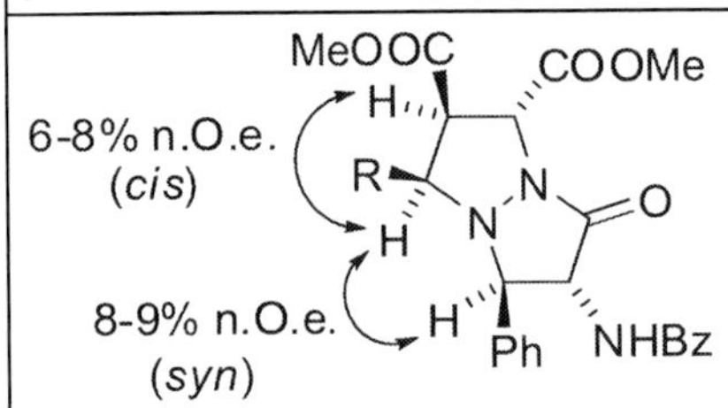

33a,r</td>
<td>28, 32</td>
<td>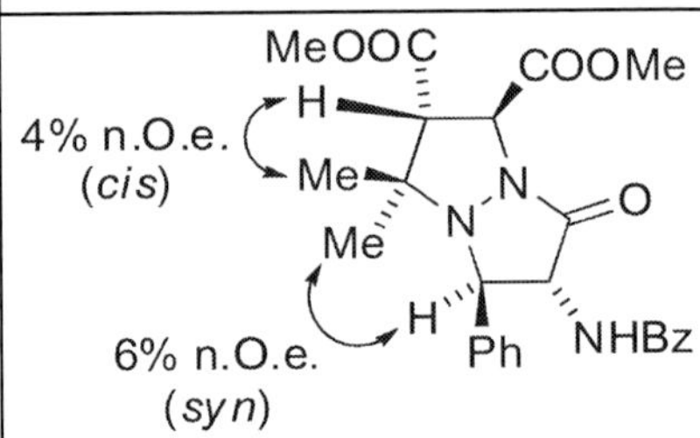

36q</td>
<td>32</td>
</tr>
<tr>
<td>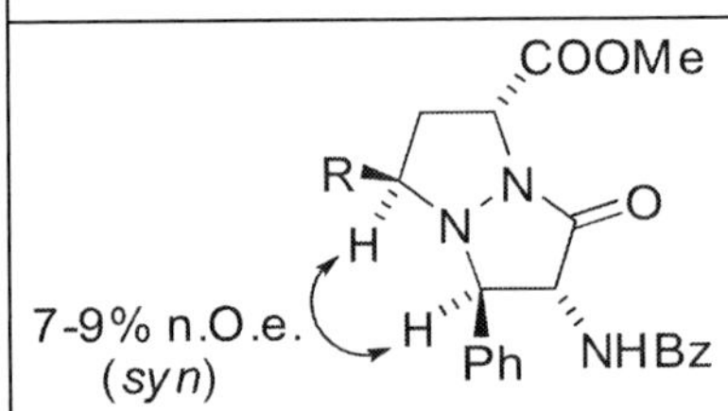

38a,r</td>
<td>28, 32</td>
<td>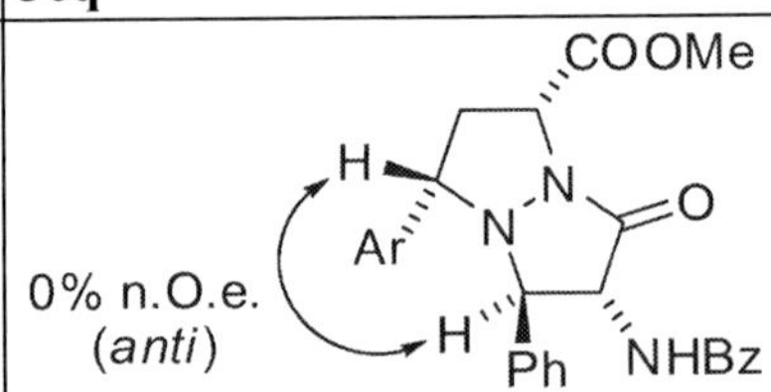

38n</td>
<td>28</td>
</tr>
<tr>
<td>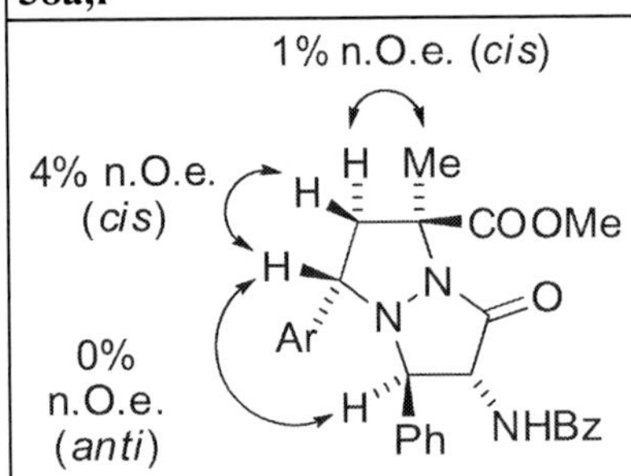

43a</td>
<td>31</td>
<td>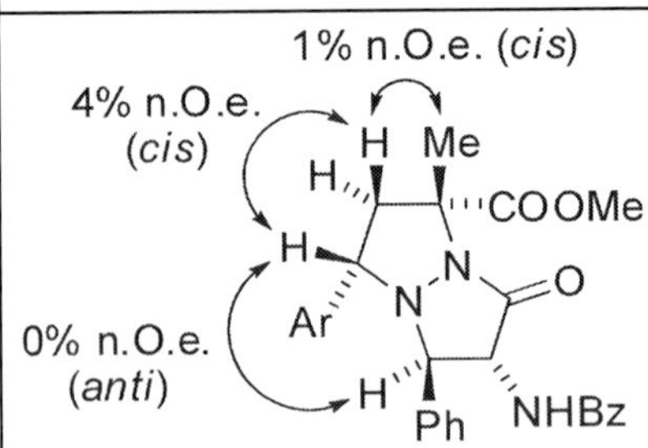

43'm,n</td>
<td>31</td>
</tr>
</table>

**Table 8. Continued**

| | | | |
|---|---|---|---|
| 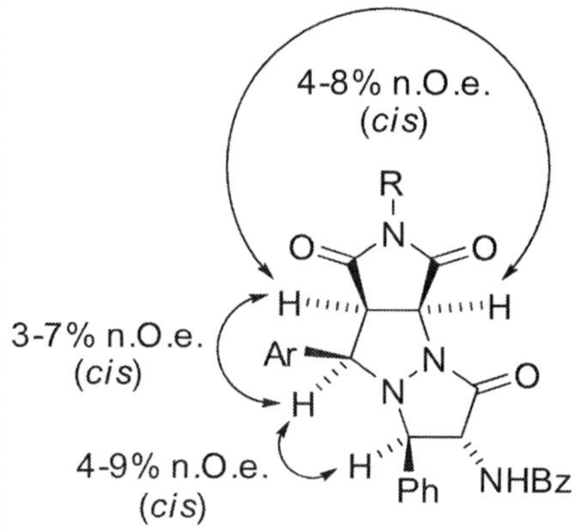<br>**44**{*1–3; 1–3*} | 29 | 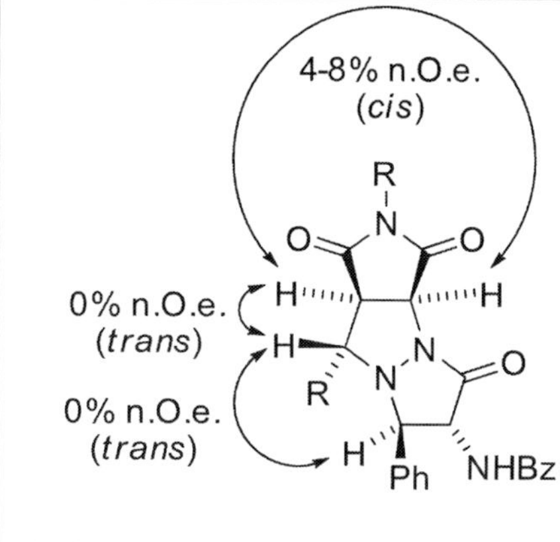<br>**44'**{*4,5; 1–3*}, **42'r** | 29, 32 |
| 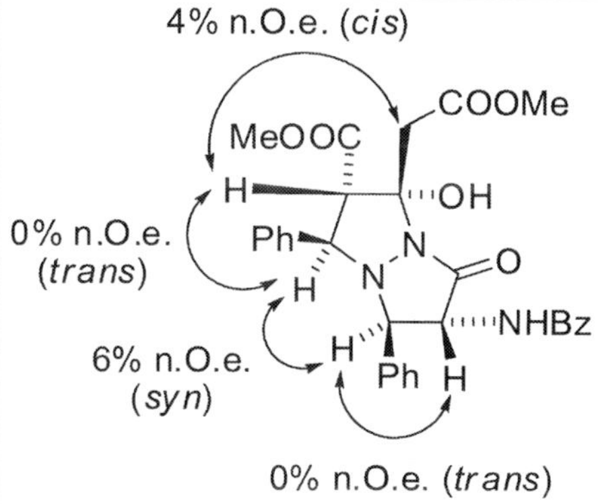<br>**43**{*1; 4*} | 30 | 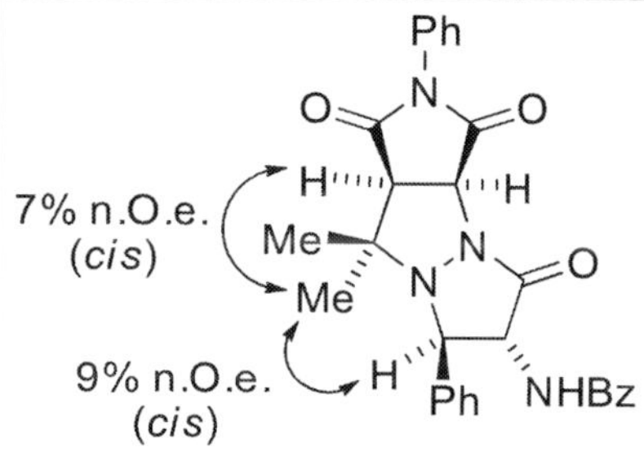<br>42q | 32 |

On the basis of typical values of vicinal coupling constants, $^3J_{H\text{-}H}$, and following Karplus equation [64], dihedral angles *H*–C–C–*H* between vicinal protons can be calculated. In the 4,5-*trans*-pyrazolidinone *1*, a large coupling constant $^3J_{H\text{-}H} \sim 12$ Hz, corresponds to the dihedral angle $\Phi_{H4\text{-}H5} \sim 180°$, *i.e.* to the envelope conformation of compound *1* with axial protons 4-H and 5-H. In azomethine imines *6*, increased $sp^2$ character of the pyrazolidinone ring induces planarization of the ring with a smaller dihedral angle $\Phi_{H4\text{-}H5} \sim 120°$, which is reflected in a smaller coupling constant, $J_{H4\text{-}H5} \sim 5$ Hz (Figure 5).

In addition to vicinal coupling constant, $^3J_{H\text{-}H}$, the configurations of cycloadducts were also confirmed by correlation of chemical shifts for 1-H, 2-H, 3-H, 5-H, and 6-H. Correlation between typical $^1$H NMR data and the relative configuration in all types of cycloadducts are summarized in Tables 9–11.

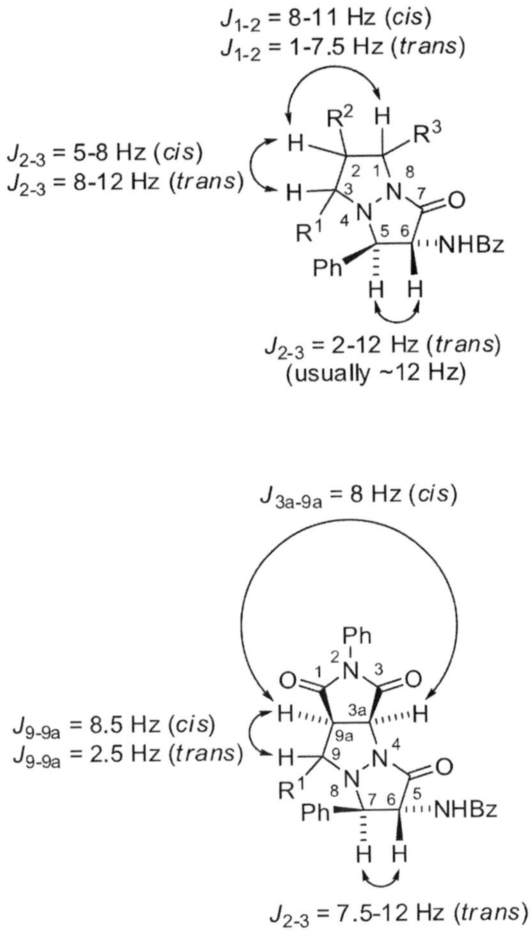


Figure 4.

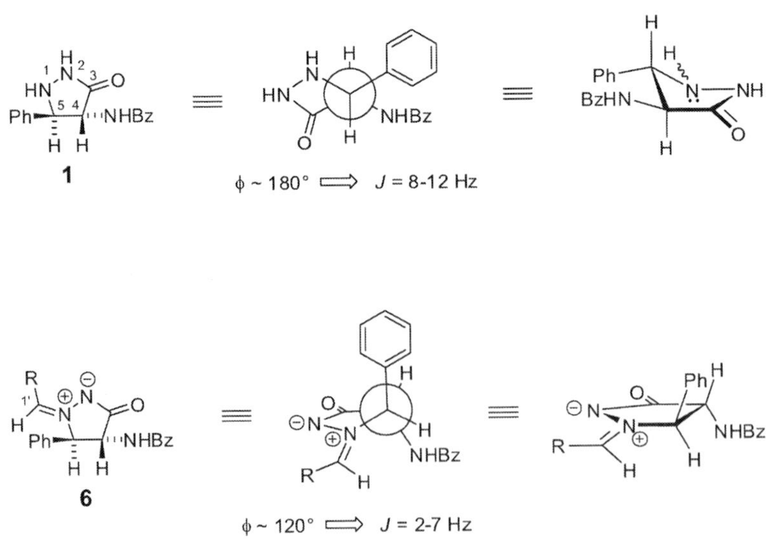


Figure 5.

**Table 9. Selected NMR data for cycloadducts obtained in cycloadditions to acetylenic dipolarophiles[a]**

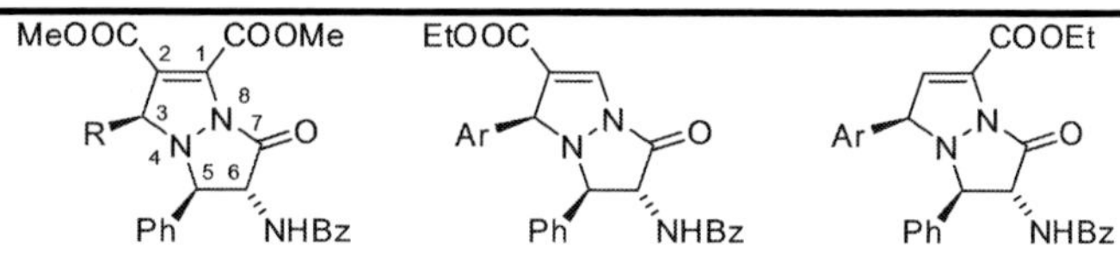


Compounds **17** Compounds **27** Compounds **28**

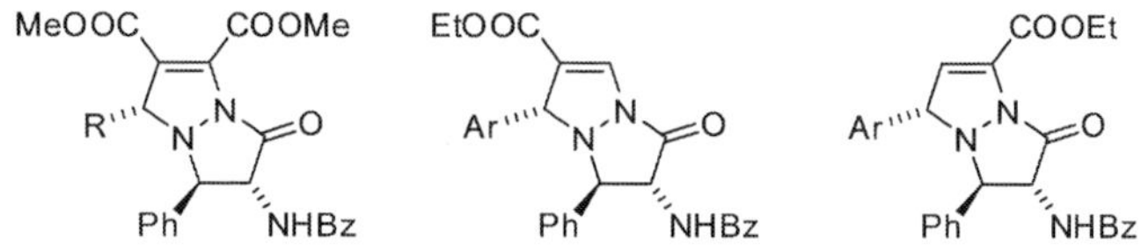


Compounds **17'** Compounds **27'** Compounds **28'**

| | $\delta_H$ (ppm) | | | | | $^3J_{H-H}$ | | | |
|---|---|---|---|---|---|---|---|---|---|
| | 1–H | 2–H | 3–H | 5–H | 6–H | 1–2 | 1–3 | 2–3 | 5–6 |
| **17** | - | - | 5.4 | 4.8 | 5.2 | - | - | - | 10.9 |
| **17'** | - | - | 5.5 | 4.4 | 5.3 | - | - | - | 10.9 |
| **27** | 7.7 | - | 5.6 | 4.8 | 4.7 | - | 1.7 | - | 11.1 |
| **27'** | 7.7 | - | 5.6 | 4.1 | 5.1 | - | 1.2 | - | 10.9 |
| **28** | - | 6.2 | 5.3 | 4.6 | 5.0 | 2.4 | - | - | 11.5 |
| **28'** | - | 6.3 | 5.3 | 3.8 | 5.1 | 2.9 | - | - | 11.5 |

[1]Typical δ and *J* values are given.

**Table 10. Selected NMR data for cycloadducts obtained in cycloadditions to dialkyl maleates, dimethyl fumarate, and maleimides[a]**

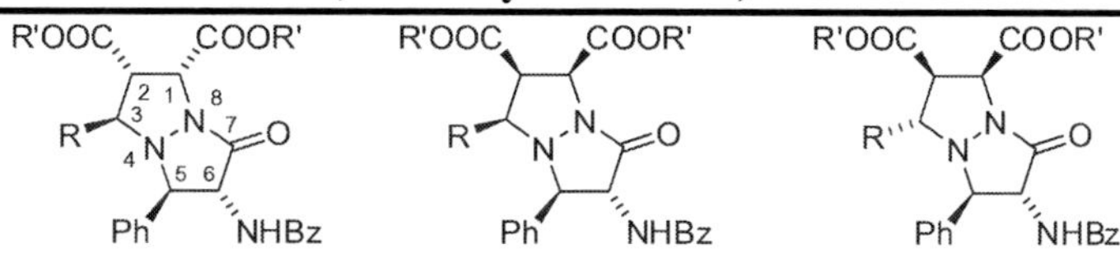


Compounds **29,31/31'** Compounds **32/32'** Compounds **30, 33/33'**

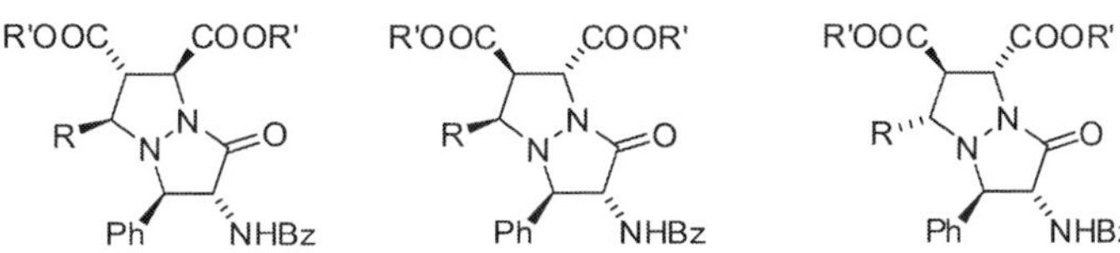


Compounds **34** Compounds **35** Compounds **36**

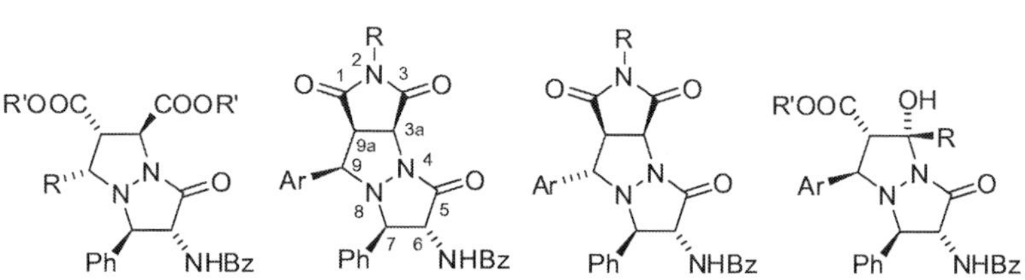


Compounds **37** Compounds **44** Compounds **44'** Compounds **45**

**Table 10. (Continued)**

| | $\delta_H$ (ppm) | | | | | $^3J_{H-H}$ | | | |
|---|---|---|---|---|---|---|---|---|---|
| | 1–H | 2–H | 3–H | 5–H | 6–H | 1–2 | 1–3 | 2–3 | 5–6 |
| **29, 31/31'** | 4.8 | 3.8 | 4.4 | 4.4 | 5.6 | 9.0 | 1.0 | 11.0 | 12.0 |
| **32/32'** | 5.1 | 3.5 | 4.2 | 4.5 | 4.9 | 7.2 | 0 | 9.3 | 9.1 |
| **30, 33/33'** | 5.1 | 4.5 | 5.3 | 4.4 | 5.0 | 8.7 | 0 | 9.7 | 9.8 |
| **34** | 5.2 | 3.6 | 4.3 | 4.5 | 4.8 | 7.2 | 0 | 9.0 | 10.1 |
| **35** | 4.8 | 3.7 | 4.4 | 4.2 | 5.6 | 2.0 | 0 | 6.8 | 11.9 |
| **36** | 5.1 | 3.8 | 4.1 | 4.8 | 5.7 | 5.3 | 0 | 8.5 | 9.0 |
| **37** | 5.4 | 4.3 | 4.6 | 3.7 | 5.2 | 7.5 | 0 | 7.5 | 9.4 |
| **45** | - | 3.6 | 4.4 | 4.3 | 5.1 | - | 0 | 11.0 | 12.0 |
| | 3a–H | 9a–H | 9–H | 7–H | 6–H | 3a–9a | 3a–9 | 9–9a | 6–7 |
| **44** | 5.4 | 4.2 | 4.9 | 4.3 | 4.6 | 7.9 | 0 | 8.5 | 7.5 |
| **44'** | 5.5 | 4.3 | 5.2 | 4.1 | 5.0 | 8.3 | 0 | 2.5 | 12.4 |

[a] Typical δ and *J* values are given.

**Table 11. Selected NMR data for cycloadducts obtained in cycloadditions to methyl acrylate and methyl methacrylate[a]**

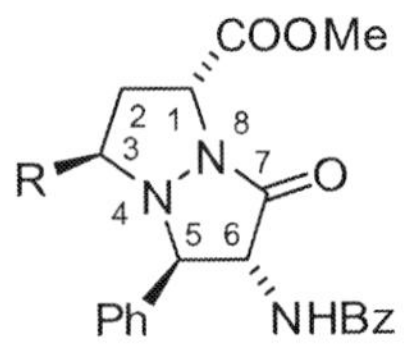

Compounds **38**

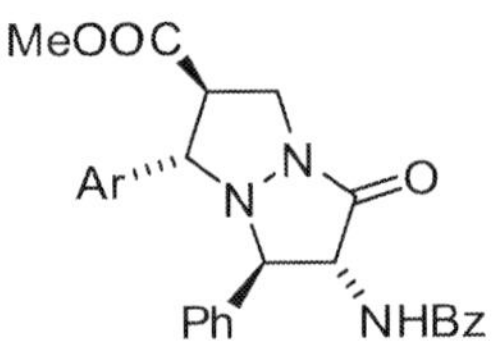

Compounds **39**

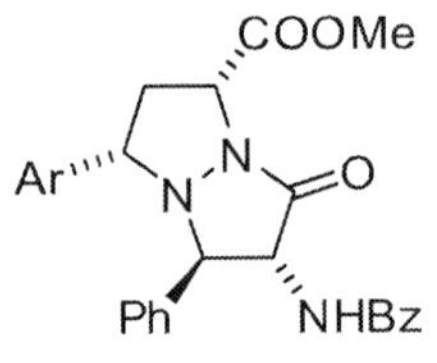

Compounds **40**

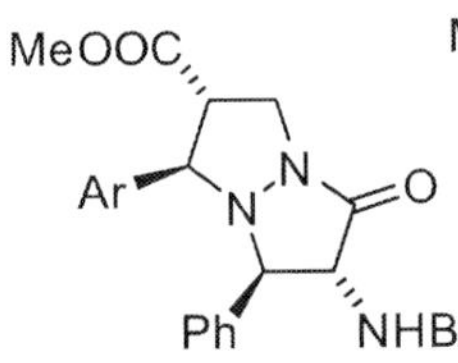

Compounds **41**

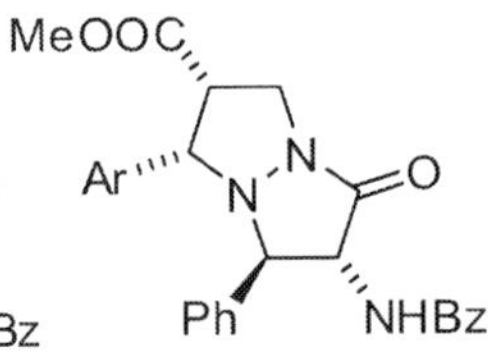

Compounds **42**

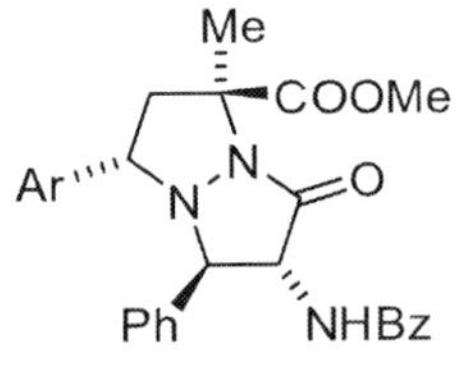

Compounds **43**

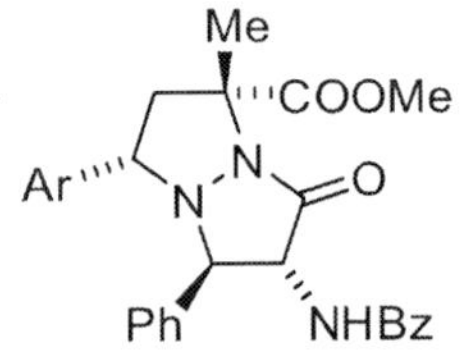

Compounds **43'**

| | $\delta_H$ (ppm) | | | | | | $^3J_{H-H}$ | | | | |
|---|---|---|---|---|---|---|---|---|---|---|---|
| | 1–H | 2–Ha | 2–Hb | 3–H | 5–H | 6–H | 1–2a | 1–2b | 2a–3 | 2b–3 | 5–6 |
| **38** | 4.6 | 2.6 | 2.8 | 4.1 | 4.3 | 5.5 | 1.5 | 9.0 | 5.3 | 11.3 | 12.2 |
| **40** | 4.7 | 2.5 | 2.8 | 4.0 | 4.7 | 5.6 | 6.4 | 10.4 | 10.6 | 7.5 | 6.4 |
| **43** | - | 2.4 | 2.8 | 4.1 | 4.3 | 5.4 | - | - | 11.3 | 5.6 | 12.0 |
| **43'** | - | 2.3 | 2.8 | 3.9 | 4.3 | 5.2 | - | - | 6.3 | 10.5 | 2.5 |
| | 1–Ha | 1–Hb | 2–H | 3–H | 5–H | 6–H | 1a–2 | 1b–2 | 2–3 | | 5–6 |
| **39** | 3.9 | 4.2 | 3.6 | 4.5 | 4.8 | 5.6 | 9.4 | 5.2 | 7.9 | | 9.4 |
| **41** | 3.9 | 4.4 | 3.8 | 4.8 | 4.3 | 4.9 | 10.0 | 4.9 | 8.5 | | 7.2 |
| **2** | 4.1 | 4.2 | 3.6 | 4.4 | 4.7 | 5.5 | 7.0 | 9.0 | 10.5 | | 6.8 |

[a] Typical δ and *J* values are given.

## 6. Conclusion

(4*R**,5*R**)-4-Benzoylamino-5-phenyl-3-pyrazolidinone (*1*) and its azomethine imines *6* are easily available and useful building blocks for the synthesis of functionalized pyrazoles. So far, pyrazolidinone *1* and its azomethine imines *6* have been used for the preparation of 3-phenyl-3-(substituted 1*H*-pyrazol-1-yl)alanine derivatives and as model compounds in cycloadditions to various dipolarophile leading to polysubstituted pyrazolo[1,2–*a*]pyrazolones as heterocyclic analogues of dipeptides. Particularly the study on stereochemistry of cycloadditions to (4*R**,5*R**)-1-alkylidene-4-benzoylamino-5-phenyl-3-pyrazolidinon-1-azomethine imines *6* gave promising results. In extension, this methodology was used in the synthesis of two small combinatorial libraries of fused pyrazolones. In conclusion, the results of this study which indicate, that by suitable choice and combination of azomethine imines *6* and dipolarophiles, these cycloaddition reactions could be used to carry out a stereoselective diversity-oriented synthesis of 2-acylamino-1*H*,5*H*-pyrazolo[1,2–*a*]pyrazolone-7-carboxylates with variable, yet predictable, configurations in search for novel bioactive compounds.

## 7. References


[1] Kitamura, R. *Rept. Japan. Assoc. Advancement Sci.* 1942, *16*, 541–544.

[2] Spasov, A.; Kurtev, B. *Annuaire univ. Sofia, Faculte phys.-math* 1947, *43(Livre 2)*, 37–51.

[3] Stodola, F. H. *J. Org. Chem.* 1948, *13*, 757–762.

[4] Clarke, H. T. *Chemistry of Penicillin*; Clarke H. T.; Ed.; Oxford University Press: Oxford, 1949, pp. 688–848.

[5] Gagnon, P. E.; Boivin, J. L.; Jones, R. N. *Can. J. Research* 1949, *27B*, 190–204.

[6] Otis, C. V. *J. Soc. Motion Picture Engrs.* 1949, *52*, 534–539.

[7] For a review on 3-pyrazolidinones see: (a) Dorn, H. *Chem. Heterocycl. Compd. USSR* 1981, 3–31. (b) Claramunt, R. M.; Elguero, *J. Org. Proc. Prep. Int.* 1991, *23*, 273–320.

[8] (a) Sabat, M.; VanRens, J. C.; Brugel, T. A.; Maier, J.; Laufersweiler, M. J.; Golebiowski, A.; De, B.; Easwaran, V.; Hsieh, L. C.; Rosegen, J.; Berberich, S.; Suchanek, E.; Janusz, M. J. *Bioorg. Med. Chem. Lett.*, 2006, *16*, 4257–4261. (b) Holan, G.; Virgona, C. T.; Watson, K. G.; Finkelstein, B. L. *Bioorg. Med. Chem. Lett.*, 1996, *6*, 77–80. (c) Kosower, E. M.; Hershkowitz, E. *Israeli IL 94,658*, 07 Oct. 1994, Appl. 07 Jun. 1990, 23 pp.; *Chem. Abstr.*: 1995, *122*, 214077s.

[9] Chen, J.-H.; Venkatesham, U.; Lee, L.-C.; Chen, K. *Tetrahedron* 2006, *62*, 887–893.

[10] Cusan, C.; Spalluto, G.; Prato, M.; Adams, M.; Bodensieck, A.; Bauer, R.; Tubaro, A.; Bernardi, P.; Da Ros, T. *Il Farmaco* 2005, *60*, 327–332.

[11] Ternansky, R. J.; Draheim, S. E. *Tetrahedron* 1992, *48*, 777–796.

[12] Hanessian, S.; McNaughton-Smith, G.; Lombart, H.-G.; Lubell, W. D. *Tetrahedron* 1997, *53*, 12789.

[13] Chuang, T.-H.; Sharpless, K. B. *Helv. Chim. Acta* 2000, *83*, 1734–1743.

[14] Panfil, I.; Urbanczyk-Lipkowska, Z.; Suwinska, K.; Solecka, J.; Chmielewski, M. *Tetrahedron* 2002, *58*, 1199–1212.

[15] Sibi, M. P.; Stanley, L. M.; Nie, X.; Venkatraman, L.; Liu, M.; Jasperse, C. P. *J. Am. Chem. Soc.*, 2007, *129*, 395–405.

[16] Nakana, H.; Tsugawa, N.; Takahashi, K.; Okuyama, Y.; Fujita, R.; *Tetrahedron*, 2006, *62*, 10897–10887.

[17] Sibi, M. P.; Soeta, T. *J. Am. Chem. Soc.*, 2007, *129*, 4522–4523.

[18] Sibi, M. P.; Prabagaran, N. *Synlett*, 2004, 2421–2424.

[19] Sibi, M. P.; Liu, M. *Org. Lett.*, 2001, *3*, 4181–4184.

[20] Lin, C.-H.; Yang, K.-S.; Pan, J.-F.; Chen, K. *Tetrahedron Lett.*, 2000, *41*, 6815–6819.

[21] Stanovnik, B.; Svete, J. *Targets in Heterocyclic Systems* 2000, *4*, 105–137.

[22] Svete, J. *J. Heterocycl. Chem.* 2002, *39*, 437–454.

[23] Svete, J. *ARKIVOC*, 2006, *Part vii*, 35–46.

[24] Budovskii, E. I.; Khomutov, R. M.; Karpeiskii, M. Ya.; Severin, E. S.; Kochetkov, N. K. *Zh. Obshch. Khim.* 1960, *30*, 2569–2573.

[25] Iyengar, D. S.; Prasad, K. K.; Venkataratnam, R. V. *Aust. J. Chem.* 1974, *27*, 2439–2446.

[26] Svete, J.; Prešeren, A.; Stanovnik, B.; Golič, L.; Golič Grdadolnik, S. *J. Heterocycl. Chem.* 1997, *34*, 1323–1328.

[27] Zupančič, S.; Svete, J.; Stanovnik, B. *J. Heterocycl. Chem.* 1999, *36*, 607–610.

[28] Pezdirc, L.; Jovanovski, V.; Bevk, D.; Jakše, R.; Pirc, S.; Meden, A.; Stanovnik, B.; Svete, J. *Tetrahedron* 2005, *61*, 3977–3990.

[29] Pezdirc, L.; Cerkovnik, J., Pirc, S.; Stanovnik, B.; Svete, J. *Tetrahedron*, 2007, *63*, 991–999.

[30] Pezdirc, L.; Bevk, D.; Grošelj, U.; Meden, A.; Stanovnik, B.; Svete, J.; *J. Comb. Chem.* 2007, *9*, 717–723.

[31] Pezdirc, L.; Stanovnik, B.; Svete, J.; to be published.

[32] Pezdirc, L.; Grošelj, U.; Meden, A.; Stanovnik, B.; Svete, J. *J. Heterocycl. Chem.* 2008, *45*, 181–188.

[33] Turk, C., Stanovnik, B.; Svete, J. to be published.

[34] Grošelj, A.; Rečnik, S.; Meden, A.; Stanovnik, B.; Svete, J. *Acta. Chem. Slov.* 2006, *53*, 245–256.

[35] Pezdirc, L.; Grošelj, U.; Meden, A.; Stanovnik, B.; Svete, J. *Tetrahedron Lett.* 2007, *48*, 5205–5208.

[36] Svete, J.; Stanovnik, B. *Heterocycles* 1999, *51*, 2073–2078.

[37] Prešeren, A.; Svete, J.; Stanovnik, B. *J. Heterocycl. Chem.* 1999, *36*, 799–801.

[38] *1,3-Dipolar Cycloaddition Chemistry*; Padwa, A.; Ed.; Vol 1; John Wiley & Sons, Inc.: New York, 1984.

[39] *1,3-Dipolar Cycloaddition Chemistry*; Padwa, A.; Ed.; Vol 2; John Wiley & Sons, Inc.: New York, 1984.

[40] Grashey, R. In *Azomethine Imines* in *1,3-Dipolar Cycloaddition Chemistry*; Padwa, A.; Ed.; Vol 1; John Wiley & Sons, Inc.: New York, 1984, pp 733–814.

[41] Schantl, J. G. In *Azomethine Imines* in *Science of Synthesis, Houben-Weyl Methods of Organic Transformations*; Georg Thieme Verlag: Stuttgart, 2004, Vol. 27, pp 731–824.

[42] Stanovnik, B. *Tetrahedron* 1991, *47*, 2925–2945.

[43] Žličar, M.; Stanovnik, B.; Tišler, M. *Tetrahedron* 1992, *48*, 7965–7972.

[44] Žličar, M.; Stanovnik, B.; Tišler, M. *J. Heterocycl. Chem.* 1993, *30*, 1209–1211.

[45] Stanovnik, B.; Jelen, B.; Žličar, M. *Il Farmaco* 1993, *48*, 231–242.

[46] Stanovnik, B.; Jelen, B.; Turk, C.; Žličar, M.; Svete, J. *J. Heterocycl. Chem.* 1998, *35*, 1187–1204.

[47] Roussi, F.; Bonin, M.; Chiaroni, A.; Micouin, L.; Riche, C.; Husson, H.-P. *Tetrahedron Lett.* 1999, *40*, 3727–3730.

[48] Roussi, F.; Chauveau, A.; Bonin, M.; Micouin, L.; Husson, H.-P. *Synthesis* 2000, 1170–1179.

[49] Chauveau, A.; Martens, T.; Bonin, M.; Micouin, L.; Husson, H.-P. *Synthesis* 2002, 1885–1890.

[50] Chung, F.; Chauveau, A.; Seltki, M.; Bonin, M.; Micouin, L. *Tetrahedron Lett.* 2004, *45*, 3127–3130.

[51] Elguero, J. In *Pyrazoles* in *Comprehensive Heterocyclic Chemistry II*; Katritzky, A. R.; Rees, C. W.; Scriven, E. F. V.; Eds.; Vol 3; Elsevier Science Ltd., Oxford, 1996, pp. 1–75; and references cited therein.

[52] Chen, J.-H.; Venkatesham, U.; Lee, L.-C.; Chen, K. *Tetrahedron* 2006, *62*, 887–893.

[53] Cusan, C.; Spalluto, G.; Prato, M.; Adams, M.; Bodensieck, A.; Bauer, R.; Tubaro, A.; Bernardi, P.; Da Ros, T. *Il Farmaco* 2005, *60*, 327–332.

[54] Pezdirc, L.; Stanovnik, B.; Svete, J. *Z. Naturforsch.* 2008, *63b*, 375–383.

[55] H. C. Kolb, M. G. Finn, and K. B. Sharpless, *Angew. Chem. Int. Ed.*, 2001, *40*, 2004–2021.

[56] V. D. Bock, H. Hiemstra, J. H. Van Maarseveen, *Eur. J. Org. Chem.* 2006, 51–68.

[57] V. V. Rostovtsev, L. G. Green, V. V. Fokin, K. B. Sharpless, *Angew. Chem. Int. Ed.*, 2002, *41*, 2596–2599.

[58] F. Himo, T. Lovell, R. Hilgraf, V. V. Rostovtsev, L. Noodelman, K. B. Sharpless, V. V. Fokin, *J. Am. Chem. Soc.*, 2005, *127*, 210–216.

[59] V. O. Rodinov, V. V. Fokin, M. G. Finn, *Angew. Chem. Int. Ed.*, 2005, *44*, 2210–2215.

[60] Shintani, R.; Fu, G. C. *J. Am. Chem. Soc.* 2003, *125*, 10778–10779.

[61] Huisgen, R. In *1,3-Dipolar Cycloadditions – Introduction, Survey, Mechanism* in 1,3-Dipolar Cycloaddition Chemistry; Padwa, A.; Ed.; Vol 1; John Wiley & Sons, Inc.: New York, 1984, pp 1–176.

[62] Houk, K. N.; Yamaguchi, K. In *Theory of 1,3-Dipolar Cycloadditions* in *1,3-Dipolar Cycloaddition Chemistry*; Padwa, A.; Ed.; Vol 2; John Wiley & Sons, Inc.: New York, 1984, pp 407–450.

[63] Turk, C.; Svete, J.; Stanovnik, B.; Golič, L.; Golič Grdadolnik, S.; Golobič, A.; Selič, L. *Helv. Chim. Acta* 2001,*84*,146–156.

[64] Karplus, M. *J. Chem. Phys.* 1959, *30*, 11–15.

In: Stereochemistry Research Trends
Editor: M. A. Horvat and J. H. Golob
ISBN 978-1-60456-825-7


*Chapter 6*

# Stereochemical Aspects of Interaction between Some Monosaccharides and L-α-amino Acids in Aqueous Solutions

***Elena V. Parfenyuk*[*] *and Galina A. Kulikova***
Institute of Solution Chemistry of Russian Academy of Sciences,
1 Akademicheskaya Str., Ivanovo, 153045, Russia

## Abstract

Specific binding of carbohydrates by proteins has been the focus of increased study during the last decades because it plays a key role in important biological processes. The specificity of the carbohydrate-protein interactions is determined in many cases by stereochemical diversity of small sugars which are constituents of oligosaccharide chains of glycoproteins and glycolipids. Most monosaccharides are stereoisomers which only differ in the spatial configuration of specific OH groups. However it is enough for amazing diversity in their chemical and biological activity. As oligosaccharides and proteins are particular complex molecules, one useful approach that can be used to understand the nature and mechanism of protein-carbohydrate interactions is to study model compounds. Monosaccharides are monomer units of the oligosaccharide chains of glycoproteins and glycolipids. Amino acids are basic components of proteins and can be considered as their model compounds. The aim of this article is to report the data on interactions between some isomeric monosaccharides and L-α-amino acids in aqueous solutions. It has been found that stereochemistry of the monosaccharides has significant influence on their selective binding with the amino acids.

[*] Fax: +7 4932 336237; E-mail: evp@isc-ras.ru

# Introduction

The carbohydrate-protein interactions are basic in very important processes in living organisms. They appear to be essential in the process of infection by certain pathogenic species [1-2]; they determine the immunochemistry of blood [3], they have an important role in cancer pathology [4], *etc.* Besides it was found that saccharides and polyols help in stabilizing the native conformation of globular proteins, enzymes [5-7]. Some studies show that certain carbohydrates protect proteins by binding to the dried protein, thus serving as a "water substitute" when the hydration shell of the protein is removed [7]. Therefore specific binding of carbohydrates by proteins has been the focus of increased study during the last decades.

However in spite of numerous experimental and theoretical studies of protein-carbohydrate interactions [1-13], the mechanism by which proteins bind carbohydrates remains a much debated area of inquiry. It is now accepted that, in general, the specificities and affinities of combining sites are determined mainly by hydrogen bonds, electrostatic interactions with promoting to these interaction der Waals contacts and hydrophobic interactions. Taking into account directivity of hydrogen bond in space, it is expected that stereochemical structure of saccharides should play an important role in these interactions.

Yet little is known about the relative roles of these interactions in selectivity of binding. One useful approach that can be used to understand the nature and mechanism of protein-carbohydrate interactions is to study model compounds. Amino acids are basic components of proteins. A determination of thermodynamic parameters characterizing solute-solute and solute-solvent interactions for low molecular weight solutes that incorporate some of structural features of such complex molecules can provide information on the various non-covalent interactions and their structural details in the molecular recognition processes. The aim of this work is to study the interactions between some L-amino acids (alanine (Ala), phenylalanine (Phe), leucine (Leu)) and D-galactose and β-D-fructose in water. Analysis of the effect of strreochemical structure of monosaccharides on their interaction with a series of amino acids on the basis of combination of the data obtained in this work with those reported earlier [14] is carried out.

# Experimental

## Materials

L-alanine , L-phenylalanine (from Merck), L-leucine, D-galactose (from Fluka) and β-D-fructose (from ICN Biomedicals) were used without further purification. The samples of the reagents were dried in a vacuum at 353 K during several days before use. Solutions were prepared by weight using doubly distilled degassed water.

## Methods

### *Isothermal Titration Calorimetry (ICT)*

The heats of interactions were measured on an automatic titration differential calorimeter [15] at 298.15 K. The initial concentration of an aqueous solution of amino acid placed into the calorimetric cell was 2-3 mM. The initial concentration of an aqueous solution of monosaccharide placed into the syringe was 0.20-0.25 M. The volume of an injection was $7{,}644 \cdot 10^{-5}$ $dm^3$. Primary treatment of the experimental data consists of determination of single heat effects corresponding to response of the system after injection of separate doses. The heat effect is defined as the product of the peak area and cell constant $\alpha$. Control experiments were performed by identical injections of the monosaccharide solution into the cell containing water. Appropriate corrections of the heats of interactions relative to the heats of dilution were made.

At first approach the determined thermodynamic characteristics can be considered as standard because all the data are obtained in dilute solutions at the temperature 298.15 K and no concentration dependences of these characteristics for the range of monosaccharide and amino acids studied were observed. At the treatment of the calorimetric results, 1:1, 2:1 and 1:2 models of binding between amino acid and monosaccharide were considered. The best fitting of the experimental data was observed using 1 : 1 model.

For a 1:1 binding between monosaccharides (S) and amino acid (AA)

$$(S)_{hydr} + (AA)_{hydr} \leftrightarrow (S \cdot AA)_{hydr}$$

the stability constant can be calculated as

$$K = \frac{[S \cdot AA]_i}{[S]_i \cdot [AA]_i} \quad (1)$$

or

$$K = \frac{[S \cdot AA]_i}{([S_0]_i - [S \cdot AA]_i) \cdot ([AA_0]_i - [S \cdot AA]_i)} \quad (2),$$

where $[S \cdot AA]_i$, $[S]_i$ , $[AA]_I$ are the equilibrium concentrations of associate, monosaccharide and amino acid, respectively; $[S]_0$ $[AA]_0$ are the initial concentrations of monosaccharide and amino acid after each injection.

Heat evolved (positive peaks, exothermic) or absorbed (negative peaks, endothermic) after *i* th injection is given as

$$\Delta Q_i = \pm \Delta H \cdot V_i \cdot [S \cdot AA]_i \quad (3)$$

where $V_i$ is the solution volume after each injection. Resolution of the system of equations (2) and (3) allows calculation of enthalpy change and stability constant values by continuously changing K and $\Delta H$ to minimize the function $\Sigma(Q_i(exp)-Q_i(cal))^2$. Optimization process is interrupted when minimum of the function $\Sigma(Q_i(exp)-Q_i(cal))^2$ is found. To exclude possible local minima, for each system under study profile of the function $\Sigma(Q_i(exp)-Q_i(cal))^2 = f(\Delta H_i, K_i)$ was analyzed. For all the studied systems the function has one minimum. Besides, the minimum of dispersion characterized the greatest correspondence of experimental and calculated single heat effects. This testifies to the global character of the minimum and confirms the reliability of the calculated thermodynamic characteristics.

Titration curves for the systems in which associate formation has been found are presented in Figure 1. For Leu – D-galactose, Leu – D-fructose and Ala –D-fructose systems, no difference between the titration heat effects and heat effects of dilution was detected.

***UV Absorption Spectroscopy***

UV absorption spectra at room temperature were measured with a double beam spectrophotometer Cary 100 using 10mm quartz cuvettes. The spectra were recorded in the range 190-300 nm. The concentration of Phe was $1\times10^{-2}$mol·dm$^{-3}$. The concentration of D-galactose was verified in the range $0 \leq C_{Gal} < 0.2$ mol·dm$^{-3}$.

## DISCUSSION

Monosaccharides are unique class of organic compounds. Each monosaccharide molecule (pentose or hexose) has stereoisomers which differ from each other by spatial arrangement of OH groups (equatorial or axial). This arrangement or stereochemical configuration determines diversity in physical-chemical properties of crystal isomer monosaccharides and their interaction with certain solutes or solvents in solution and as consequence biological activity. First of all stereoselectivity of the monosaccharide molecules is connected with their ability to hydrogen bonding because hydrogen bonds are highly directional. Moreover, spatial orientation of OH groups has influence on hydrophobicity of the monosaccharide molecule. Hydrophobic patches on the α- and β-faces of the molecule originate from the hydrophobic surface formed by CH, $CH_2$ groups and can be a target for van der Waals interactions. Thus stereochemical configuration of isomer monocchararides should affect on their interaction with another solutes including amino acids.

Analysis of the theoretical and experimental data reported in literature showed that the interactions between monosaccharides and amino acids are accompanied by small energy changes. This is because the interactions in the systems are the confluence of many weak polar and nonpolar intermolecular forces which often compensate one another. Moreover, in water the interactions between such hydrophilic solutes as monosaccharides and amino acids have to compete with solute-solvent interactions.

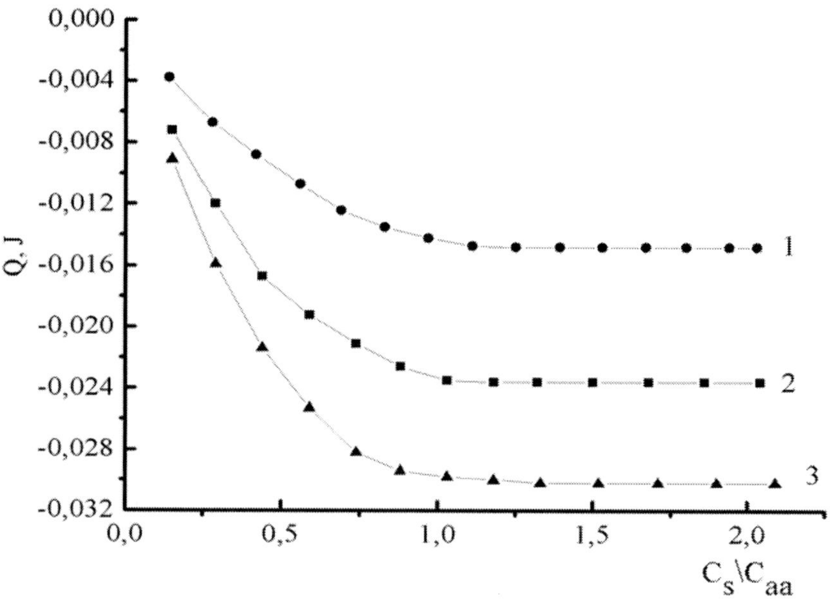


Figure 1. Calorimetric titration curves of aqueous solution of L-phenylalanine (2.75 mM) (curve 1) and L-alanine (2.60 mM) (curve 2) by aqueous solution of D-galactose (0.2418 M) and of aqueous solution of L-phenylalanine (3.00 mM) (curve 3) by aqueous solution of D-fructose (0.2444 M) at 298.15 K.

In general the interactions between amino acids and monosaccharides in aqueous solutions have been studied by means of the thermodynamic parameters of transfer of the amino acids from water to aqueous solutions of monosaccharides of different compositions [17-22]. The reported values of enthalpies $(\Delta H_{tr})$, partial molar heat capacities $(\overline{C}^0{}_{p2,tr})$ and partial molar volumes $(\overline{V}^0{}_{2,tr})$ of transfer of series of amino acids and peptides from water to aqueous glucose are positive and increase with increasing glucose concentration. These results indicate that the ionic-hydrophilic and hydrophilic- hydrophilic interactions are dominant. But with increasing alkyl group size in amino acid molecules, the contribution of hydrophilic-hydrophobic interactions with sugar increases. All these effects are associated with differences of water structure in hydration shells of the solutes and in solvent bulk. The same conclusions have been made on the basis of the data on enthalpic pairwise interaction coefficients [19] and ultrasonic measurements [22] of some amino acids in aqueous glucose solution.

However the main objects of the studies mentioned above were amino acids but not monosaccharides and their stereochemistry. Moreover, the interactions between solutes were studied at high concentrations of monosaccharide. During our studies concerning selective interactions between saccharides and amino acids [16, 23-24], the stability constants and thermodynamic parameters of interaction of some L-α amino acids with D-galactose and D-fructose were obtained by isothermal titration calorimetry. They are listed in Table 1. The same data for D-glucose reported earlier [14] are also presented in the Table 1.

It should be noted that all studied amino acids are in zwitterionic form in aqueous solutions. They contain amino-acid residues of different nature and size. The mentioned monosaccharides are isomers. According to literature data, in aqueous solution they exist in equilibrium of a few tautomeric forms [25-27]. However, as is well known, the preponderant tautomer in the equilibrium mixture is the β-pyranose because this tautomer is the most thermodynamically favorable in water [26]. The percentage ratio of α/β anomers for D-glucose and D-galactose is approximately equal [25,27]. D-glucose and D-galactose are aldohexoses which differ from each other only by spatial orientation of the OH group at the fourth carbon atom. D-fructose is ketohexose (Figure 1).

The data show that Ala forms the same stability associates with D-glucose and D-galactose. However the associate formation of Ala with D-fructose is not observed. This may be result from different extent of hydration of the monosaccharide molecule. The stronger the hydration of the monosaccharide molecule, the weaker their interaction with the amino acid. However literature data on $g_{xx}$ coefficients [27], volumetric properties [28], B-viscosity coefficients [29], listed in Table 2 indicate that the extent of hydration of the monosaccharide molecules decreases in the following order: glucose > fructose > galactose. Hence, the lack of the associate formation of Ala with D-fructose is not caused by the hydration state of the monosaccharide. Another reason may be that the principle of complementarity or mutual fitting of Ala and D-fructose molecules is not kept. It is well known that for efficient binding, the structures of interacting particles must be complementary to each other according to form, size and positions of binding groups. It is likely that the binding sites of Ala and D-fructose are not complementary to each other.

The associate formation between the studied amino acids and monosaccharides is accompanied by small exothermic effects. It should be noted that the interactions in the systems under study are water-mediated and most of the observed phenomena are probably due to changes in the state of the solvent. However, solute-solute direct interactions can not be excluded. A main source of negative enthalpy changes are specific interactions. Several hydroxyl groups of the monosaccharide molecules indicate on their hydrophilic character. As has been mentioned above, in aqueous solutions the amino acid molecules occur as zwitterions. Thus, such specific interactions as hydrogen bonding or ion-dipole interactions can dominate at the associate formation. The substitution of a hydrogen atom in $CH_3$-group of Ala molecule by a hydroxyl group in the molecule Ser results in a 2 times increase of the stability constant value for the associate of D-glucose with the latter [14]. This means that the specific interactions contribute significantly to the stability of the formed associates.

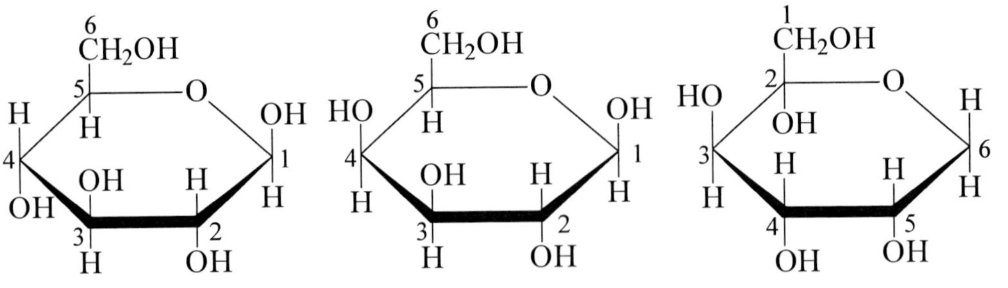

Glucose (1e2e3e4e6e)

Galactose (1e2e3e4a6e)

Fructose (1e2a3e4e5a)

Figure 2. Schematic representation of stereochemical configurations of the monosaccharide molecules (confirmations of dominant confirmer in aqueous solution, a= axial; e – equatorial hydroxyl group).

**Table 1. Stability constants and thermodynamic parameters of interaction (kJ·mol$^{-1}$) between some monosaccharides and L-α-amino acids in aqueous solutions**

| Amino acid | D-glucose[a] | | | D-galactose | | | D-fructose | | |
|---|---|---|---|---|---|---|---|---|---|
| | K | -ΔH | TΔS | K | -ΔH | TΔS | K | -ΔH | TΔS |
| Ala | 3744±171 | 2.65±0.12 | 17.7 | 2660±553 | 0.36±0.08 | 19.2 | No associate formation | | |
| Ser | 6340±860 | 2.84±0.33 | 18.9 | - | - | - | - | - | - |
| Leu | No associate formation | | | No associate formation | | | No associate formation | | |
| Phe | 14003±1670 | 1.85±0.22 | 21.8 | 3873±640 | 0.19±0.03 | 20.3 | 2386±209 | 0.39±0.08 | 18.9 |

Stoichiometry of the formed associates is 1 : 1.

All amino acids studied are in isoelectric point (The measured pH values of the solutions are 6.04, 5.65, 5.98, 5.98, respectively).

[a] Ref [14].

**Table 2. Properties of aqueous monosaccharide solutions at 298, 15 K.**

| Monosaccharide | Stereochemical configuration [a] | $\overline{V}_2^0 \cdot 10^{22}$, [b] cm$^3$·molecule$^{-1}$ | $g_{xx}$, [c] J·mol$^{-1}$ (mol·kg$^{-1}$)$^{-1}$ | B-coefficient [d] (308.15K) | $\left(\delta\overline{V}_2^0/\delta T\right)_p$, [b] cm$^3$·K$^{-1}$·mol$^{-1}$ |
|---|---|---|---|---|---|
| D-glucose | 1e2e3e4e6e | 1.858±0.001 | 92 | 0.4428 | 0.084±0.008 |
| D-fructose | 1e2a3e4e5a | 1.844±0.001 | 83 | 0.4370 | 0.09±0.012 |
| D-galactose | 1e2e3e4a6e | 1.831±0.001 | 65 | 0.3935 | 0.078±0.024 |

[a] Dominant configuration in aqueous solution.

[b] Ref [28]; [c] Ref [27]; [d] Ref [29].

At the same time the monosaccharide and amino acid molecules are amphiphilic. They contain hydrophobic groups which form hydrophobic fragments. The hydrophobicity of the monosaccharide molecules is determined by their stereochemical structure. Spatial orientation of OH groups of monosaccharide molecules promotes the formation of the hydrophobic fragments on their surfaces that provides their ability to van der Walls contacts. Such interactions also give a negative contribution to the ΔH values.

As can be seen from Table 1, Phe forms the associates both with D-galactose and D-glucose but the stability constant of the associate with D-glucose is more than 3 times greater than that with D-galactose. Such selectivity in the binding may be explained by difference in the stereochemical structure of the monosaccharide molecules.

It is well known that carbohydrates can interact with proteins through CH/π interactions [30-32]. The interactions between hydrophobic C-H groups of carbohydrate and π-electron systems of aromatic amino-acid residues are characterized by parallel orientation of the pyranose or furanose ring with the aromatic ring of amino-acid residue and governed by dispersive and charge transfer interactions. In contrast to D-galactose, all hydroxyl groups of D-glucose are equatorial (Figure 2). It is likely that an axial orientation of the OH group at the fourth (C4) carbon atom of D-galactose molecule prevents or weakens CH/π interaction between Phe and this monosaccharide. Authors of work [32] showed that the aromatic residue is important for discriminating galactose from glucose in sugar- lectin interactions. It should be noted that the effect of the spatial orientation of the OH group at C4 on monosaccharide - crown ether interactions has been found in works [33,34]. The key role of this factor in the interactions between proteins and carbohydrayes has been marked in different studies [35-

37]. In work [14] the contribution of CH/π interaction to the stability constant is confirmed by UV absorption spectra of Phe in aqueous solutions of D-glucose of different compositions. According to literature data [38,39], the bands at 205 and 257 nm are associated with $\pi \rightarrow \pi^*$ transition of benzene ring of Phe. These bands are often used for quantification of the amino acid in solution. Analogous spectral measurements were performed for Phe in aqueous D-galactose solutions. The spectra are presented in Figure 3. As can be seen from Figure 3, no appreciable changes in the spectra are observed at additions of D-galactose. Thus, it may be concluded that this type of interaction does not contribute significantly to the stability of the Phe – D-galactose associate.

The stability constants and thermodynamic parameters of interactions between Phe and D-fructose and D-galactose are close each other. It is likely that the nature of forces responsible for associate formation in these systems is similar.

No associate formation was found between Leu and the studied monosaccharides. Possibly, this is because Leu molecule has long alkyl side chain that can hinder sterically from the associate formation. Besides, Leu is strongly hydrated. The hydration enthalpies of Ala, Phe and Leu are -137.3, -145.7 and -147.2 kJ·mol$^{-1}$, respectively [40]. Thus, the strong hydration of Leu is another factor that may prevent formation of the associate between the amino acid and the monosaccharides under study.

The formed associates are preferentially entropy stabilized. As has been mentioned above, the reagents under study are strongly hydrated. The interactions between them are accompanied by significant redistribution of water molecules between the bulk and the hydration shells. Because water in the bulk and the hydration shells is thermodynamically unequal, the solvent reorganization should have a significant effect on the associate formation. The entropy term values testify about it.

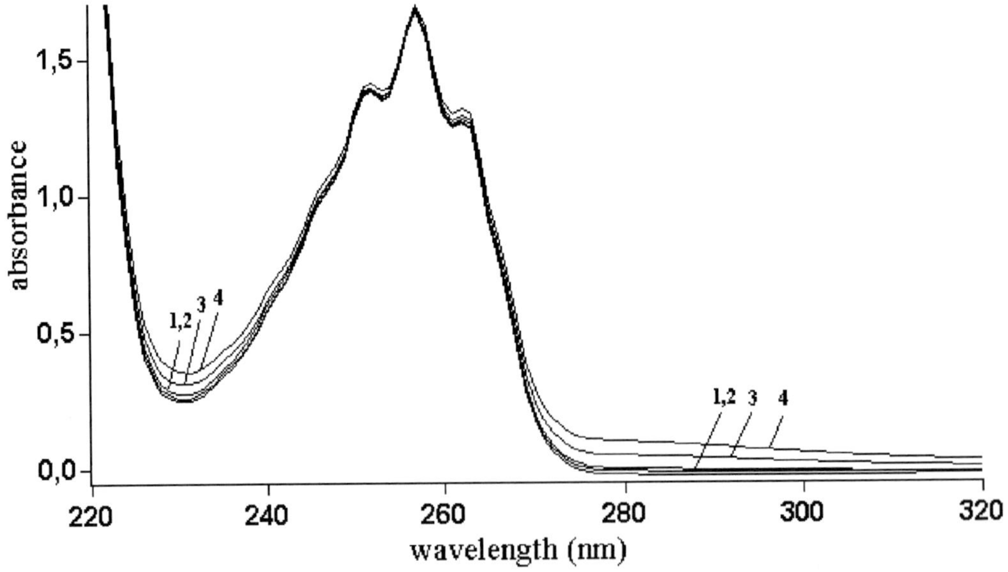


Figure 3. UV spectral results upon addition of D-galactose to $10^{-2}$ mol·dm$^{-3}$ aqueous Phe: 1, [Glc] = 0.20 mol·dm$^{-3}$; 2, [Glc] = 0.10 mol·dm$^{-3}$; 3, [Glc] = 0.05 mol·dm$^{-3}$; 4, [Glc] = 0.00 mol·dm$^{-3}$.

## Conclusion

The results reported in this work show that there is a remarkable selectivity in the binding of the studied monosaccharides by the amino acids in dilute aqueous solution. The interaction between some amino acids and monosaccharides results in formation of thermodynamically stable associates. The selectivity is affected by stereochemical structure of the monosaccharides as well as the nature of the amino acids. Spatial orientation of OH groups in the monosaccharide molecules has influence (i) on formation of hydrophobic patches on their surfaces and determines their ability to van der Waals contacts; (ii) on specific interactions between the monosaccharides and amino acids. The solvent reorganization plays an important role in the associate formation.

## References

[1] Olofsson S., Bergstrom T. Glycoconjugate glycans as viral receptors. *Ann. Med.,* 2005, v.37, N 3, p.154-172.

[2] Karlsson K.A. Pathogen-host protein-carbohydrate interactions as the basis of important infections. *Adv. Exp. Med. Biol.*, 2001, N.491, p.431-443.

[3] Lemieux R.U. The origin of the specificity in the recognition of oligosaccharides by proteins. *Chem. Soc. Rev.,* 1989 v.18, N 3, p.347-374.

[4] Gorelic E., Raz A. On the role of cell surface carbohydrates and their binding proteins (lectins) in tumor metastasis. *Cancer and Metastasis Rev.,* 2001, v.20, p.245-277.

[5] Chang L.L., Shepherd D., Sun J., Quellette D., Grant K.L., Tang X.C., Pikal M.J. Mechanism of protein stabilization by sugars during freeze-drying and storage: native structure, preservation, specific interaction, and/or immobilization in a glass matrix? *J. Pharm. Sci.,* 2005, v.94, N 7, p.1427-1444.

[6] Sampedro J.D., Uribe S. Trehalose-enzyme interactions results in structure stabilization and activity inhibition. The role of viscosity. *Mol. Cell Biochem.,* 2004, v.256. N.1, p.319-327.

[7] Arakawa T., Kita Y. and Carpenter J.F. Protein –solvent interactions in pharmaceutical formulations. *Pharm. Res.,* 1991, v.8, N 3, p.285-291.

[8] Ambrosi M., Cameron N.R., Davis B.G. Lectins: tools for the molecular understanding of the glycocode. *Org. Biol. Chem.,* 2005, v.3, N 9, p.1593-1608.

[9] Poveda A, and Jiménez-Barbera J., NMR studies of carbohydrate-protein interactions in solution. *Chem. Soc. Rev.,* 1997, v.27, p.133-143.

[10] Neumann D., Lehr C.-M., Lenhof H.-P., Kohlbacher O. Computational modeling of sugar-lectin interaction. *Adv. Drug Deliv.Rev.*, 2004, v.56, N 4, p.437-457.

[11] Johnson M.A., Pinto B.M.. NMR spectroscopic and molecular modeling studies of protein-carbohydrate and protein-peptide interactions. *Carbohydr. Res.,* 2004, v.339, N 5, p.907-928

[12] Lypez-Dhez E.C., Bone S. An investigation of the water-binding properties of protein+sugar systems. *Phys. Med. Biol.* , 2000, v.45, N12, p.3577-3588.

[13] Srinivas V.R., Bhanuprakash R., Surolia A. A predominantly hydrophobic recognition of H-antigenic sugars by winged bean lectin: a thermodynamic study. *FEBS Lett.*, 1999, N 450, p.181-185.

[14] Kulikova G.A., Parfenyuk E.V. Influence of side chain of L-α-amino acids on their interactions with D-glucose in dilute aqueous solutions. *J. Solut. Chem.*, 2008, (in press)

[15] Lebedeva N.Sh., Mikhailovsky K.V., Vyugin A.I. A differential automatic titration calorimeter. *Russ. J. Phys. Chem.*, 2001, v.75, N 6, p.1140-1142.

[16] Parfenyuk E.V., Davydova O.I., Lebedeva N.Sh. Interactions of D-maltose and sucrose with some amino acids in aqueous solutions. *J. Solut. Chem.*, 2004, v.33, N 1, p.1-10, (2004).

[17] Bhat R., Kishore N. and Ahluwalia J.C. Thermodynamic studies of transfer of some amino acids and peptides from water to aqueous glucose and sucrose solutions at 298.15 K. *J. Chem. Soc., Faraday Trans. I,* 1988, v.84, N 8, p.2651-2665.

[18] Lou Y., Lin R. Enthalpy of transfer of amino acids from water to aqueous glucose solutions at 298.15 K. *Thermochim. Acta,* 1998, v.316, N 2, p.145-148.

[19] Li Sh., Hu X, Lin R., Zong H. Enthalpic interaction of glycine in aqueous glucose and sucrose solutions at 298.15 K. *Thermochim. Acta,* 1999, v.342, N 1, p.1-5.

[20] Li Sh., Sang W., Lin R. Partial molar volumes of glycine, L-alanine, and L-serin in aqueous glucose solutions at T = 298.25 K. *J. Chem. Thermodyn.*, 2002, v. 34, p.1761-1768.

[21] Zhaoa Ch., Mab P., Lia J. Partial molar volumes and viscosity B-coefficients of arginine in aqueous glucose, sucrose and L-ascorbic acid solutions at 298.15 K. *J. Chem. Thermodyn.* 2005, v.37, N 1, p.37-42.

[22] Banipal T.S., Sehgal G. Partial molal adiabatic compressibilities of transfer of some amino acids and peptides from water to aqueous sodium chloride and aqueous glucose solutions. *Thermochim. Acta.*, 1995, v.262, p.175-183.

[23] Parfenyuk E.V., Lebedeva N.Sh., Davydova O.I. Thermodynamic characteristics of interactions between L-serine and DL-alanine and sucrose in water. *Russ. J. Phys. Chem.*, 2003, v.77, N 4, p. 757-758.

[24] Parfenyuk E.V. and Davydova O.I. Stability constants for associate of saccharose and raffinise with glycine and DL-alanine in aqueous solutions. *Russ. J. Phys. Chem.*, 2004, v.78, N 5, p……. .

[25] Schmidt R.K., Karplus M., and Brady J.W. The anomeric equilibrium in D-xylose: Free energy and the role of solvent structure. *J. Am. Chem. Soc.*, 1996, v.118, N 2, 541-546.

[26] Tvarovska I.: "The structure and conformational properties of carbohydrates", in Náray-Szabó G. (ed.) *Theoretical chemistry of biological systems,* Elsevier, 1986, pp.283-348.

[27] Barone G. Physical chemistry of aueous solutions of oligosaccharides. *Thermochimica Acta,* v.162, N 1, p.17-30.

[28] Banipal P.K., Banipal T.S., Lark B.S. and Ahluwalia J.C. Partial molar heat capacities and volumes of some mono-, di- and tri-saccarides in water at 298.15, 308.15 and 318.15K. *J. Chem. Soc.Faraday Trans.*, 1997, v.93, N 1, p.81-87

[29] Dey P.C., Motin M.A., Biswas T.K. and Huque E.M. Apparent molar volume viscosity studies of some carbohydrates in solutions. *Monatsh. Chem.*, 2003, v.134, p.797-809

[30] Murakami M. The influence of CH/pi interactions to the function of carbohydrate binding proteins. *Protein Pept. Lett.,* 2002, v.9, N3, p.195-209

[31] Spiwok V., Lipovova P., Skalova T., Buchtelova E., Hasek J., Kralova B. Role of CH/pi interactions in substrat binding by Escherichia coli beta-galactosidase. *Carbohydr. Res.,* 2004, v.339, N 13, p.2275-2280.

[32] Sujatha M.S., Sasidhar Y.U., Balaji P.V. Insights into the role of the aromatic residue in galactose-binding sites: MP2/6-311G++** study on galactose- and glucose-aromatic residue analogue complex. *Biochem.,* 2005, v.44, N 23, p.8854-6662.

[33] Volkova N.L., Parfenyuk E.V. Selective interactions of 18-crown-6 with D-glucose and D-galactose in aqueous solutions: Titration calorimetry, densimetry, viscosimetry. Thermochim. *Acta,* 2005, v.435, N1, p.108-112.

[34] Volkova N.L., Parfenyuk E.V. Parameters of the formation of molecular complexes in D-glucose –(D-galactose)-15-crown-5-water ternary systems. *Russ. J. Phys. Chem. A,* 2007, v.81, N7, p.1151-1155.

[35] Cheong Y., Shim G., Kang D., Kim Y. Carbohydrate binding specificity of pea lectin studied by NMR spectroscopy and molecular dynamic simulations. *J. Mol. Struct.,* 1999, v.475, N 2, p.219-232/

[36] Weis W.I., Drickmer K. Structural basis of lectin-carbohydrate recognition. *Annu. Rev. Biochem.* 1996, v.65, p.441-473.

[37] Gabius H.-J. Biological information transfer beyond the genetic code: *The sugar code. Naturwissensch.,* 2000, v.87, p.108-121

[38] Lo Shui-Yin, Li W. Nanostructures in very dilute aqueous solutions. *Russ. Chem. J.,* 1999, v.43, n 5, p.40-48.

[39] Rodante F. Thermodynamics of the “standard” α-amino acids in water at $25^0$C. Thermochim. *Acta,* 1989, v. 149, N 1, p.157-171.

In: Stereochemistry Research Trends
Editor: M. A. Horvat and J. H. Golob

ISBN 978-1-60456-825-7


*Chapter 7*

# CHALLENGES IN CARBOHYDRATE CHEMISTRY: A SURVEY OF MODERN APPROACHES FOR L-HEXOSE STEREOSELECTIVE SYNTHESIS

***Annalisa Guaragna, Daniele D'Alonzo and Giovanni Palumbo***
Dipartimento di Chimica Organica e Biochimica, Universita` di Napoli Federico II
Complesso Universitario Monte Sant'Angelo, Via Cinthia 4, I-80126 Napoli, Italy

## ABSTRACT

In the last two decades a new key role of carbohydrates in living organisms has emerged: they have demonstrated to be the major information carriers between cells and their surroundings, whether free or in the form of glycoconjugates with proteins or lipids. This finding has rapidly brought increasing interest in several aspects of carbohydrate chemistry, driven by their potential use in biochemical and pharmaceutical research. The need for specific saccharides in significant amounts, in order to thoroughly understand their biological functions, has inspired modern synthetic approaches. The demand for compounds which can mimic natural substances by way of similar structure and/or biological function has stimulated elaboration of innovative preparative procedures in carbohydrate synthesis. The search for analogues of mono- and oligosaccharides resistant to hydrolytic/enzymatic cleavage has been carried out as well. All these efforts in chemistry and biology have converged in establishing a well defined discipline, called glycoscience.

Especially with respect to investigations on mono- and oligosaccharides, an interest on mirror-image carbohydrates has emerged as an extremely intriguing topic, since they have been found to be key components of several bioactive compounds, such as oligosaccharides, antibiotics, glycopeptides, terpene glycosides, steroid glycosides and other clinically useful agents. However, while most D-sugars are available in large quantity from natural materials, their L-enantiomers are rather rare in nature and then not easily accessible from inexpensive sources. As demand for their synthesis in considerable amount and high purity is more and more pressing, in recent years intense efforts have been focused on the development of new and general methodologies for the construction of L-hexoses.

In this chapter we present a survey of the synthetic routes to L-sugars, mainly the ones coming from the new century, although some former and creative approaches will also be reviewed. These methodologies include specific or general approaches and range from carbon chain elongation (homologation), epimerization of D-sugars by chemical or enzymatic methods, up to the most recent amino acid-catalyzed stereoselective aldol reactions.

## 1. Introduction

### 1.1. Emerging Roles of Carbohydrates in Drug Discovery: from Natural Glycans to Synthetic Glycomimetics

Because of their enormous structural and functional diversity, carbohydrates have been found to play crucial roles in numerous fields, whether they are in the form of monomers, oligomers, polymers, or as components of biopolymers and other naturally occurring substances [1]. Although up to few decades ago most carbohydrates were thought to be able to play roles just as structural and food storage elements, whereas control of the organism functioning was left to the other main natural polymers, such as DNA and proteins, nowadays the increased awareness of vital biological roles exerted by oligosaccharides has led to a renewed interest in carbohydrate field, which ranges from the elucidation of their structural complexity, their structure-function relationships up to the exploitation of their vast potential in drug development. With a re-evaluation of the importance of carbohydrates, analogously to Proteomics and Genomics, the term "Glycomics" [2] (or "Glycobiology" [3] or "Glycoscience" [4]) has been coined to refer to the role of carbohydrates in biological events, [5] including carbohydrate-carbohydrate, carbohydrate-protein, and carbohydrate-nucleic acid interactions (Figure 1).

In the form of glycopeptides, glycolipids, glycosaminoglycans, proteoglycans, or other glycoconjugates, carbohydrates have been found on cell surfaces to coordinate biological events including signal transduction, [6] inflammation processes, [7] cell-cell interactions, [8] bacteria-host interactions, [9] viral entry, [10] fertility and development. [11]

As our understanding of carbohydrate-mediated biological recognition processes advances to the molecular level, new carbohydrate-based therapeutics are becoming readily available and widely utilized for the treatment of a broad range of inflammatory and infectious diseases. [12] In the search for chemical tools able to block glycoconjugate formation and/or function, carbohydrate mimetics (glycomimetics) have emerged as a group of artificially or naturally derived compounds with shapes and polarities resembling sugars, but offering improved biological, pharmacological and drug-like properties.

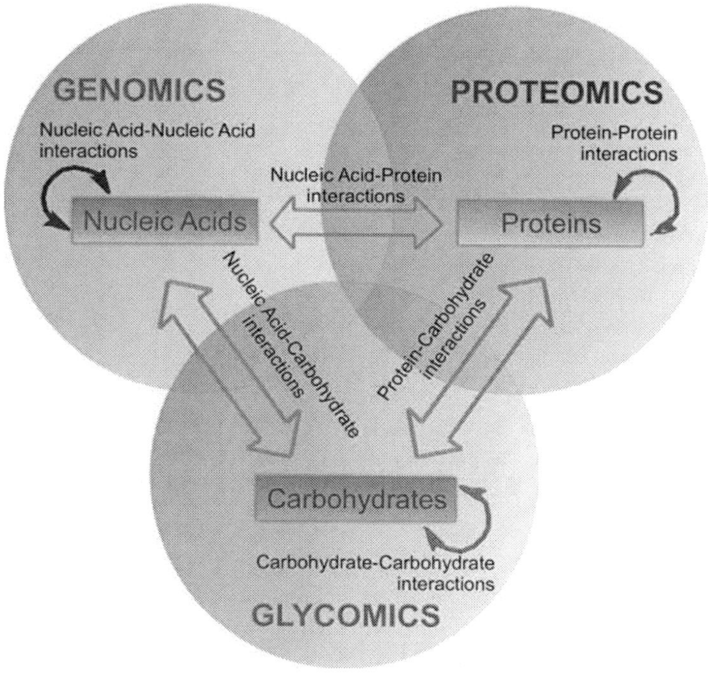


Figure 1. Biopolymer interactions.

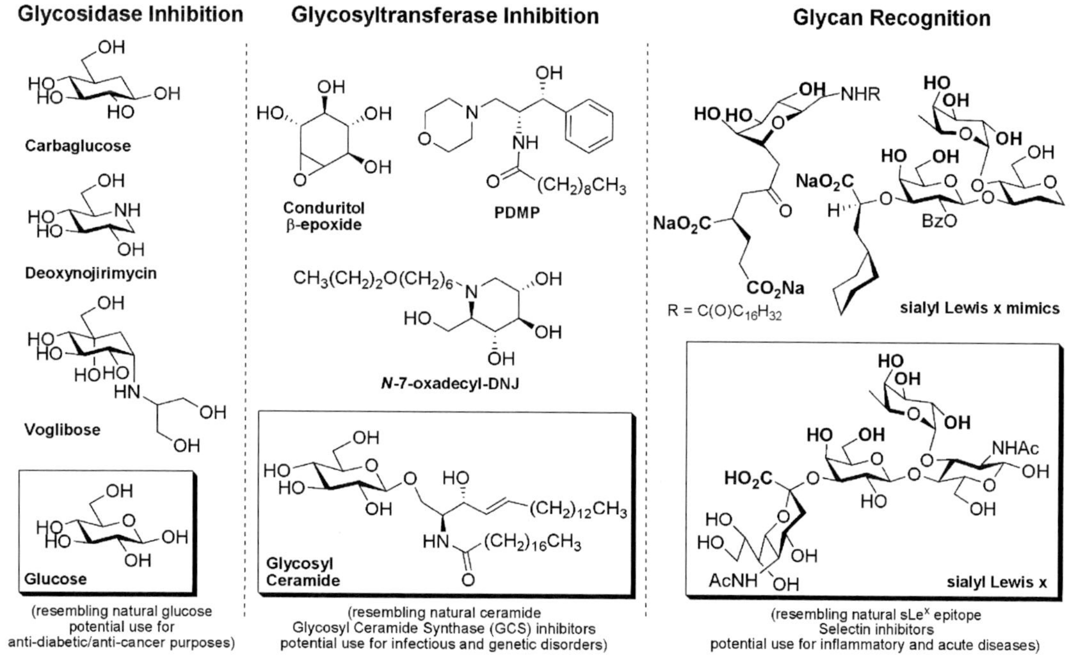


Figure 2. Use of glycomimetics for several therapeutic purposes.

Glycomimetic research has so far progressed in numerous fields of investigation (Figure 2); indeed, innovative strategies have been developed for the design and synthesis of new sugar-like agents able to interfere in carbohydrate-receptor binding events, whether as antagonists of carbohydrate-protein interaction (such as glycan-selectin recognition) [13] or as inhibitors of carbohydrate processing enzymes (glycosidases and glycosyltransferases). [14,15]

## 1.2. Role of Chirality at Chemistry-Biology Interface: Unnatural Carbohydrates as Tools for Glycobiology

In the emerging field of functional glycomics, quite a few innovative research lines are focusing on the employ of unnatural mono- and oligosaccharides, driven by the possibility to explore the function of molecular chirality in carbohydrate recognition. [16]

The role of chirality at chemistry-biology interface is a topic of enormous fascination and importance in glycobiology as well as in other fields of investigation. [17] Indeed, optically active molecules often hold specific conformational and high order structures associated with their chiral features; their structures are deeply connected with their properties and functions, including the ability for chiral recognition. Nevertheless, the relationships between stereochemistry of a potentially active molecule and its effective biological function are not always predictable. Particularly, even if it is often assumed that biomolecular recognition processes are stereospecific, there are many notable exceptions. For instance, mammalian taste bud receptor recognizes D-sugars *1* and L-sugars *ent-1* with equal avidity [18] (Figure 3). In a different context, human deoxycytidine kinase (dCK), the enzyme responsible for the phosphorylation of cytidine nucleosides, identified as a target enzyme for HIV, HSV and HBV inhibition, recognizes and phosphorylates D-series nucleoside analogues, as well as their L-enantiomers (as reported [19] in case of BCH 189 (*2*) and 3TC (*ent-2*), Figure 3). Furthermore, although glycosides belonging to D-series are the natural substrates for all glycosidases (with the exception L-fucosidases), unnatural deoxy-L-allonojirimycin (*3*) is a much better inhibitor (IC50 = 30 $\mu$M) of $\alpha$-D-mannosidase than deoxy-D-mannojirimycin (*4*, IC50 = 840 $\mu$M) (Figure 3) [20].

These data demonstrates fairly well that enantiospecificity, although important in many cases, could not be a universal trait related to biological function. To further probe this fundamental aspect, a number of research groups have already engaged in the synthesis and evaluation of mirror-image biopolymers [21]. Similarly, mirror-image carbohydrates may be useful for addressing significance of molecular chirality in glycobiology, particularly in circumstances where biomolecular interactions are not well defined [22].

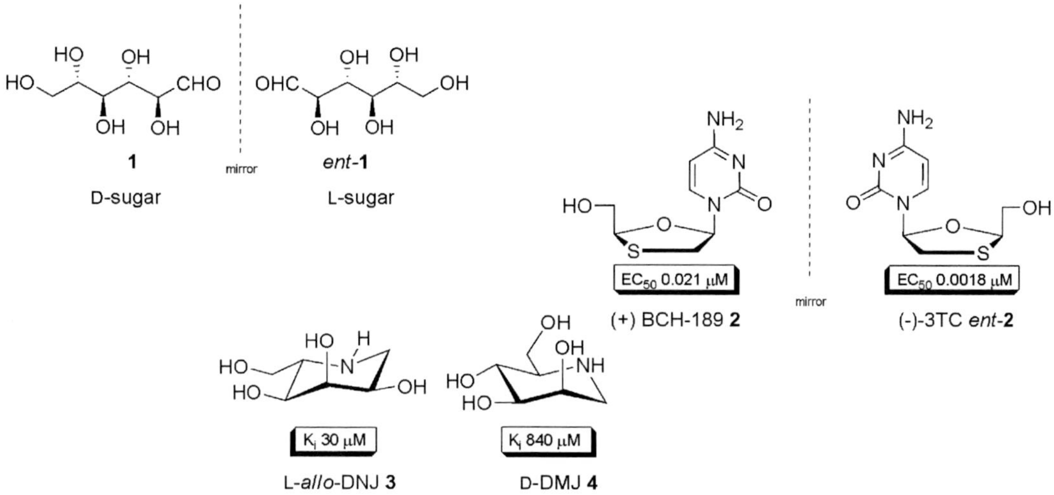


Figure 3. The relationships between molecular stereochemistry and biological function are not always predictable.

## 1.3. Rare Sugars: a Chemically Available Source of Building Blocks for the Synthesis of Bioactive Molecules

In view of the various biochemical pathways and disease processes in which glycoconjugates are crucially engaged – such as angiogenesis, [23] cancer, [24] tissue repair, cardiovascular diseases, [25] immune-system functions, [26] microbial and viral pathogenesis [27] – the possibilities for their employ as therapeutics and diagnostics are numerous and exciting. However, while numerous common D-sugars are available in huge quantities from natural products and therefore constitute a convenient source of starting materials for organic synthesis, their L-enantiomers [28] result to be much less available; nonetheless they have been often recognized as remarkable components of biologically relevant compounds. Especially, L-hexoses (mainly in their pyranosidic form) are key constituents of numerous bioactive [29] oligosaccharides, antibiotics, glycopeptides and terpene glycosides, as well as steroid glycosides and other clinically useful agents. [30] Some significant examples are represented by L-gulopyranoside-containing compounds such as the nucleoside antibiotic Adenomycin (*5*) [31] and the antitumor drug Bleomycin A2 (*6*) [32] (Figure 4). Moreover, L-altrose (*7*) has been found to be a typical component of extracellular polysaccharides from *Butyrivibrio fibrisolvens* strain CF3. [33] On the other hand, L-mannose has been discovered in some steroid glycosides (*8*), [34] and its phenol derivatives have been used as potent substrates for measuring the α-L-mannosidase activity of commercial naringinase. [35]

With the heightened biological and medicinal interest in unnatural carbohydrates, in recent years many methodologies for their construction have been described. [36] Indeed, almost all L-hexoses are not commercially available; this feature, together with the practical difficulties in obtaining these compounds from natural sources has led chemists to develop new and convenient methods for their production. Although the pursuit of approaches towards this kind of molecules already comprises a large amount of protocols, the area of investigation is still active, as new methodologies, new concepts and the development of innovative and general strategies are taking root. Considering the modern approaches for the synthesis of natural and unnatural carbohydrates, quite a few ways to generate monosaccharide units in selective manner have emerged:

- diastereoselective syntheses, wherein new stereocentres are introduced into compounds bearing preexisting ones;
- enantioselective syntheses from achiral substrates by introduction of a new stereogenic centre which fixes the steric series;

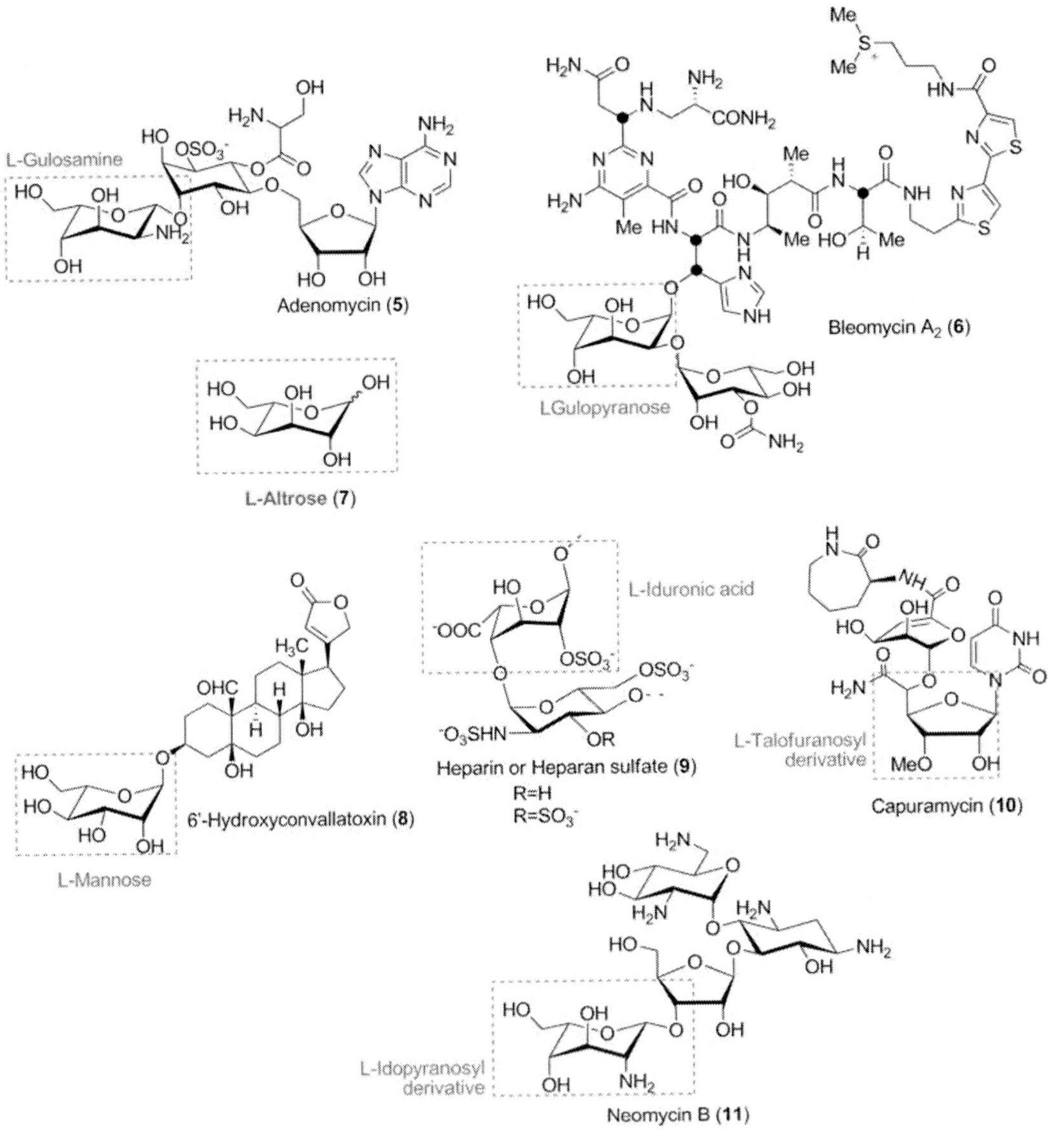


Figure 4. L-Hexoses and their related biomolecules.

- enantioselective syntheses with simultaneous introduction of multiple stereocentres by coupling of two prochiral substrates, which results in the direct construction of target molecules via highly enantioselective procedures;
- monosaccharide epimerizations by chemical or enzymatic methods, which relies on the availability of appropriate D-series sugars as starting materials.

This report covers the most important synthetic approaches developed during the new century, with the exception of some pioneering and creative methodologies, which in spite of the new emerging technologies remain among the most intriguing and elegant approaches to the preparation of such molecules. Farther on, a short account of some successful applications of these methodologies for the synthesis of biologically relevant oligosaccharide sequences will be briefly discussed, in order to show the remarkable potential of these procedures to create useful tools for studies in glycobiology and glycomedicine.

# 2. Diastereoselective Synthesis

## 2.1. General Route to L-Hexoses from L-Ascorbic Acid

Quite a few *de novo* synthetic approaches can be found in the recent literature using chiral synthons which exploit their inherent chirality to set the configuration of final products. Following such strategy, a novel stereodivergent approach to rare hexoses has been successfully reported using L-ascorbic acid as common starting material. [37]

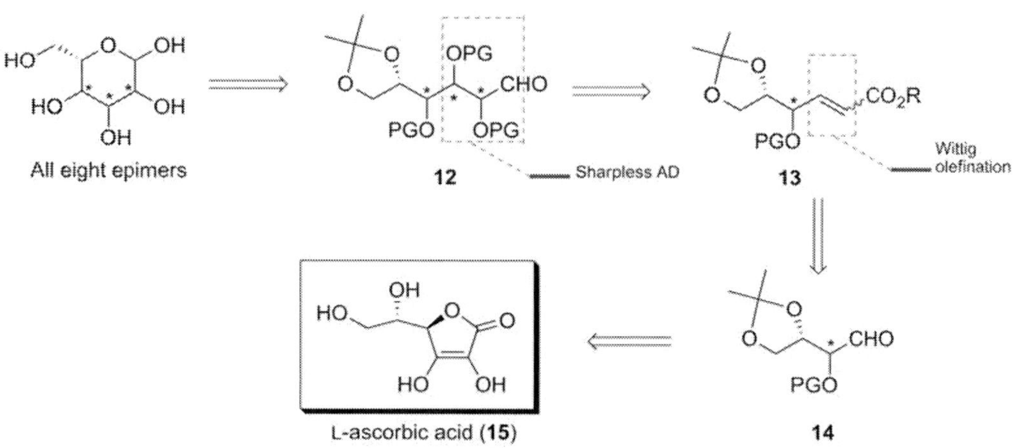


Scheme 1. L-Hexoses from ascorbic acid: retrosynthetic path.

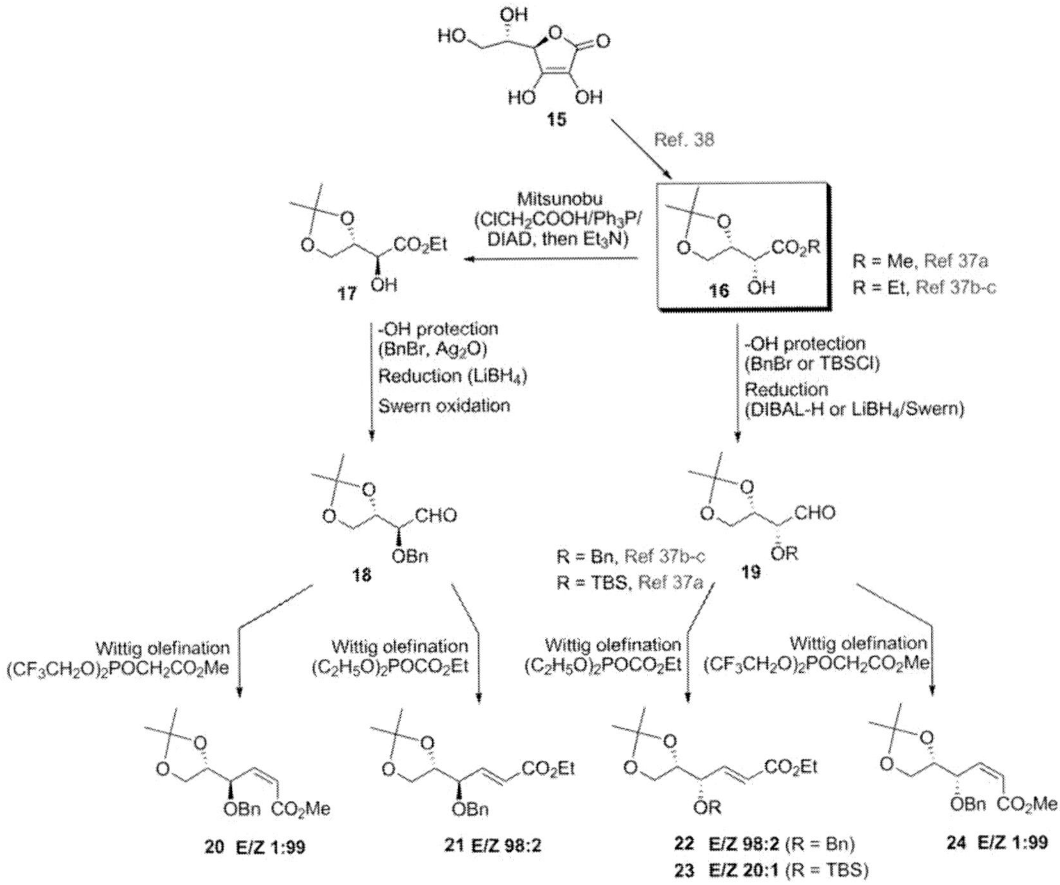


Scheme 2. Preparation of the key intermediates from ascorbic acid.

As depicted in Scheme 1, the strategy comprised the following major steps: (i) preparation of chiral aldehyde *14*, readily available from L-ascorbic acid (*15*); (ii) synthesis of α,β-unsaturated ester *13* with the specified *E* or *Z* configuration *via* Wittig olefination reaction; (iii) stereoselective dihydroxylation of ester *13* to give protected polyhydroxylaldehydes *12*; (iv) six-membered cyclization of aldehydes *12* to provide desired L-hexopyranoses.

Key compound *16* was prepared from L-ascorbic acid in two steps by a known procedure [38] (Scheme 2); conversely, inversion of configuration at *C*-2 position in *16* by the convenient chloroacetate modification [39] of Mitsunobu reaction provided α-hydroxy ester *17* after selective chloroacetate group removal in mild basic conditions. Then, ester group reduction in *16* and *17* to give diastereomeric aldehydes *18* and *19* followed by Wittig-type olefination reaction afforded four intermediates *20-24* (diastereomers in pairs at *C*-4 position).

Once synthesis of olefin intermediates *20-24* was achieved, Sharpless asymmetric dihydroxylation represented the key step required to provide desired polyhydroxylated compounds *25-32*, [40] demonstrating to be a powerful tool to overcome the inherent facial selectivity of such substrates and to gain access to stereoisomeric diols in good yields and excellent stereoselectivities. Use of (DHQD)2PHAL and (DHQ)2PHAL, as ligands in AD-mix reagents, respectively addressed osmylation towards the β- and the α-face of the olefin, giving diol intermediates *25-32*. Finally, conversion of the ester function into formyl group and six-membered ring closure afforded target molecules *41-48* (Scheme 3).

As highlighted in Scheme 3, the general feature of such a kind of methodology has been demonstrated by the synthesis of all eight epimers of L-aldohexoses in stereoselective manner.

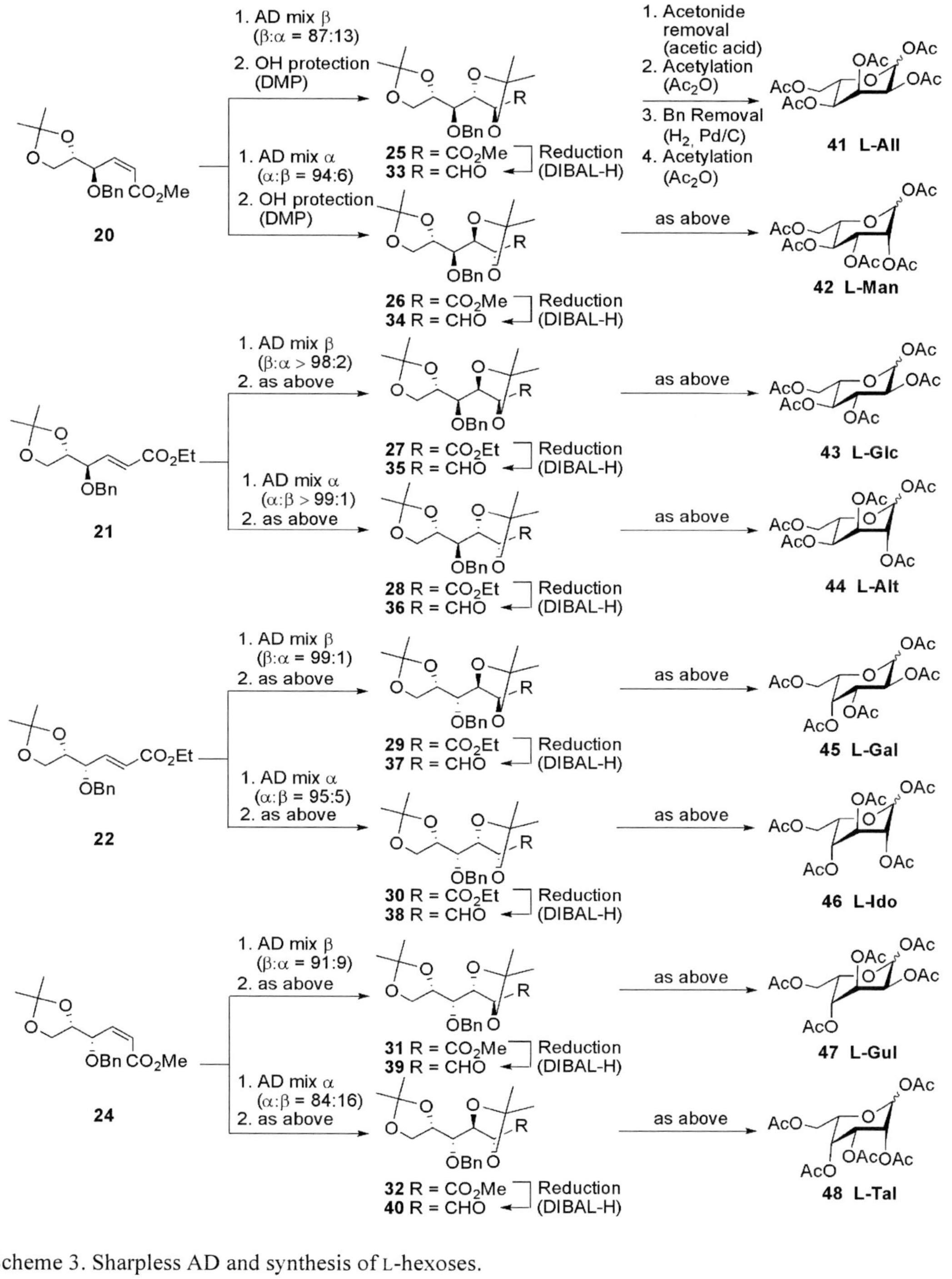

Scheme 3. Sharpless AD and synthesis of L-hexoses.

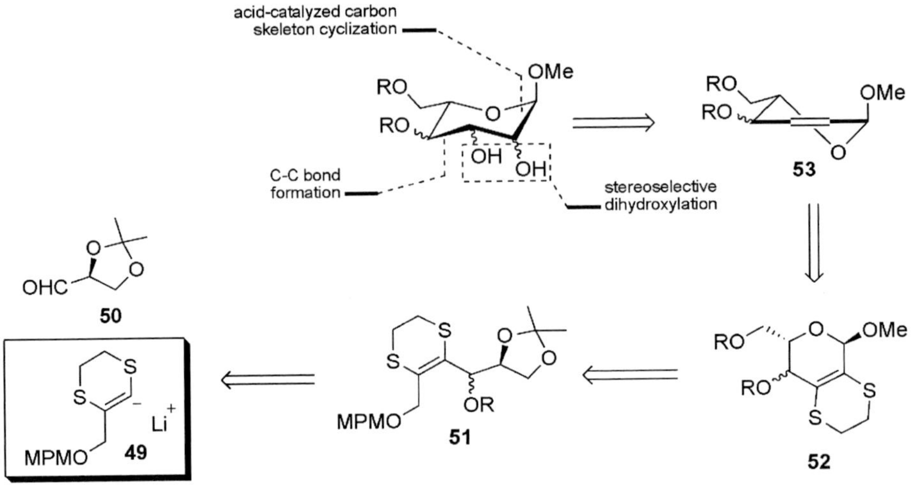


Scheme 4. L-Hexoses by a synthetic equivalent of an allylic alcohol: retrosynthetic path.

## 2.2. Versatile Access to Hexopyranoses by Homologation of Chiral Substrates with an Allylic Alcohol Moiety

Thanks to continuous efforts in the search of new reagents devoted to the construction of molecular building blocks via efficient and stereoselective carbon-carbon bond-forming reactions, many brilliant solutions have been developed for the synthesis of polyhydroxylated molecules, such as carbohydrates and their analogues. [41] Among numerous examples of homologating agents developed for this purpose, the opportunity of introducing a synthetic equivalent of an allylic alcohol on a chiral substrate, which has been demonstrated to hold great potential for synthetic aims, is nowadays restricted only to a few examples. In this context, the 3-*C*-lithiated (5,6-dihydro-1,4-dithiin-2-yl)[(4-methoxybenzyl)oxy]methane (*49*), which versatility has been already displayed in many examples, [42] has been recently exploited to open up a general procedure for L-hexopyranose synthesis. [43]

As outlined in retrosynthetic path, this route comprised: (a) elongation of a chiral electrophile (e.g. 2,3-*O*-isopropylidene-L-glyceraldehyde, *50*) by the 3-*C* lithiated building block *49*; (b) carbon skeleton cyclization, to generate bicycle intermediate *52*; (c) dithioethylene bridge removal under reducing conditions, to obtain a 2,3-unsaturated sugar moiety *53*; (d) double bond functionalization by stereoselective dihydroxylation of *53* (Scheme 4).

As the coupling reaction between *49* and glyceraldehyde *50* was achieved, two diastereomers *54* and *55* were produced, each one being employed throughout the synthesis (Scheme 5). MPM group removal under mild conditions (DDQ) followed by oxidation with PCC furnished aldehydes *56-57*. Then, exposure to Amberlyst 15 in methanol allowed, in a *one pot* simple procedure, conversion of formyl group into its di-*O*-methyl acetal, acetonide deprotection, and then the intramolecular transacetalation to give bicyclic compounds *58-59*. Finally, primary hydroxyl function protection and dithioethylene bridge removal (Raney-Ni) led to key unsaturated pyranosyl derivatives *60-61*. Hence, four L-epimers (L-*manno*-, L-*altro*-, L-*ido*- and L-*gulo*-hexopyranosides *62-65*) have been synthesized, with a high degree of stereoselectivity, by appropriate *syn*- and *anti*-dihydroxylation reactions of the double bond in olefins *60-61*.

The opportunity to have a strategy of general applicability arose from the employ of 1,6-anhydro derivatives *70-71* (Scheme 6), obtained by a further cyclization of bicycle intermediates *58-59* or more conveniently by a *domino* procedure starting from coupling products *66-67*. [44] Particularly, it was exploited a peculiar feature of 1,6-anhydro derivatives, according to which all stereocentres are in opposite orientation with respect to the corresponding classical pyranosides. Therefore, an inverse stereoselectivity was attained in *syn*- and *anti*-dihydroxylation reactions, affording L-*allo*-, L-*gluco*- and L-*galacto*-hexopyranosides *72*, *75* and *76*.

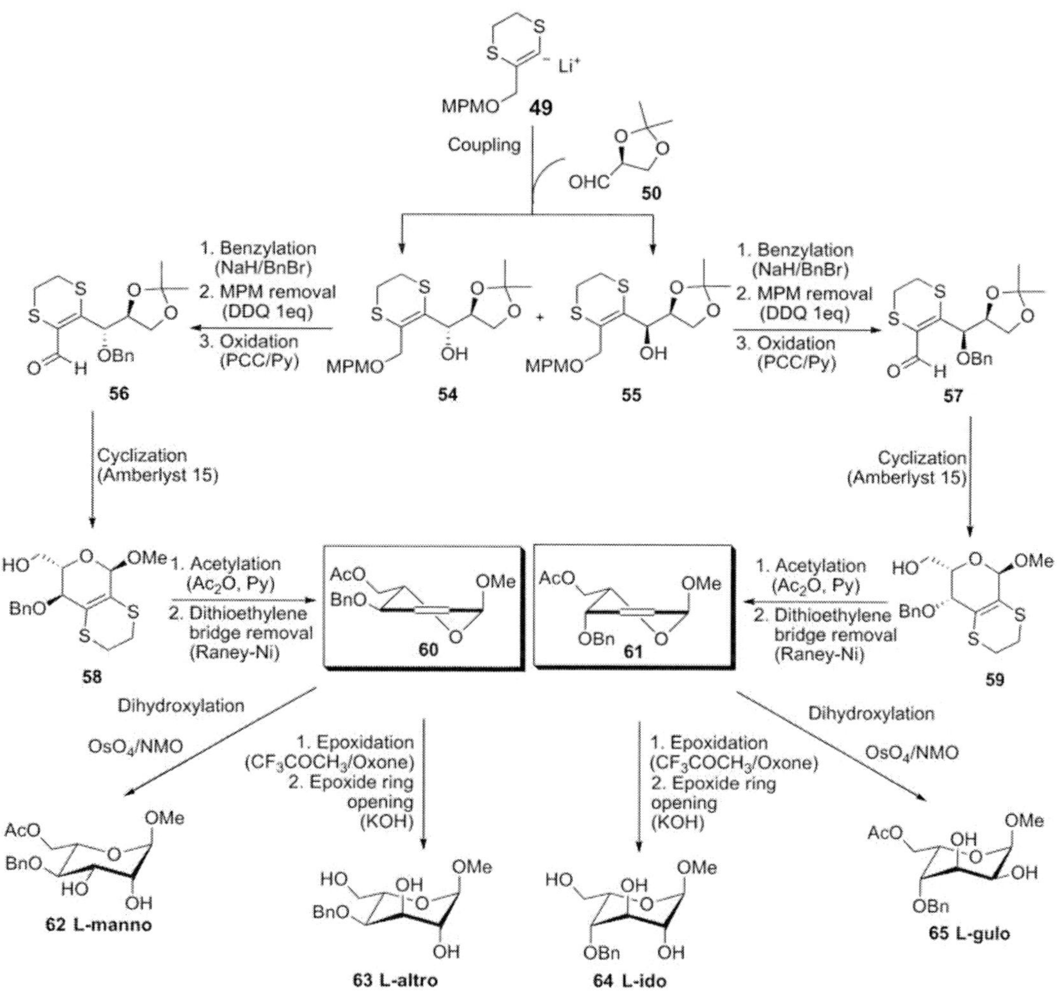


Scheme 5. Synthesis of L-hexoses with *manno*-, *altro*-, *ido*- and *gulo*-configuration.

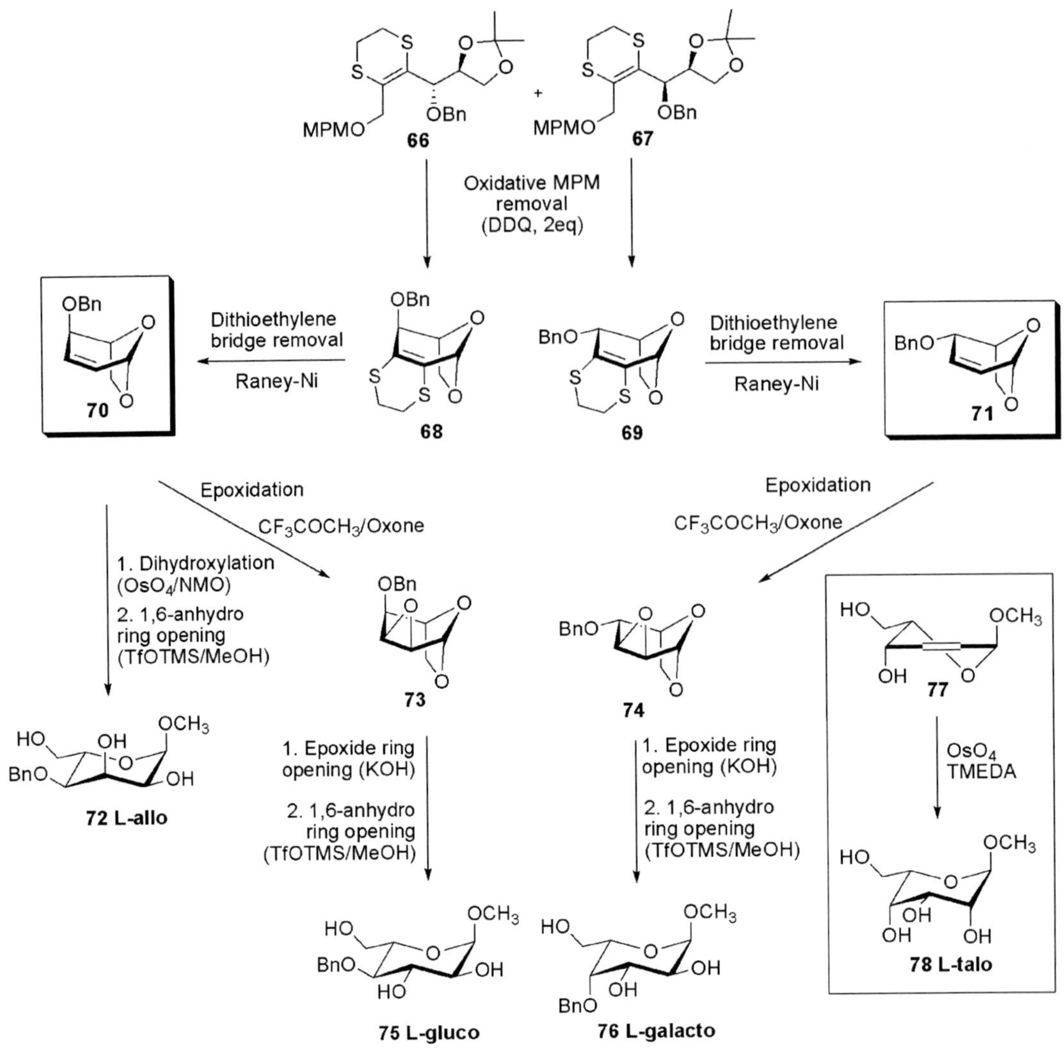


Scheme 6. Synthesis of L-hexoses with allo-, gluco-, galacto- and talo-configuration.

Finally, stereocontrolled dihydroxylation of free diol *77* afforded the remaining L-*talo*-hexopyranoside *78* (Scheme 6).

As shown, the current approach enables the synthesis of all eight epimers; on the other hand, replacement of aldehyde *50* with its D-enantiomer allows to prepare D-series hexoses. Moreover, as final compounds are orthogonally protected, they are promptly utilizable as building blocks for further synthetic manipulations.

As proof of the versatility of this methodology, several carbohydrate-related compounds were stereoselectively prepared, such as 4-deoxy D- and L-hexoses, [42a,45] deoxy-iminosugars, [46] and six-membered nucleoside analogues. [47]

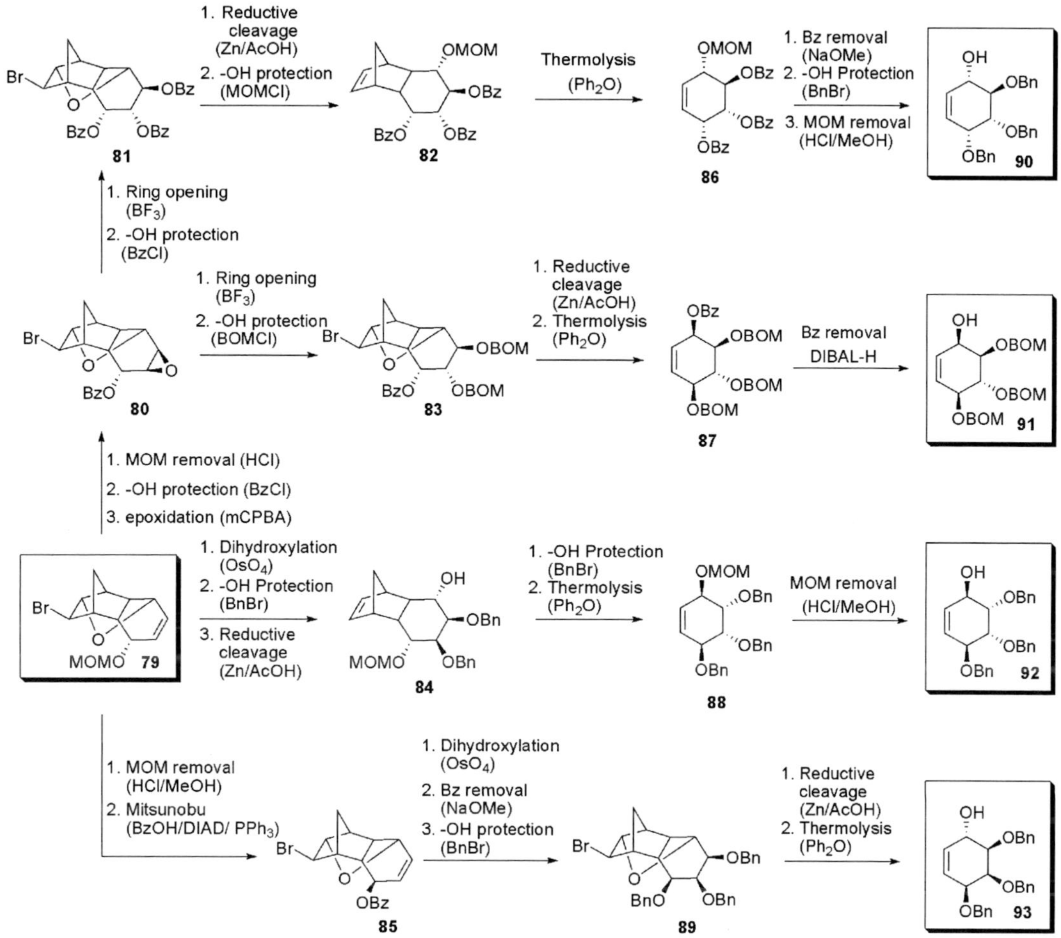

Scheme 7. Hexoses from a common cyclohexanoid chiral starting material: synthesis of key intermediates *90-93*.

## 2.3. Convergent Route to Enantiopure Aldohexoses from a Common Cyclohexanoid Chiral Building Block

An alternative substrate-controlled synthesis, virtually leading to all eight enantiopure aldohexoses, has been developed starting from a common cyclohexanoid chiral building block *79* (Scheme 7). [48] This key building block, which has been prepared by both enzymatic [49] and chemical [50] procedures, serves as chiral equivalent of a *cis*-1,4-dihydroxycyclohexane-2,5-diene, allowing diastereocontrolled construction of a variety of natural products on the basis of its molecular bias. [51] In a preliminary study, synthesis of four L- and D-aldohexose diastereomers (D-Glc, D-Gal, D-Man, D-All, L-Ido, L-Tal, L-Gul, L-Alt) has been reported *via* protected conduritols *90-93* (Schemes 7 and 8). More specifically, suitable dihydroxylation (OsO4/NMO) and epoxidation (*m*CPBA) reactions carried out on bromo-ether *79* enabled the synthesis of four cyclohexenol derivatives *90-93*, considering the reductive cleavage (Zn/AcOH) and the subsequent cyclohexene unmasking (Ph2O, reflux) on intermediates *79*, *81*, *83* and *89* as the key steps.

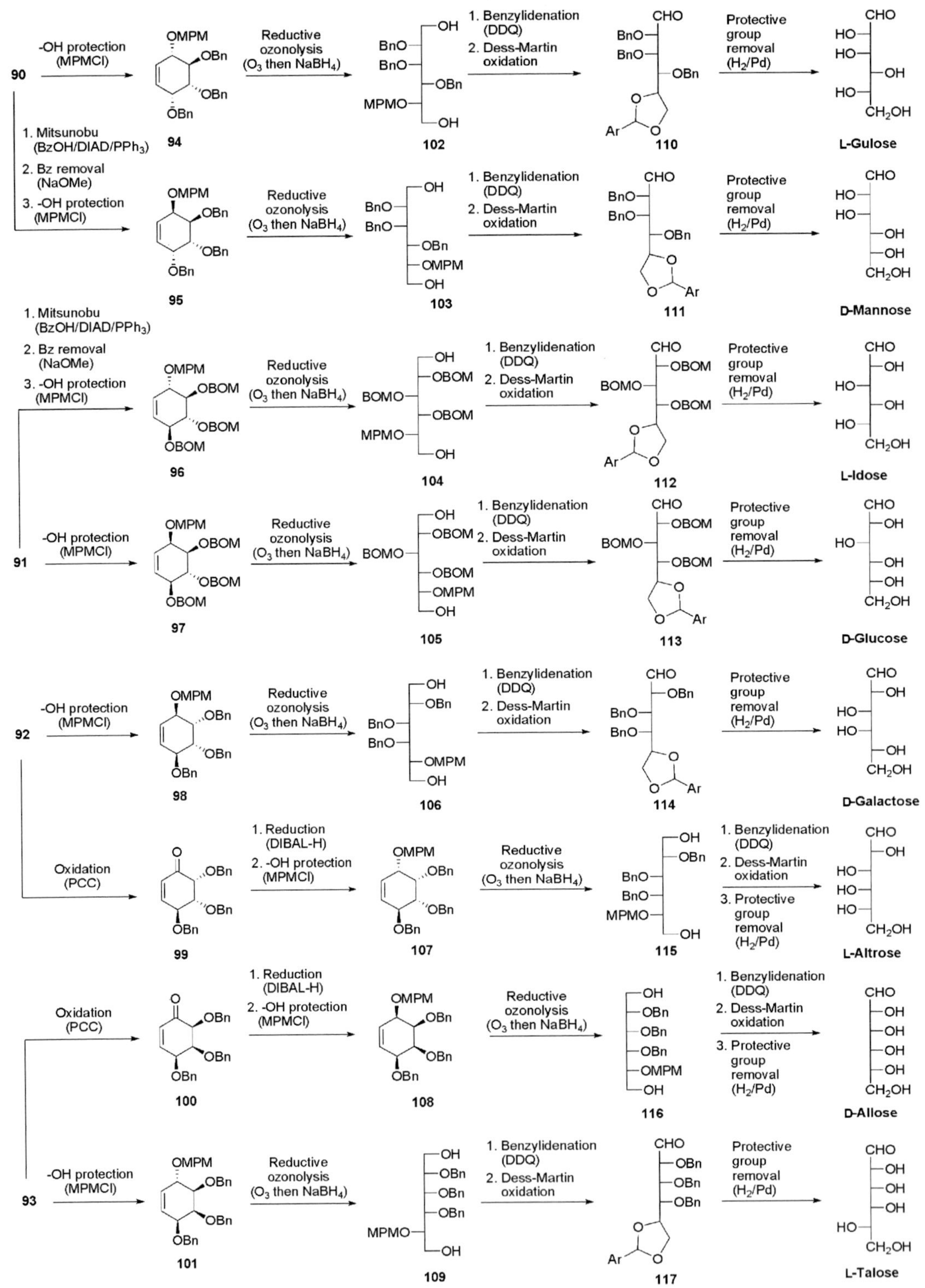


Scheme 8. Synthesis of D- and L-aldohexoses from conduritols *90-93*.

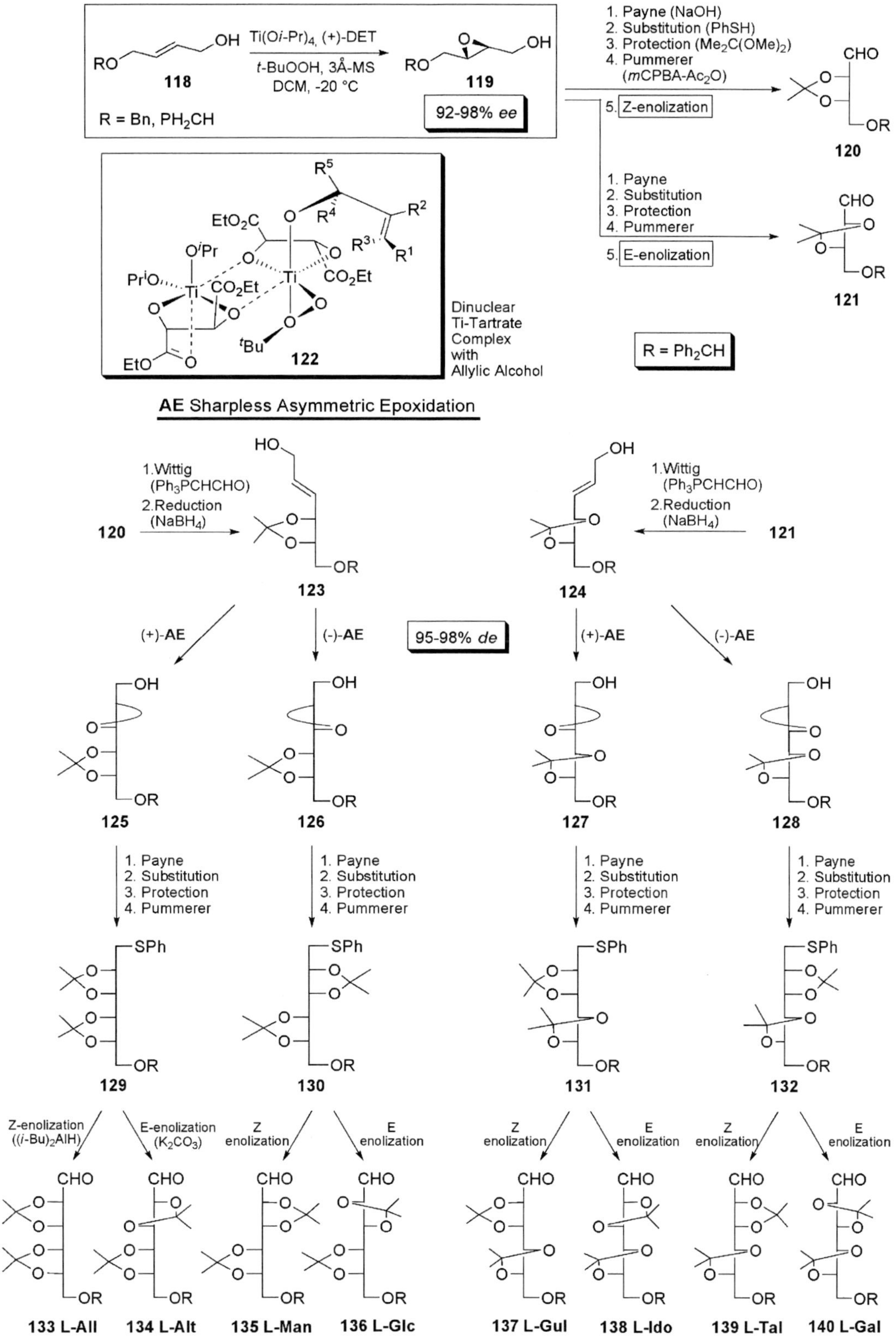


Scheme 9. The Sharpless approach to the synthesis of L-hexoses.

Conversion of conduritols *90-93* into aldohexoses was then carried out in a straightforward way (Scheme 8). Hydroxyl protection of compounds *90-93* afforded MPM derivatives *94*, *97*, *98*, *101*; on the other hand, inversion of configuration at the free hydroxyl function in intermediates *90*, *91* by Mitsunobu reaction (BzOH/DIAD/PPh3) followed by hydroxyl protection (NaOMe, then MPMCl) gave olefins *95* and *96*; conversely, oxidation/reduction procedure (PCC, then DIBAL-H) carried out on conduritols *92* and *93* enabled to prepare derivatives *107* and *108*. Reductive ozonolysis on protected cyclohexenols (O3 then NaBH4) generated diols *102-106*, *109*, *115* and *116*, then *C*-1 Dess-Martin oxidation and protective group removal afforded four L- and four D- free aldohexoses (L-Gul, L-Ido, L-Alt, L-Tal, D-Man, D-Glc, D-Gal and D-All).

Although versatility of the cyclohexanoid chiral building block *79* for the preparation of enantiopure aldohexoses has been herein demonstrated, a more general application of this method has been suggested, since enantiomers of molecules *90-93* may be also accessible, as the authors already reported, starting from the enantiomer of building block *79*. [49-50]

## 3. Enantioselective Synthesis

### 3.1. L-Sugars by Asymmetric Carbon Chain Elongation: the Early Approach

Masamune and Sharpless have provided a seminal and brilliant illustration of the potential of catalyst-induced double stereodifferentiation in their asymmetric approach to L-hexoses. [52] They accomplished the synthesis of the eight L-sugars by means of a reiterative reaction sequence involving the Katsuki-Sharpless asymmetric epoxidation (AE), the Payne rearrangement (NaOH), the Pummerer rearrangement (Ac2O) and the Wittig reaction as key steps (Scheme 9). The titanium-tartrate-oxidant-allylic alcohol complex *122* was firstly used [53] to generate the key chiral starting material *119* in high optical purity (92-98% *ee*) and subsequently for the diastereoselective elaboration (95-98% *de*) of the late-stage intermediates *123-124* [(+)-AE and (–)-AE were alternatively used]. Ultimately, all four contiguous stereocentres have been controlled independently, providing access to each of the eight hexoses *133-140* by essentially the same route [54] (Scheme 9).

The appeal of this methodology lies in the ready syntheses, in high enantiomeric excess, of sugars belonging to both series, depending on the use as chiral auxiliary of either the (-) or (+) enantiomer of diethyl tartrate (DET) in the first asymmetric epoxidation reaction.

### 3.2. Access to Hexoses via Sharpless Asymmetric Dihydroxylation of Furfural Derivatives

As a great deal of new reagents and methods has been built up, as consequence of profitable studies in organometallic chemistry, [55] many meaningful approaches have been developed for gaining access to enantiopure compounds containing a polyhydroxylated chain embedded in their structural framework.

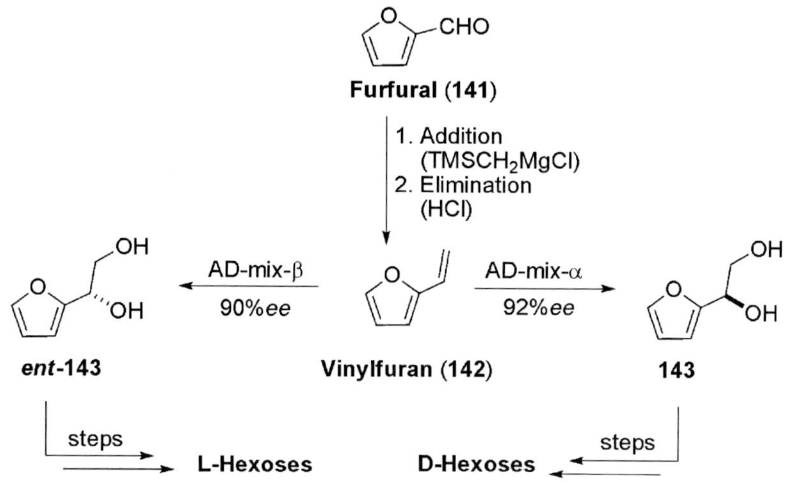


Scheme 10. Enantioselective approach to hexoses via Sharpless AD of vinylfuran (*142*).

Among the numerous examples of reagents reported so far, use of furfural (*141*) and its derivatives (Scheme 10) deserves a special attention because of its great synthetic potential originated from the appropriate manipulation of the five-membered ring. Compound *141* has already demonstrated to be a useful synthon for the assembly of numerous biologically relevant enantiopure complex molecules. [56]

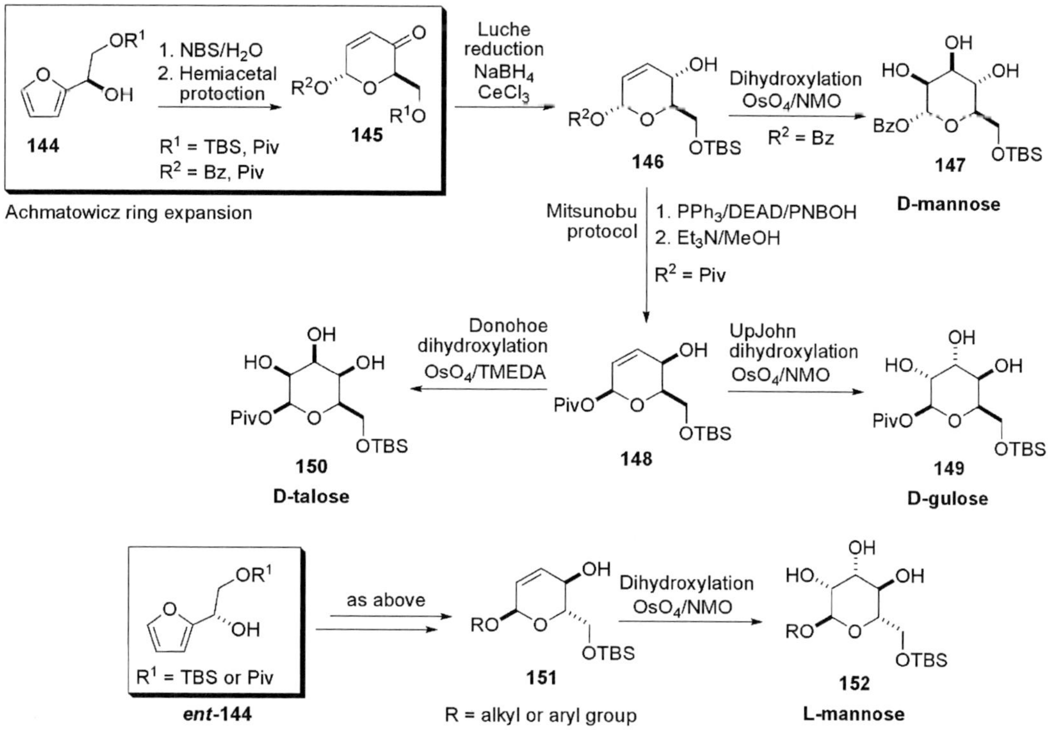


Scheme 11. Stereoselective synthesis of hexoses with talo-, manno- and gulo-configuration.

With the purpose to achieve a more efficient synthetic methodology for both D- and L-hexoses rather than traditional strategies, which start from available sugars and are based on a great number of tedious and unfavourable protection/deprotection steps, a new and convenient route has been devised. More specifically, the Sharpless asymmetric dihydroxylation (AD) reaction has been successfully applied to develop an intriguing and expeditious enantioselective approach for the *de novo* synthesis of natural and unnatural carbohydrates. [57] This approach relied upon the use of osmium chemistry on a furfural derivative, vinylfuran *142*, [58] to create all stereocenters of the desired monosaccharides (Scheme 10).

With the aim to test the breadth of such methodology, early experiments have been carried out starting from diol *143*, gaining entry to the less expensive D-sugars. Crucial to this approach was also the two-step route toward pyranone *145* from alcohol *144* (Scheme 11). Pyranone *145* was prepared by Achmatowicz ring expansion reaction followed by hemiacetal protection. In this case, the more reactive axial anomeric alcohols can be acylated selectively (>20:1) at -78 °C, whereas at room temperature a 1:1 mixture of anomers was produced.

As two key stereocentres were established on *145*, further synthetic modifications allowed introducing the remaining hexose functionalities (Scheme 11). Pyranones *145* were stereoselectively reduced under Luche conditions [59] (NaBH4/CeCl3) to give allylic alcohols *146* as the only observable stereoisomers. Then, alcohols *146* were diastereoselectively dihydroxylated to give D-*manno*-triols *147* with complete stereocontrol. Similarly, D-gulose and D-talose derivatives *149* and *150* were prepared from diastereomer intermediate *148* by respectively applying the Upjohn [60] and Donohoe's [61] conditions for the stereoselective dihydroxylation reaction (Scheme 11).

It's noteworthy to underline that the same synthetic path, carried out from *ent-144*, enabled preparation of L-mannose derivative *152* (Scheme 11); [62] similarly L-talose and L-gulose can be synthesized.

The versatility of this methodology has been further tested by synthesizing several monosaccharide-related molecules, such as deoxy-sugars and amino-sugars, [63,64] nucleoside analogues [65] and iminosugars. [66]

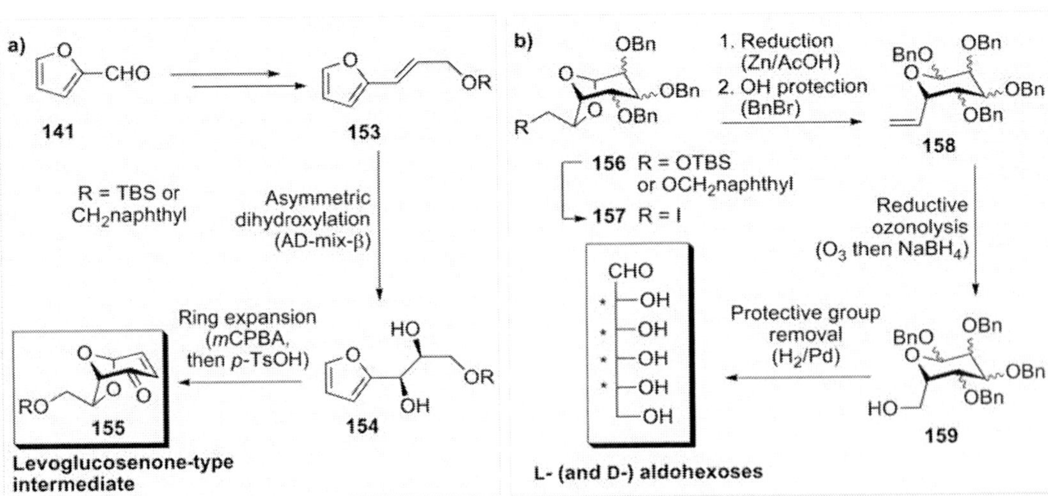


Scheme 12. Key stages of L- (and D-) aldohexose synthesis based on furfural.

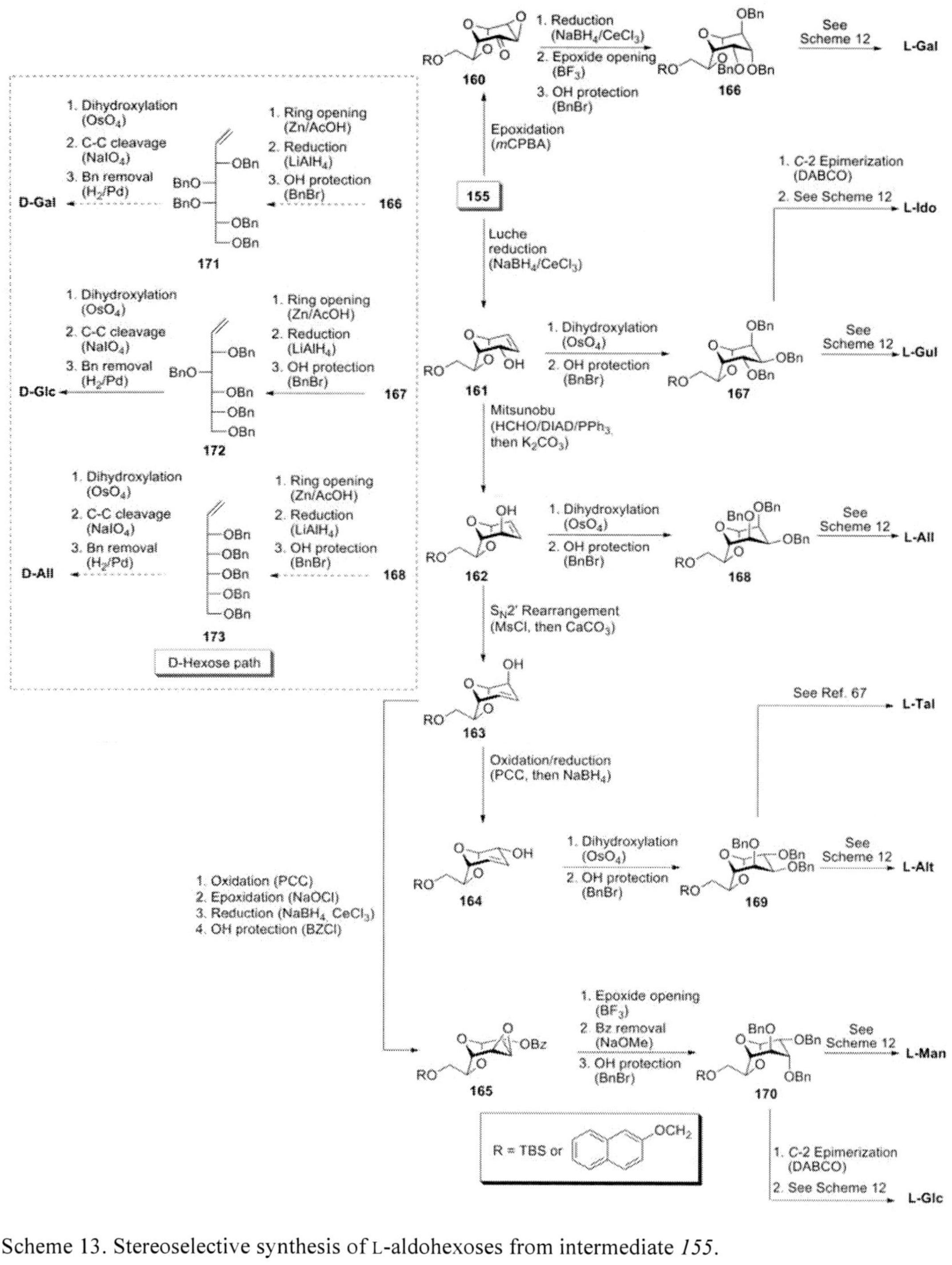


Scheme 13. Stereoselective synthesis of L-aldohexoses from intermediate *155*.

As further application of furfural-based building blocks, an integrated enantio- and diastereocontrolled synthesis of all eight L-aldohexoses (as well as their D-enantiomers) has been explored, [67] relying on two key stages: a) Sharpless AD reaction on furfural derivative *153*, then ring expansion to obtain the key enone *155* (Scheme 12, *left side*); b) conversion of compound *156* into intermediate *158* via iodide *157*, then reductive ozonolysis/hydrogenation to afford desired aldohexoses (Scheme 12, *right side*).

The main steps of this route are outlined in Scheme 13. The synthetic utility of levoglucosenone derivative *155* having an extra hydroxymethyl functionality lies in the opportunity to control the regioselective cleavage of bicyclic system and to discriminate the two terminal functionalities of the substrate. Enone *155* was used in two ways, since *C*-1 and *C*-6 positions can be inverted, with the aim to produce all eight hexoses, as well as their D-counterparts, from the same precursor (Scheme 13). In major detail, stereoselective epoxidation (*m*CPBA) of intermediate *155* and dihydroxylation (OsO4) of *161* allowed to respectively prepare compounds *166* and *167*, which led to L-galactose, L-gulose and L-idose. On the other hand, Mitsunobu inversion of *C*-4 hydroxyl group in *161* brought to the formation of *162*, leading to L-allose.

Moreover, regio- and stereoselective SN2′ type rearrangement on compound *162*, followed by an oxidation/reduction step gave derivative *164*, which *syn* and *anti* dihydroxylation respectively afforded compounds *169* and *170*, from which remaining L-talose, L-altrose, L-mannose and L-glucose were obtained.

As anticipated (Scheme 12), access to D-aldohexoses was also possible, as demonstrated by the synthesis D-glucose (D-galactose and D-allose can be prepared as well), obtained by sequential ring-opening/reduction/dihydroxylation/C-C bond cleavage procedure carried out on intermediates *167* (Scheme 13).

### 3.3. Hetero Diels Alder Cycloadditions: Useful Tools for Stereoselective Monosaccharide Construction

[4+2] Hetero Diels-Alder (HDA) reactions, particularly those carried out in an asymmetric environment, provide an efficient way to stereoselectively approach to the *de novo* synthesis of monosaccharides. This matter has been extensively reviewed in the past years. [68] From a theoretical standpoint, application of asymmetric HDA reactions, focused to prepare optically active carbohydrate derivatives, offers some essential advantages, compared to the manipulation of naturally occurring monosaccharides, as it allows synthesizing each carbohydrate enantiomer with a very short number of steps. [68b]

Among the existing approaches in [4+2] asymmetric HDA reactions, [69] couplings have been studied between oxa-1,3-diene derivatives *174* and homodienophiles such as *175*, leading to dihydropyrans *176*; alternatively, reactions have been examined between substituted 1,3-dienes, such as *177*, and heterodienophiles, like aldehyde *178*, giving adducts *179*; in both cases, *176* and *179* can be easily functionalized to give sugars in the desired manner (Scheme 14). The required asymmetry can be furnished by performing reactions in presence of chiral Lewis acids, or by using dienes or dienophiles equipped with appropriate chiral auxiliaries, which are removed in a late stage of the synthesis.

Several pioneering routes towards carbohydrates using HDA protocols have included furan, which can be considered as chemically equivalent to an 1,4-oxygenated 1,3-butadiene. [70] Reactions with dienophiles bearing an appropriate chiral auxiliary in presence of Lewis acids afforded diastereomeric cycloadducts in optically pure form, affording enantiopure carbohydrates (such as D- or L-allose, D- or L-talose [71]) in a few steps.

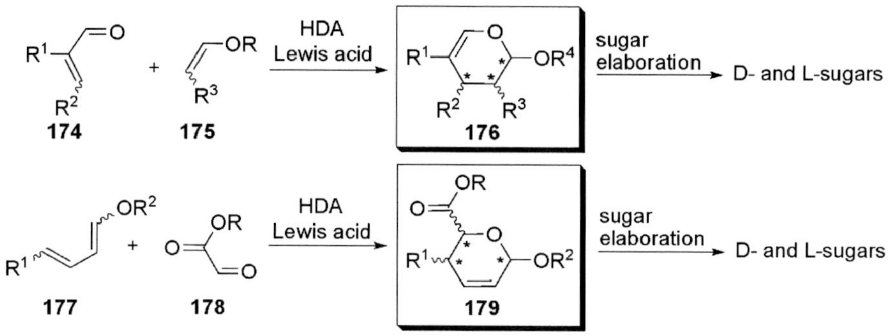

Scheme 14. Monosaccharide synthesis by hetero Diels Alder (HDA) cycloaddition.

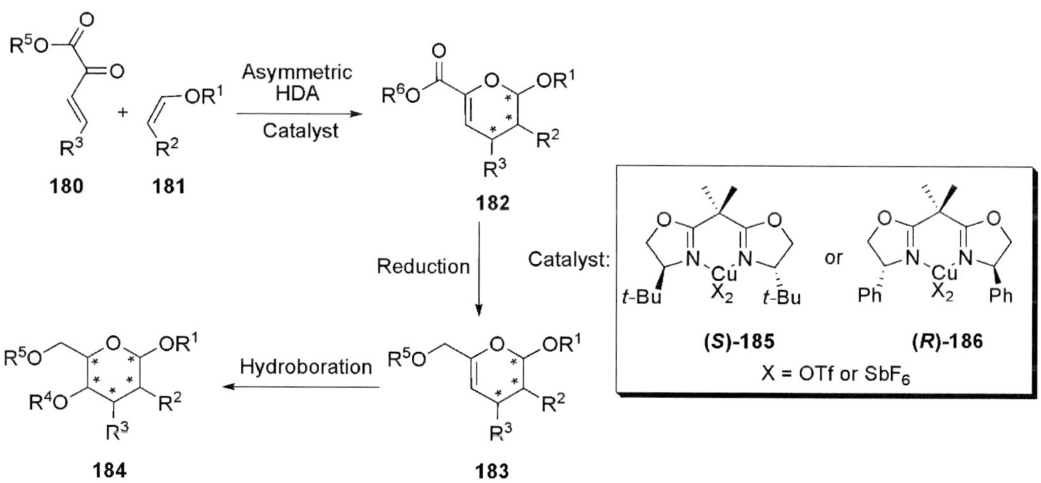

Scheme 15. Enantioselective carbohydrate synthesis by chiral Lewis acid catalysts *185* and *186*.

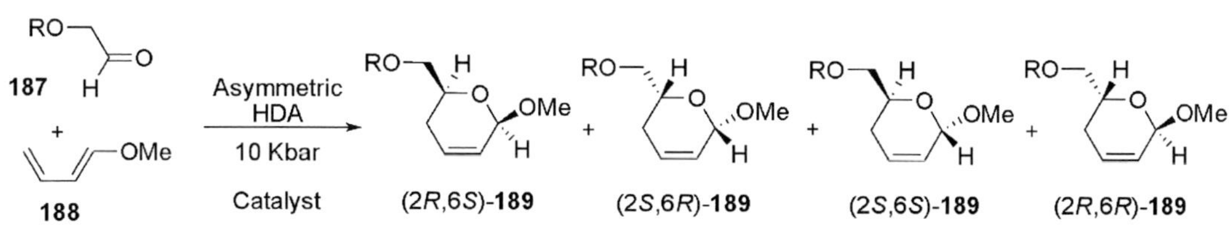

| R = $SiBu^tMe_2$ or COMe | Catalyst | (2*R*,6*S*)-**189** | (2*S*,6*R*)-**189** | (2*S*,6*S*)-**189** | (2*R*,6*R*)-**189** |
|---|---|---|---|---|---|
| | **190** | 29-32% | 24-27% | 20-22% | 21-55% |
| | **191** | 3-6% | 2-6% | 38-45% | 45-55% |

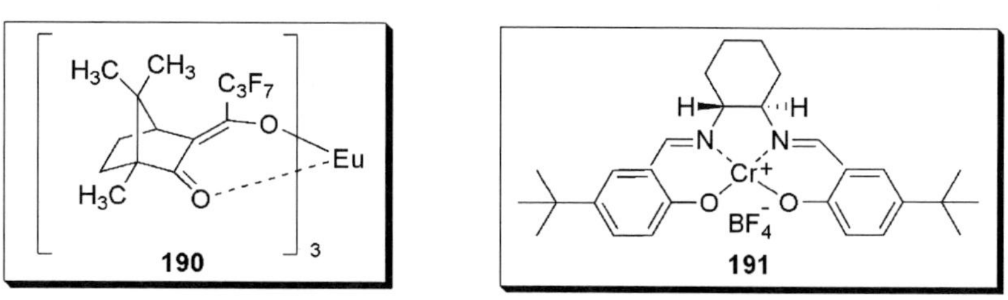

Scheme 16. Asymmetric cycloadditions by $Eu(hfc)_3$ *(190)* and Salen *(191)* catalysts.

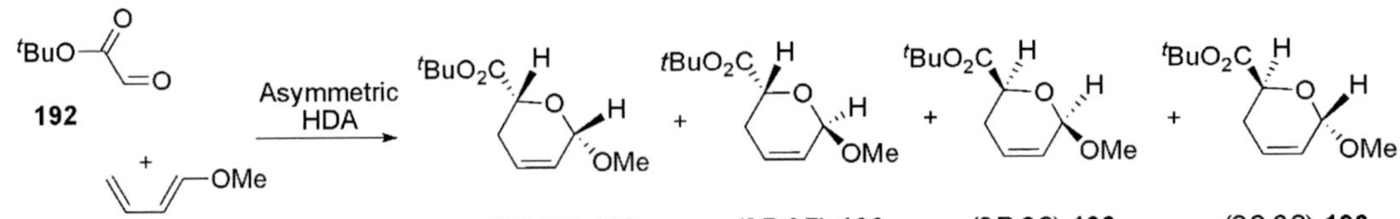


(2*S*,6*R*)-**193** (2*R*,6*R*)-**193** (2*R*,6*S*)-**193** (2*S*,6*S*)-**193**

| Catalyst | *cis:trans*-**193*** | *ee* for *cis*-**194*** | *ee* for *trans*-**194*** |
|---|---|---|---|
| **194a** | 28:72 | 24-30 (6*R*) | 16-20 (6*R*) |
| **194b** | 86:14 | 62-80 (6*R*) | 2-20 (6*S*) |
| **194c** | 87:13 | 76 (6*R*) | 18 (6*R*) |
| **194d** | 84:16 | 74 (6*R*) | 9 (6*R*) |
| **194e** | 8:2 | 47 (6*R*) | 28 (6*R*) |
| **195a** | 93:7 | 53-88 (6*S*) | 16-68 (6*S*) |
| **195b** | 64:36 | 73 (6*S*) | 52 (6*S*) |

* best reported results

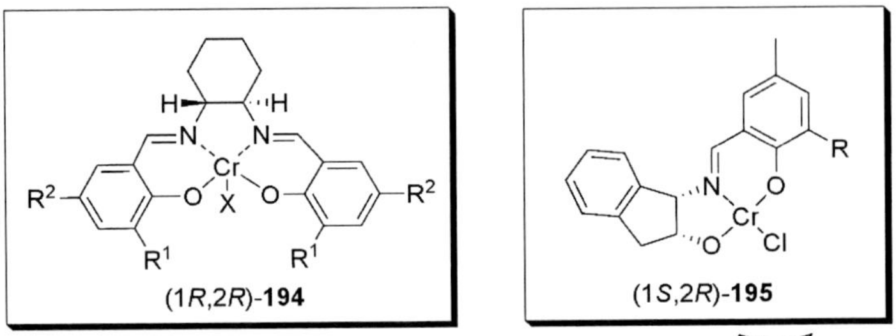

(1*R*,2*R*)-**194** (1*S*,2*R*)-**195**

**a**: $R^1 = R^2 = {}^tBu$, $X = BF_4$
**b**: $R^1 = R^2 = {}^tBu$, X = Cl
**c**: $R^1 = {}^tBu$, $R^2$ = Me, X = Cl
**d**: $R^1 = {}^tBu$, $R^2$ = OMe, X = Cl
**e**: $R^1$ = Me, $R^2 = {}^tBu$, X = Cl

**a**: R = 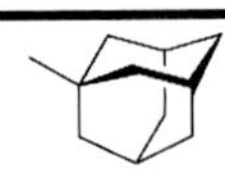

**b** R = *t*-Bu

Scheme 17. Diastereo- and enantioselective preparation of 2-enopyranosides by chiral chromium complexes.

In more recent times, endeavours have been carried out to develop catalytic enantioselective HDA reactions using chiral Lewis acids to induce the formation of chiral centres in one step. Independently, the groups of Evans [72] and Jørgensen [73] showed that β,γ-unsaturated α-ketoesters *180* reacted with ethyl vinyl ethers *181* in presence of enantiomerically pure bisoxazoline copper(II) catalysts *185* and *186*, leading to enantiomerically enriched dihydropyranes *182*, which after ester group reduction and double bond hydroboration afforded enantiopure hexoses *184* (Scheme 15).

Latest reports on this argument have explored several new ligands and reaction conditions to prepare convenient carbohydrate precursors. Resuming previous results carried out to reveal the influence of high pressure (10 kbar) in HDA reactions involving 1-methoxybuta-1,3-diene and various heterodienophiles, [74] Jurczak and coworkers have lately investigated [75] this topic in the [4+2] cycloaddition of 1-methoxybuta-1,3-diene (*188*) to glycoaldehyde derivatives *187* in an asymmetric environment induced by chiral catalysts such as Eu(hfc)3 (*190*) and (1*R*,2*R*)-(salen)Cr(III)BF4 (*191*) (Scheme 16). Application of both Eu and Salen catalysts afforded cycloadducts with unsatisfying enantiomeric excess. Indeed, Eu catalyst afforded, at relatively elevated temperature (50 °C), cycloadduct *189* with *endo*-selectivity and low enantiomeric excess. The same [4+2] cycloaddition, carried out in presence of Salen catalyst, gave cycloadduct *189* with very high *exo*-selectivity and much better, however still not satisfactory, enantiomeric excess.

In the effort to improve such selectivity, modifications of Cr(III) ligands have been developed. [76] Particularly, Salen-type complexes *194-195* (Scheme 17) were employed in reactions of 1-methoxybuta-1,3-diene (*188*) with alkyl glyoxylates such as *192*. Application of the less-active chloride complex *194b* and chiral tridentate Cr(III)Cl *195a* as catalysts afforded, under mild conditions, 2-enopyranoside cycloadducts *193* as convenient precursors of many natural products including D- and L-sugars, in variable yields (50-96%) along with high enantioselectivity (up to 88% ee).

When (1*R*,2*R*)-*194b-e* were used, predominance of cycloadduct (2*S*,6*R*)-*193* (L-sugar precursor) was observed, conversely when tridentate complex (1*S*,2*R*)-*195a* was used, the product with opposite chirality (2*R*,6*S*)-*193* (D-sugar precursor) was formed.

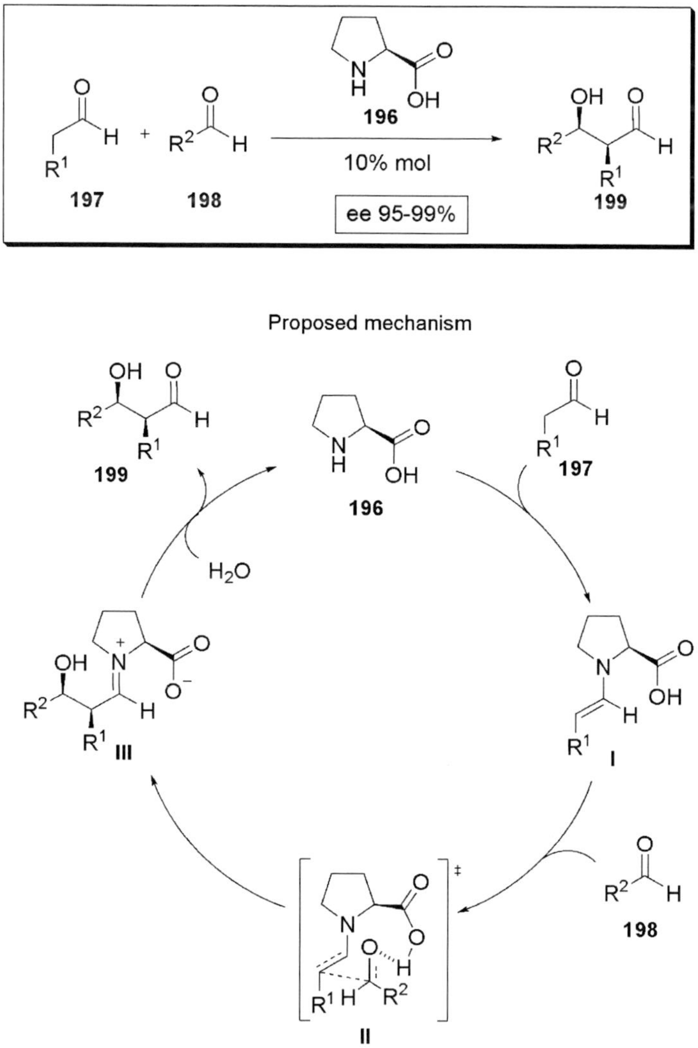


Scheme 18. Proposed mechanism for proline-catalyzed enantioselective aldol reaction.

### 3.4. Hexoses by Proline-Mediated Enantioselective Aldol Reaction

The concept of asymmetric organocatalysis by using simple amino acids has overwhelmingly emerged in the late years, as it can be applied to the immediate assembly of up to five stereogenic centers, and therefore devoted to the construction of a plethora of chemically different chiral structures. [77] With regard to the synthesis of polyhydroxylated compounds, unprotected (*S*)-proline *196* (and their derivatives) has been found to catalyze aldol reactions between non-chiral enolizable aldehydes *197* and *198* in a highly enantioselective way (Scheme 18), allowing asymmetric *de novo* synthesis of L- (and D-) hexoses and their analogues with excellent overall stereoselectivity (in some cases, the new stereocentres were assembled with almost absolute stereocontrol).

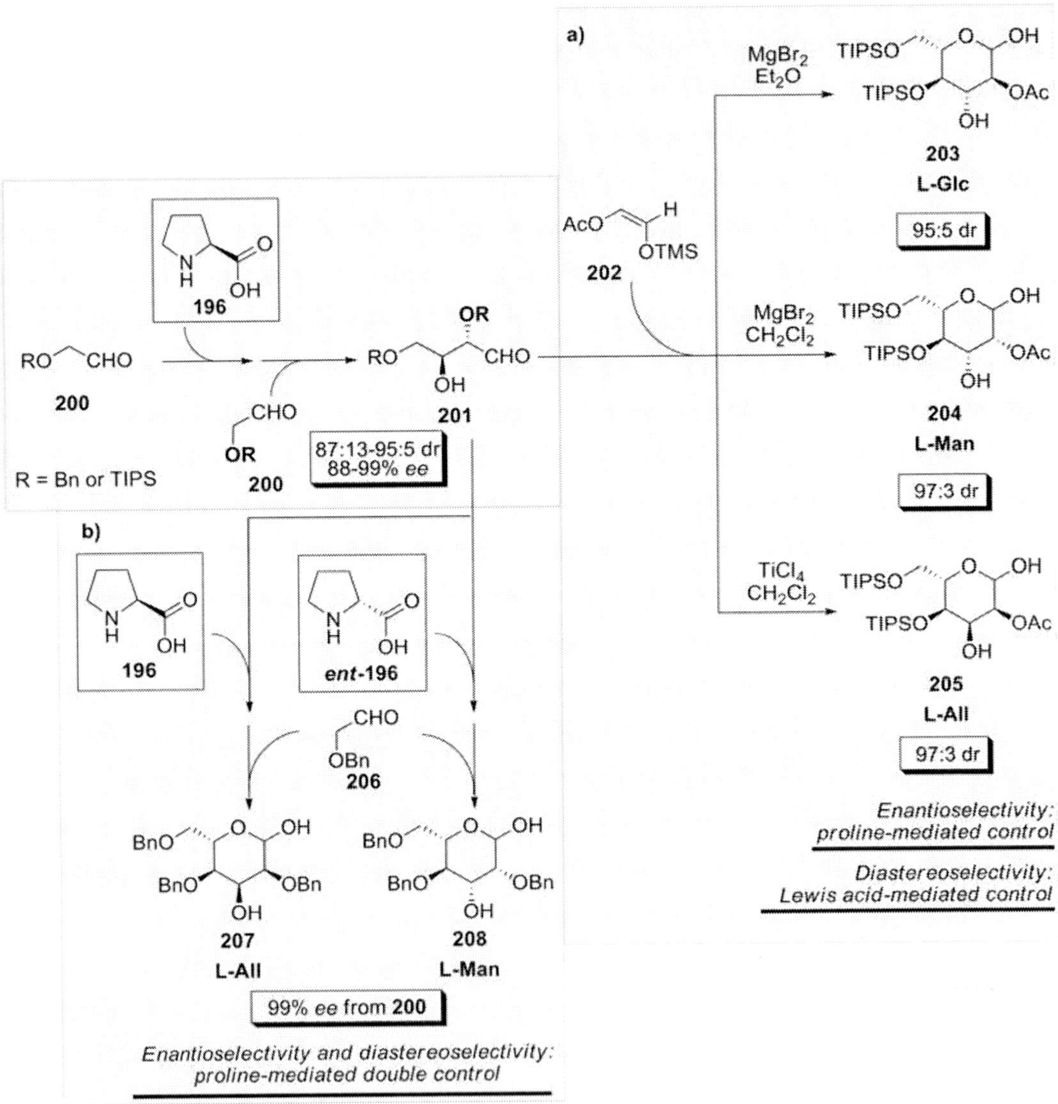


Scheme 19. Proline-catalyzed carbohydrate synthesis: (a): AA-mediated enantiomeric control and Lewis acid-mediated diastereomeric control. (b): synthesis of carbohydrates under AA-mediated enantio- and diastereomeric control.

From a conceptual standpoint, monosaccharide synthesis can be envisioned as a two-step iterative aldol sequence using simple α-oxyaldehydes such as *200* (Scheme 19). In the first practical applications of this concept, [78] aldehydes *200* were converted, in presence of (*S*)-proline (*196*), into corresponding aldol products *201* with high *ee* values (Scheme 19). Diastereoselective ratios varied in the range of 87:13-95:5 in favour of the *anti*-product. Interestingly, further reaction of *201* with starting material *200* was not observed. Hence, alcohols *201* were subjected to a second aldol reaction (with another substrate). Following this route, partially protected carbohydrates were obtained from the reaction of L-tetrose *201* with protected silyl enol ethers *202* (Scheme 19). At this stage, the stereochemical outcome of the reaction was under control of reaction conditions. For instance, employ of MgBr2 in Et2O led to L-glucose derivative *203*, whereas corresponding L-mannose derivative *204* was formed in CH2Cl2. Conversely, in presence of TiCl4, protected L-allose *205* was obtained. [79]

Very recently, such two-step reaction has been realized under complete proline control. The aldol product *201* obtained in an (*S*)-proline-catalyzed reaction was submitted to a second aldol reaction, again catalyzed by (*S*)- or (*R*)-proline (*196* or *ent-196*), respectively giving carbohydrate derivatives *207-208* (Scheme 19). Independently by the aldehydes used, enantioselectivities were excellent (99% *ee* or higher). If catalysts were used in the opposite sequence, the enantiomeric products were formed. [80] Because of the inherent simplicity of this process, a catalytic prebiotic "gluconeogenesis" has been suggested, in which amino acids may have transferred their stereochemical information to sugars. [81]

## 4. D-SUGARS EPIMERIZATION BY CHEMICAL REACTIONS

One of the most expedient and inexpensive ways to prepare L-monosaccharides in high optical purity is attained by inversion of configuration at *C*-5 stereocentre of the easily and commercially available D-monosaccharides. Following this route, L-idose can be afforded from D-glucose, L-gulose from D-mannose, L-altrose from D-galactose, L-talose from D-allose and L-allose from D-talose, as depicted in Figure 5.

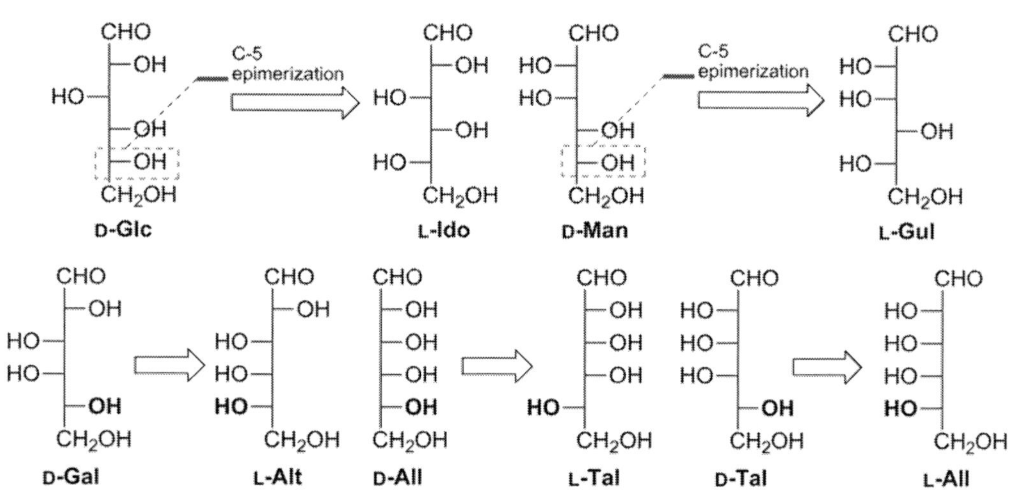


Figure 5. L-sugars via *C*-5 epimerization of D-sugars.

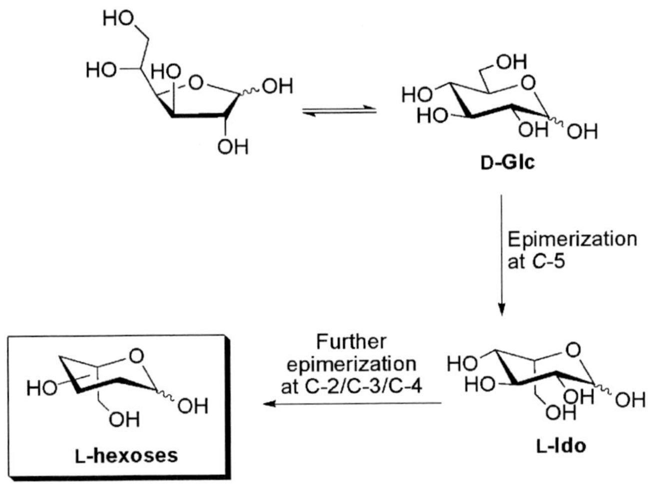


Figure 6. Regioselective epimerization of D-sugars.

### 4.1. L-Hexoses by D-Glucose Epimerization

On the basis of these findings, an attractive approach for the synthesis of L-hexoses has been achieved by sequential and specific epimerization at different positions of starting D-glucose (Figure 6). [29b,82] The method involves the early conversion of carbohydrate derivatives with D-*gluco* configuration into derivatives with L-*ido* configuration; then regioselective epimerization of L-idose at *C*-2, *C*-3 and/or *C*-4 positions furnished corresponding L-hexoses.

With this aim, D-glucose protection with double ketal fixation on the 1,2- and 3,5-dihydroxy groups afforded the *cis-anti-cis*-fused tricyclic D-gluco-furanose, namely diacetone-α-D-glucose *209*, [83] which in turn underwent elimination by consecutive treatment with triphenylphosphine (PPh3), *N*-bromosuccinimide (NBS) and 1,8-diazabicyclo[5.4.0]undecene-7 (DBU) to yield enol ether *210* with the required 5-*exo*-double bond (Scheme 20). The reaction took place in a one-pot manner via a tandem isopropylidene rearrangement, regioselective *C*-6 bromination, and β-elimination. Due to the steric hindrance on the α-face of the olefin, hydroboration proceeded with high selectivity, affording corresponding alcohol *211*. Finally, acidic hydrolysis of *211* (aq H2SO4) gave L-idose [84]. It is noteworthy to underline that compound *211* represented a common valuable intermediate for both synthetic routes depicted in Schemes 20 and 21: indeed, appropriate elaboration of its backbone led to various L-hexoses, such as L-talose, L-altrose, L-mannose, L-*allo*- and L-*gulo*-hexopyranoses.

The first required step (Scheme 20) was regioselective hydrolysis of *211* followed by 5,6-*O*-isopropylidenation to furnish alcohol *212*, which was the common precursor for the preparation of L-talose, L-altrose and L-mannose.

L-Talose is a *C*-3 epimer of L-idose, therefore a simple oxidation-reduction procedure (PDC, Ac2O, then NaBH4) for inversion at *C*-3 position of compound *212* was necessary. Then, treatment of 1,2:5,6-di-*O*-isopropylidene-β-L-talofuranose *214* with Amberlite-120 acidic resin smoothly afforded L-talose. Regarding L-altrose, the only stereochemical difference between the latter and L-idose is configuration at *C*-4 position. Epimerization of alcohol *212* to corresponding L-altrofuranosyl sugar *216* was accomplished via stereoselective hydroboration of olefin *215*. Acidic hydrolysis of isopropylidene groups in presence of Amberlite-120 resin led to L-altrose (Scheme 20).

Finally, simultaneous epimerization of *212* at both *C*-3 and *C*-4 chiral centres by stereoselective hydrogenation of enol acetate *217* afforded L-mannofuranosyl derivative *218*. Deacetylation and acidic hydrolysis of *218* gave then L-mannose.

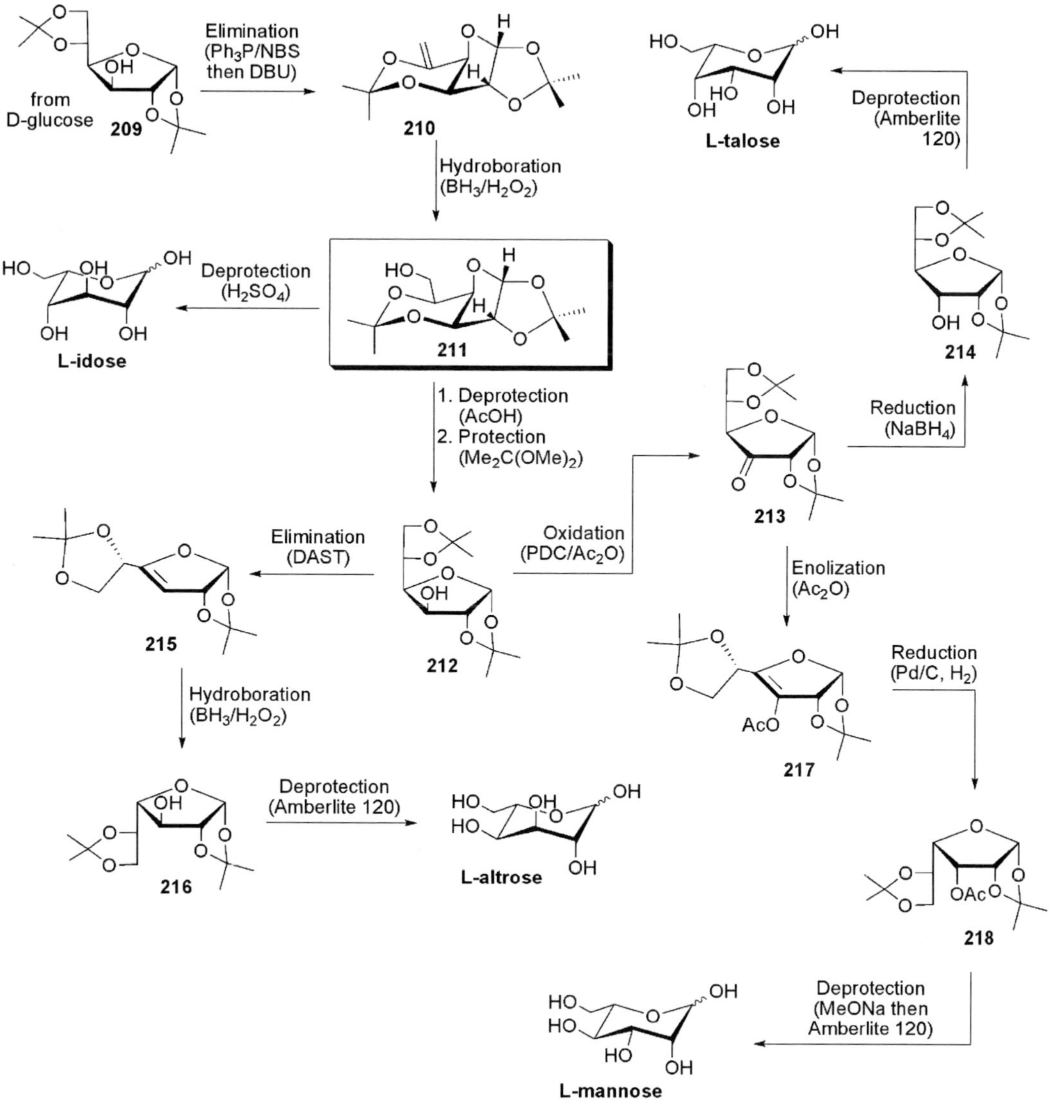


Scheme 20. Epimerization of diacetone-α-D-glucose: synthesis of L-idose, L-mannose and L-altrose.

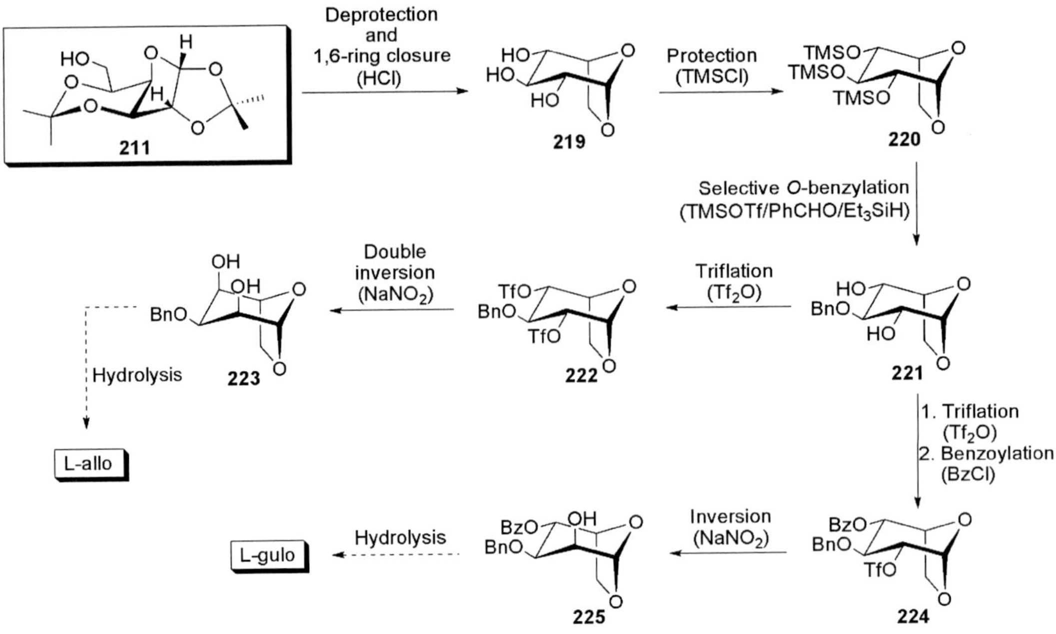


Scheme 21. 1,6-Anhydro sugars as key intermediates in the synthesis of L-gulose and L-allose.

In the endeavour to complete the series, a further interesting approach has been proposed (Scheme 21) exploiting conversion of intermediate *211* into 1,6-anhydro-β-L-idopyranose (*219*) [85] by refluxing with ethanolic HCl solution. Regioselective protection and selective epimerization of each centre in [3.2.1]bicyclic backbone have opened the way to the synthesis of two new sugars, i.e. protected 1,6-anhydro-β-L-allopyranose *223* and protected 1,6-anhydro-β-L-gulopyranose *225*.

To obtain L-*allo* derivative *223* two inversions, at *C*-2 and *C*-4 positions on the protected L-idofuranose *211* are required. Regioselective 3-*O*-benzylation of synthon *219* using TMSOTf/Et3SiH/PhCHO reductive etherification procedure of its *O*-trimethylsilylated ether *220* afforded corresponding 3-*O*-benzyl derivative *221* with good selectivity. [86] Subsequent triflation and nucleophilic substitution of diol *221* via 2,4-di-OTf *222* intermediate furnished corresponding L-*allo* 2,4-diol *223*. Alternatively, consecutive triflation and benzoylation of *221* and nucleophilic substitution of 2-OTf derivative *224* led to 1,6-anhydro-β-L-gulopyranosyl derivative *225*. Acidic hydrolysis of 1,6-anhydro derivatives *223* and *225*, as already reported for similar substrates [82], can readily afford L-*allo* and L-*gulo* hexoses.

## 4.2. L-Hexoses from D-Glycono-1,5-lactones

Taking advantage from the structural relationships between D-glucose and L-idose, D-galactose and L-altrose, and D-mannose and L-gulose, Ikegami et al. have suggested an easy method [87] to efficiently synthesize these three L-monosaccharides through a simple and practical four-step procedure. The synthesis is based on the intramolecular cyclization of δ-hydroxyalkoxamate derivatives of D-glycono-1,5-lactones by Mitsunobu reaction. In spite of its broad use in organic synthesis, this procedure represents the first application of the Mitsunobu-type cyclization devoted to the synthesis of carbohydrates.

δ-Hydroxyalkoxamic derivatives *227a-c* (Scheme 22) were prepared by reaction of D-glyconolactones *226a-c* with benzyloxyamine. Use of a Lewis acid such as trimethylaluminum (Me3Al) accelerated the reaction remarkably. Key to the successful production of δ-hydroxybenzyloxamates resulted the preparation of a glycono-1,5-lactone/benzyloxamine mixture before the addition of Me3Al.

Cyclization of *227a-c* under Mitsunobu conditions proceeded rapidly, and resulted mainly in the *O*-alkylation rather than the *N*-alkylation products (with the exception of mannose derivative *226b*, which produced the only *O*-cyclized compound) all with inversion of configuration at *C*-5 position. Mitsunobu cyclization of δ-hydroxybenzyloxamates can be rationalized assuming that formation of anion intermediates *228* and *229* were generated via deprotonation of the amide group by the reduced DEAD anion present in the reaction mixture (Scheme 22).

Acid-catalyzed hydrolysis of oximes *230a-c* (Scheme 23) to the parent L-glycono-1,5-lactones *231a-c* and their subsequent reduction provided the expected 2,3,4,6-tetra-*O*-benzyl-L-hexopyranoses, namely protected L-idose (*232a*), L-gulose (*232b*) and L-altrose (*232c*). On the other hand, intramolecular alkylation of intermediate *229* resulted in *N*-cyclization; this reaction has been recently investigated to prepare L-iminosugar precursors. [88]

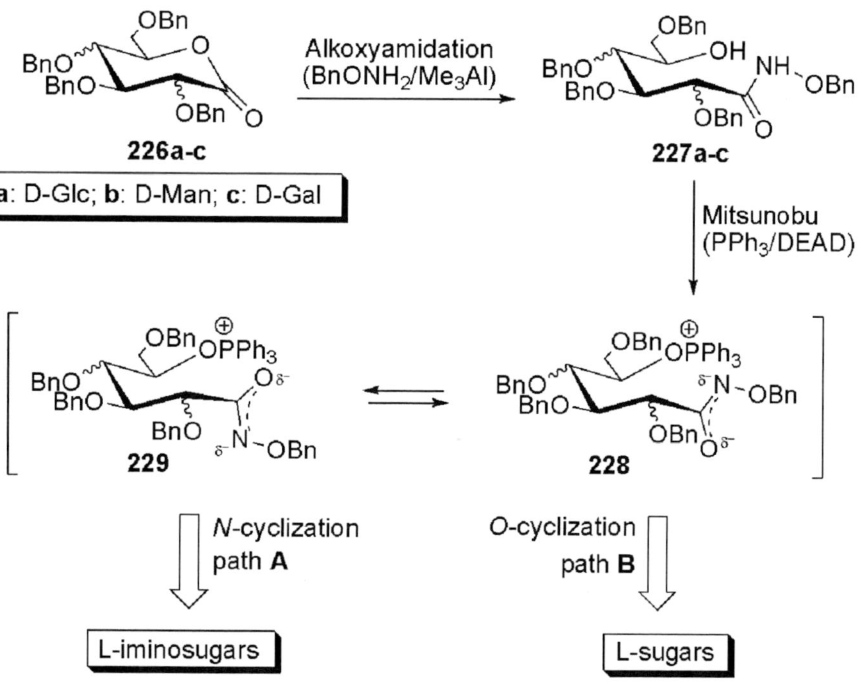


Scheme 22. Intramolecular *N*- and *O*-alkylation of alkoxamates.

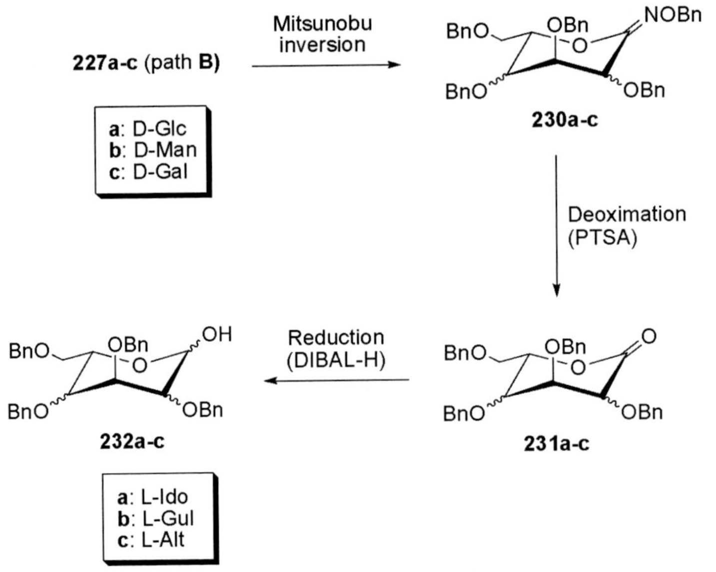


Scheme 23. Cyclization of the δ-hydroxyalkoxamates and synthesis of L-sugars.

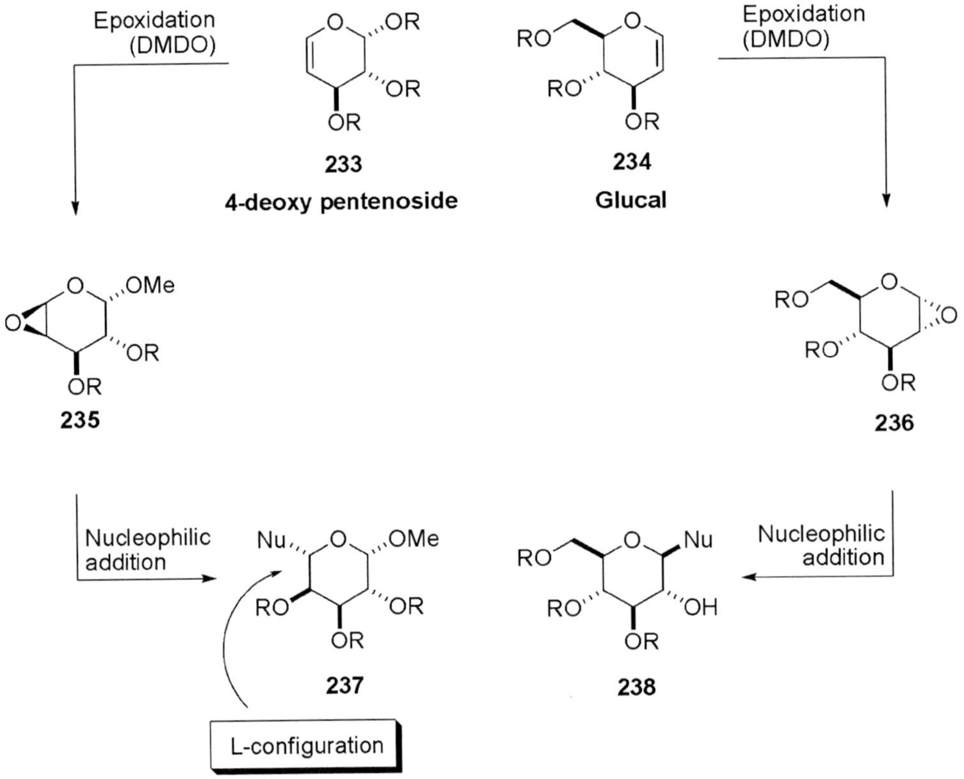


Scheme 24. Chemical resemblance between 4-deoxypentenosides and glycals.

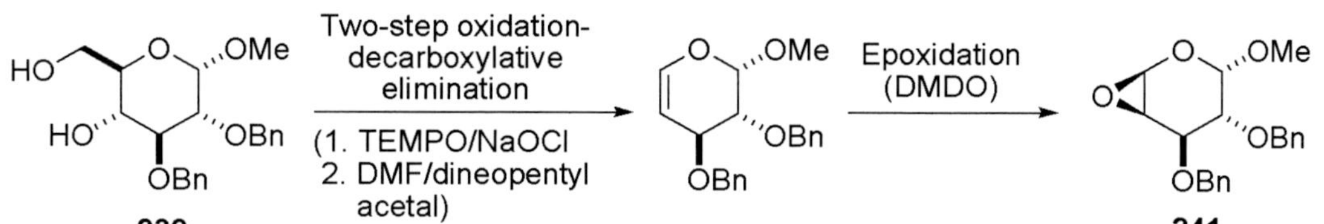


239 240 241

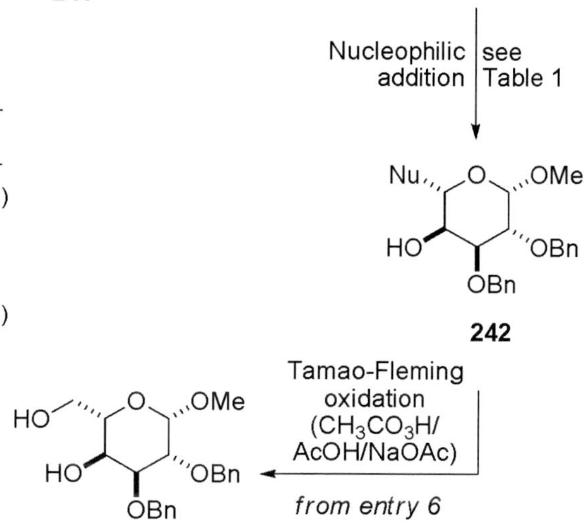


242

243 L-altro

**Table 1**. Epoxide Nucleophilic Ring-Opening

| entry | nucleophile | Reaction conditions |
|---|---|---|
| 1 | $^{13}CH_3MgI$ | Nu (3.5 eq), CuI (1.5 eq.) THF, -10 °C |
| 2 | 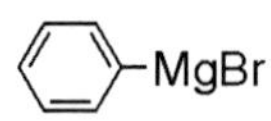 | " |
| 3 | 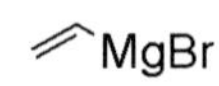 | Nu (3.5 eq), CuI (0.1 eq.) THF, -10 °C |
| 4 | 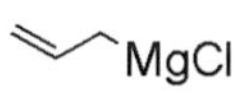 | " |
| 5 | 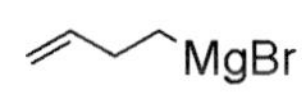 | " |
| 6 | $PhMe_2Si$⌒MgCl | " |
| 7 | KCN | Nu (10-20 eq), aq DMF, -10 °C |
| 8 | $NaN_3$ | " |
| 9 | 4-MePhSLi | Nu (9.0 eq), THF, 0 °C |
| 10 | $LiAlD_4$ | Nu (5.0 eq), $Et_2O$, rt |

Scheme 25. L-Pyranosides via 4-deoxypentenosides.

## 4.3. L-Sugars from 4-Deoxypentenosides

An intriguing approach to L-pyranosides via 4-deoxypentenosides *233* (4,5-unsaturated pentopyranosides) has been lately introduced. [89] These unsaturated sugars hold a strong resemblance to glycals such as *234*, which represent widely employed intermediates in the synthesis of a variety of natural products, [90] and generally used to provide a direct access to a broad range of natural and unnatural sugars and related glycoconjugates with medicinal value. [91]

Indeed, 4-deoxypentenosides and glycals exhibit analogous reactivity profiles; for instance, both compounds can be similarly epoxidized in stereoselective manner by dimethyldioxirane (DMDO). Then, in case of *235*, epoxide-ring opening with carbon nucleophiles furnishes sugar-like molecules with inversion of configuration (Scheme 24).

4-Deoxypentenoside *240* was prepared from the corresponding methyl α-D-glucoside *239* by a two-step oxidation-decarboxylative elimination procedure [92] (Scheme 25). Epoxidation of *240* performed by means of DMDO/acetone system proceeded with fairly good β:α selectivity (approximately 10:1). However, considerable limitations arose on the choice of the reagents and reaction media, due to the sensitivity of the resulting 4,5-epoxypyranosides to acidic hydrolysis that precluded purification by silica gel chromatography.

β-Epoxide ring-opening was explored with a broad set of nucleophiles, under SN2 conditions; in most cases it proceeded with complete regioselectivity and with inversion of stereochemistry at *C*-5 position of starting glucose, leading to corresponding alcohols *242* with L-*altro* configuration as the major products. Several compounds obtained (Table 1, *entries 3,6,7*) were amenable to be converted to fully hydroxylated L-hexopyranosides; for example, Tamao-Fleming oxidation [93] on dimethylphenylsilane *242* (Nu = CH2SiMe2Ph) yielded corresponding L-altropyranoside *243* (Scheme 25).

## 5. L-Hexoses by Enzymatic Reactions

Enzymes are nowadays recognized as valuable catalysts for organic synthesis. Considering their ability to perform reactions in a stereoselective way, the mild conditions used, the possibility to keep protecting group chemistry to a minimum, their employ in synthesis has become a valuable addition to the existing conventional methods. The high chemo-, regio-, diastereo- and enantioselectivity, as well as catalytic efficiency make enzyme-based technologies especially useful for large-scale synthesis of chiral intermediates and complex, highly functionalized molecules like carbohydrates and their analogues. [94]

In an early approach, synthesis of monosaccharides from acyclic precursors has been developed by using type I and/or type II aldolases. [95] The former class is usually found in higher plants and animals, whereas the latter is predominantly encountered in bacteria and fungi and requires a Zn2+ cofactor as Lewis acid in the active site [96] (Scheme 26). Over 30 aldolases have been identified and isolated, most of which catalyzing reversible stereoselective addition of ketone donors to aldehyde acceptors. Although aldolases were known to be quite specific for the nucleophilic donor component, a high degree of flexibility was allowed for the acceptor aldehyde component, leading to the synthesis of numerous natural and unnatural carbohydrates. [97]

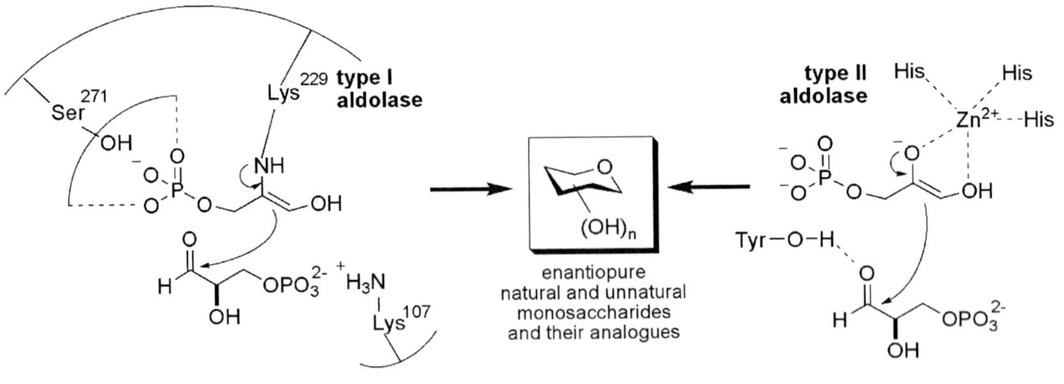


Scheme 26. Aldolase-mediated synthesis of monosaccharides.

In recent times, Izumori has developed a new concept [98] for the bioproduction of all hexose sugars (16 aldohexoses, 8 ketohexoses and 10 hexitols) starting from natural and cheap sugars as D-glucose and D-fructose and using enzymatic and microbiological reactions. The strategy involves several biotransformations between ketohexoses, hexitols and hexoses, making use of three biocatalysts necessary for the key reactions.

*D-Tagatose 3-epimerase* (DTE) was used to reverse stereochemistry at *C*-3 position of all 2-ketohexoses. [99] This enzyme afforded D-sorbose from D-tagatose, D-psicose from D-fructose [100] (Scheme 27, *left side*). In addition, L-ketohexoses were epimerized as well, giving L-fructose from L-psicose and L-tagatose from L-sorbose. [101]

*Polyol dehydrogenases (oxidoreductases)* (PD) were used to catalyze oxidation/reduction reactions between ketohexoses and the corresponding hexitols, affording D-tagatose from D/L-galactitol and L-psicose from D/L-allitol. [102] Using the reverse reactions of polyol dehydrogenases, various rare hexitols were prepared from ketohexoses: L-talitol from L-tagatose, L-allitol from L-psicose, L-talitol from L-psicose and L-iditol from L-sorbose. [103] Furthermore, considering the structural characteristics between hexitols (for instance, D-glucitol is identical to L-gulitol, D-galactitol to L-galactitol and D-gulitol to L-glucitol), several epimerization reactions between ketohexoses were carried out: for example, L-tagatose was obtained from L-psicose (Scheme 27, *in the middle*); similarly, D-sorbose was obtained from L-fructose.

*Aldose isomerases* (AI) were utilized to catalyze isomerization reactions between aldoketoses and the corresponding aldohexoses. Thus, epimerization of L-psicose furnished L-allose and L-altrose, whereas epimerization of L-tagatose gave L-talose and L-galactose (Scheme 27, *right side*).

On the basis of these data, Izumori has designed a map (Scheme 28) for the production of aldohexoses. The map enables to establish the requisite sequence to afford each desired sugar. For example, production of L-glucose from D-glucose was obtained following the appropriate shortcut (Scheme 28, *route I, II, III, IV, V and VI*).

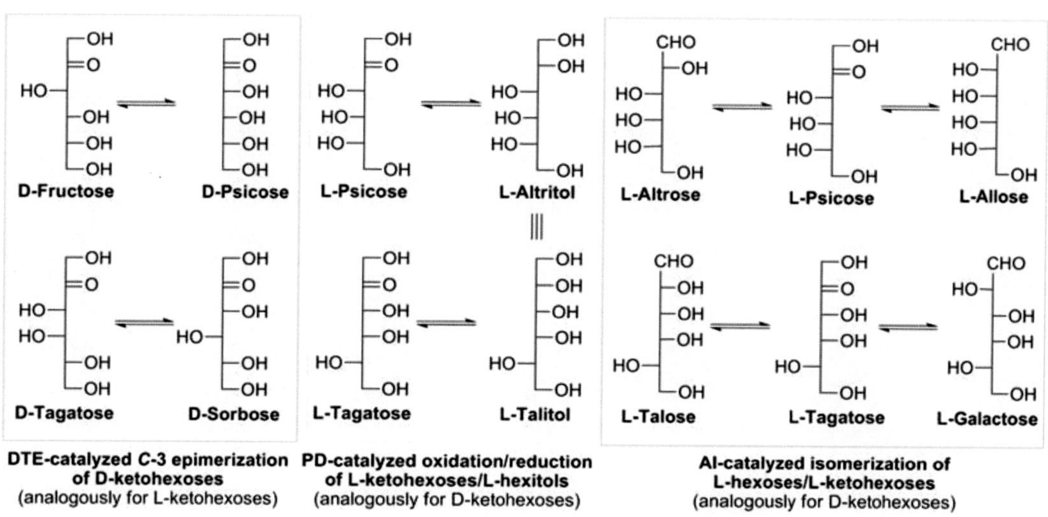


Scheme 27. Biotransformations between hexitols, ketohexoses and hexoses.

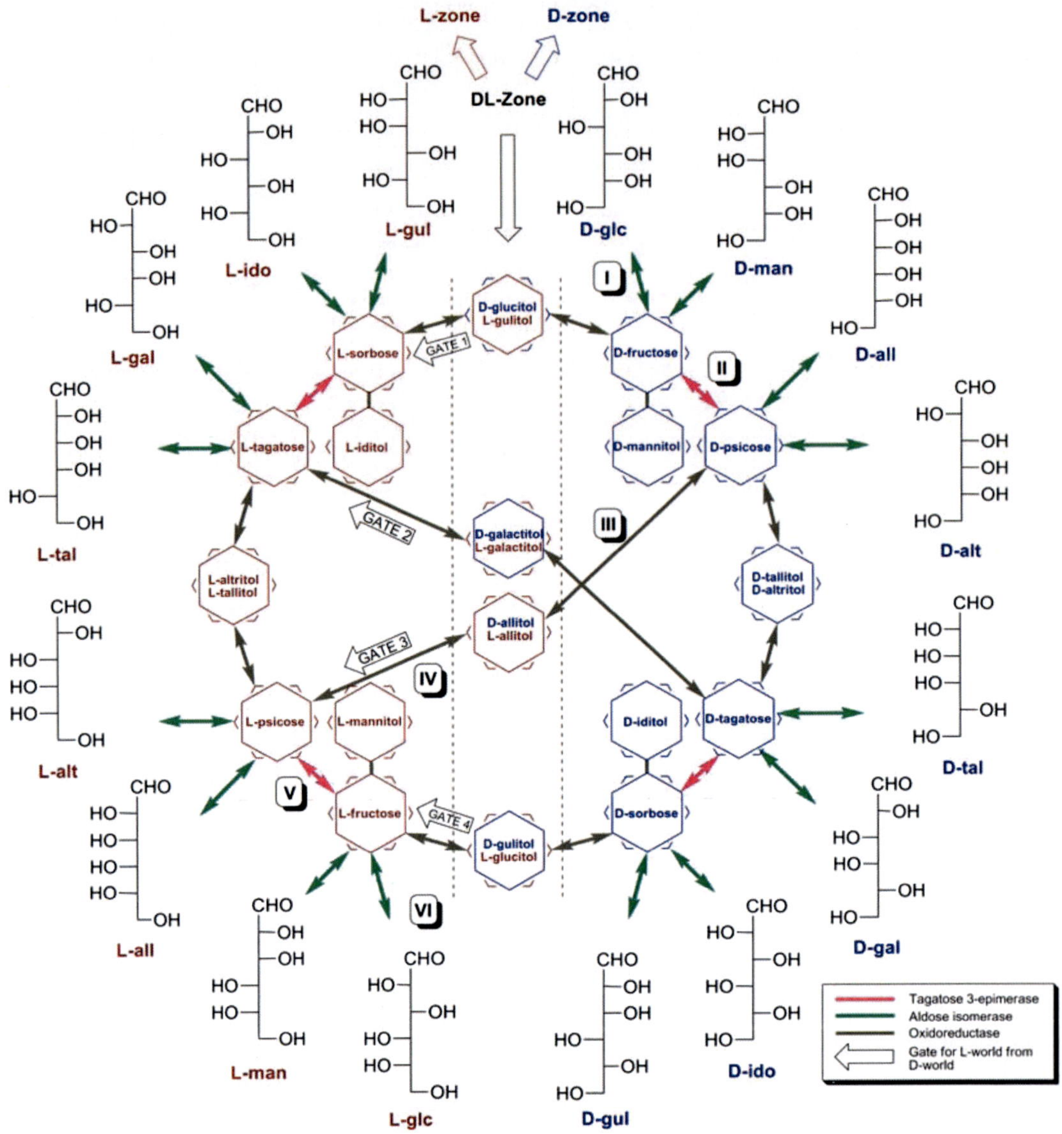


Scheme 28. "Izumoring": bioproduction of hexose sugars.

Using the same technique, a strategy for the preparation of all ketohexoses and hexitols can be easily designed as well.

## 6. Conclusions and Future Perspectives

Although the search of approaches for the synthesis of carbohydrates is now over one hundred years old, [104] the field of investigation is still dynamic, because of the entry of new technologies, new ideas and the development of structured and general approaches. Since the degree of advancement reached in organic synthesis by various synthetic methods allows preparation of almost any kind of polyfunctional system with high stereoselectivity, great versatility, and sometimes in large amount, further progresses are addressing to use such knowledge in providing suitable building blocks for the construction of more complex bioactive molecules, thus meeting and supporting chemical biology and drug discovery endeavours.

## 6.1. Application of Some Synthetic Methods for the Assembly of Biologically Relevant Oligosaccharides

A major impediment to the growth of glycobiology field is the lack of pure, widely available complex carbohydrates and glycoconjugates. In addition to the low concentrations in which such molecules are usually found from natural sources, their identification and isolation is greatly complicated by their microheterogeneity. Supply of synthetic material for biophysical and biochemical studies, therefore, represents a pressing demand and a considerable challenge for the synthetic chemist. With special reference to unnatural oligosaccharides, such as, for example, those containing L-hexose moieties, their synthesis still remains a difficult task, since approaches starting from unprotected L-hexoses are far to be practical and efficient. Thus, in recent times research has been addressed to the employ of expeditious procedures towards L-pyranoside-based building blocks for the construction of complex oligosaccharide structures.

As example of such endeavours, employ of 1,6-anhydro-β-L-hexopyranoses *247*, *252* (see Paragraph 4.1) as building blocks for the synthesis of numerous biologically potent oligosaccharides, has been highlighted [29b,86a] (Scheme 29). Indeed, 1,6-anhydrohexopyranoses are very useful synthons for carbohydrate-based syntheses of complex natural products and bioactive molecules, since their [3.2.1]bicyclic skeleton not only enables high regio- and stereocontrolled reactions, but also allows the employ of fewer protecting groups than the corresponding pyranoses. Moreover, once the requisite functionalities are properly installed, the 1,6-anhydro acetal function can be cleaved and the free hydroxyl groups can be utilized for further elaboration. For the synthetic purposes reported below, 1,6-anhydrosugars were used to obtain glycosyl donor *246* and acceptors *247* and *252*, leading to the disaccharide moiety of bleomycin A2 (*249*) as well as the disaccharide subunits of heparin and heparan sulfate (*254*, *257*).

In the frame of a project devoted to the synthesis of mirror-image oligosaccharides and glycoconjugates, an efficient route to the unnatural enantiomer of a blood group antigen has been envisaged (Scheme 30), with the aim to discover and develop novel compounds with potent biological activity. [16] Such procedure makes use of 4-deoxypentenoside building block *240* (see Paragraph 4.3) and L-glucal derivative *261*, this latter having been synthesized from D-glucal *258* (Scheme 30). [105] These unsaturated pyranosides were further elaborated to be used both as glycosyl donors (*263* and *267*) and as acceptor (*261*) in *O*-glycosidation steps to afford the H-type II blood group determinant trisaccharide *268* (D-Fuc-(α1→2)-L-Gal-(β1→4)-L-Glc*N*Ac-β-OMe).

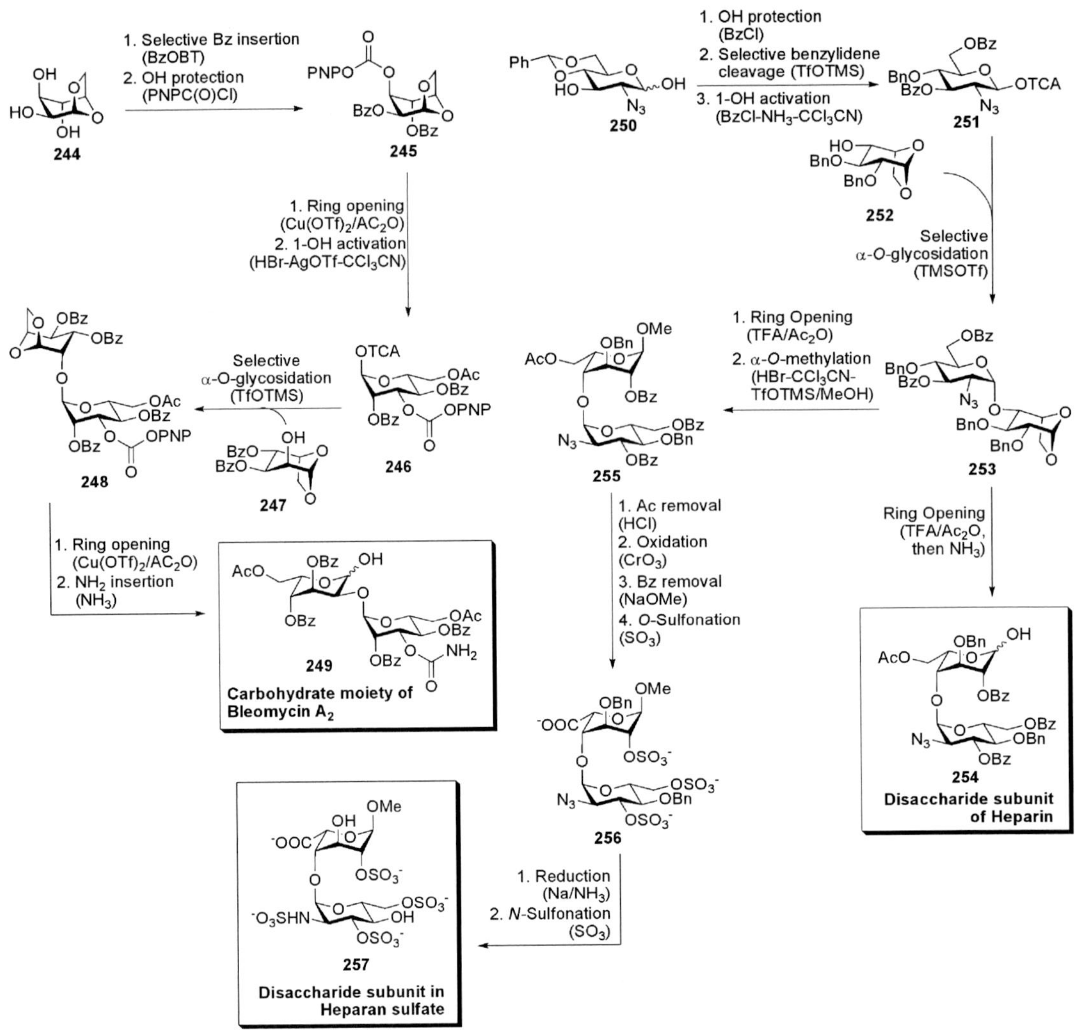


Scheme 29. Synthesis of the carbohydrate moiety of bleomycin $A_2$ *249* and of the disaccharide subunits of heparin and heparan sulfate *254* and *257*.

As further progress in monosaccharide synthesis, also *de novo* protocols for the formation of carbohydrate-like building blocks have been addressed to the synthesis of oligosaccharide sequences, often demonstrating to represent a better alternative compared with the traditional synthetic approaches starting from natural monosaccharides. Indeed in the conventional approaches up to five chemically similar hydroxyl groups need to be differentiated, requiring temporary protecting groups as well as selective masking of the anomeric hydroxyl group: this frequently results in a time consuming and expensive processes. Conversely, *de novo* strategies are convergent and minimize the number of synthetic steps. Dissection of carbohydrate building blocks into linear fragments reduces the number of OH groups and avoids temporary protection of the hemiacetal anomeric function. Moreover, it is possible to prepare several structurally-related carbohydrate building blocks from one common linear precursor by varying the conditions for stereoselective construction of the carbon skeleton.

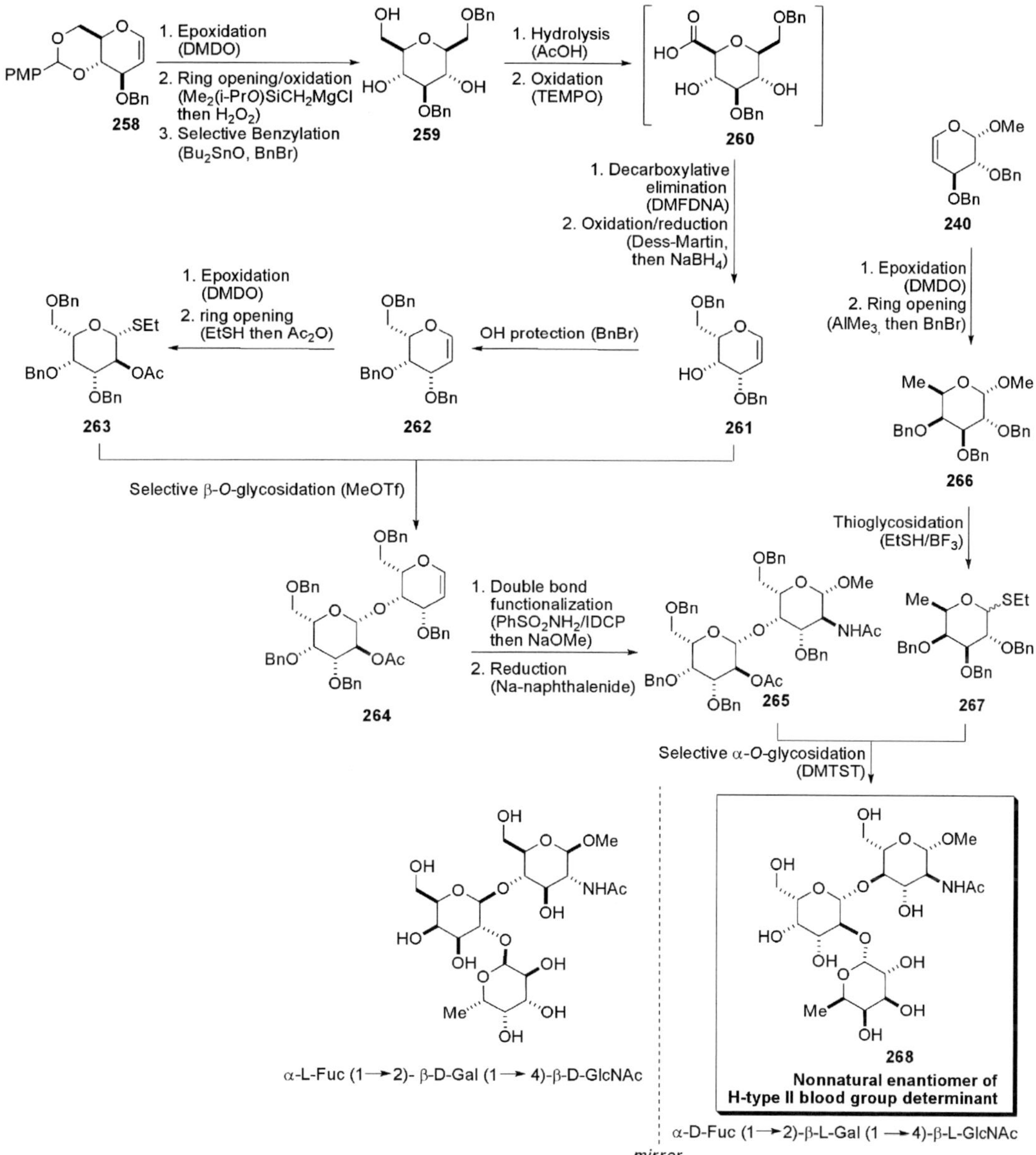


Scheme 30. Synthesis of the unnatural enantiomer of the H-type II blood group trisaccharide (*268*).

Several recent noteworthy examples demonstrate the feasibility of *de novo* strategies for oligosaccharide synthesis. In one of these approaches, mono-, di- and trisaccharide moieties of the cardio-toxin and anticancer agent digitoxin *280* [106] have been stereoselectively synthesized [107] starting from achiral acylfuran *272* by the enantioselective approach to pyranones (see Paragraph 3.2). Central for the successful outcome of this strategy was also the iterative use of a mild palladium-catalyzed glycosylation reaction, that allows the stereoselective oligomerization of pyranones *270* (Scheme 31). [108]

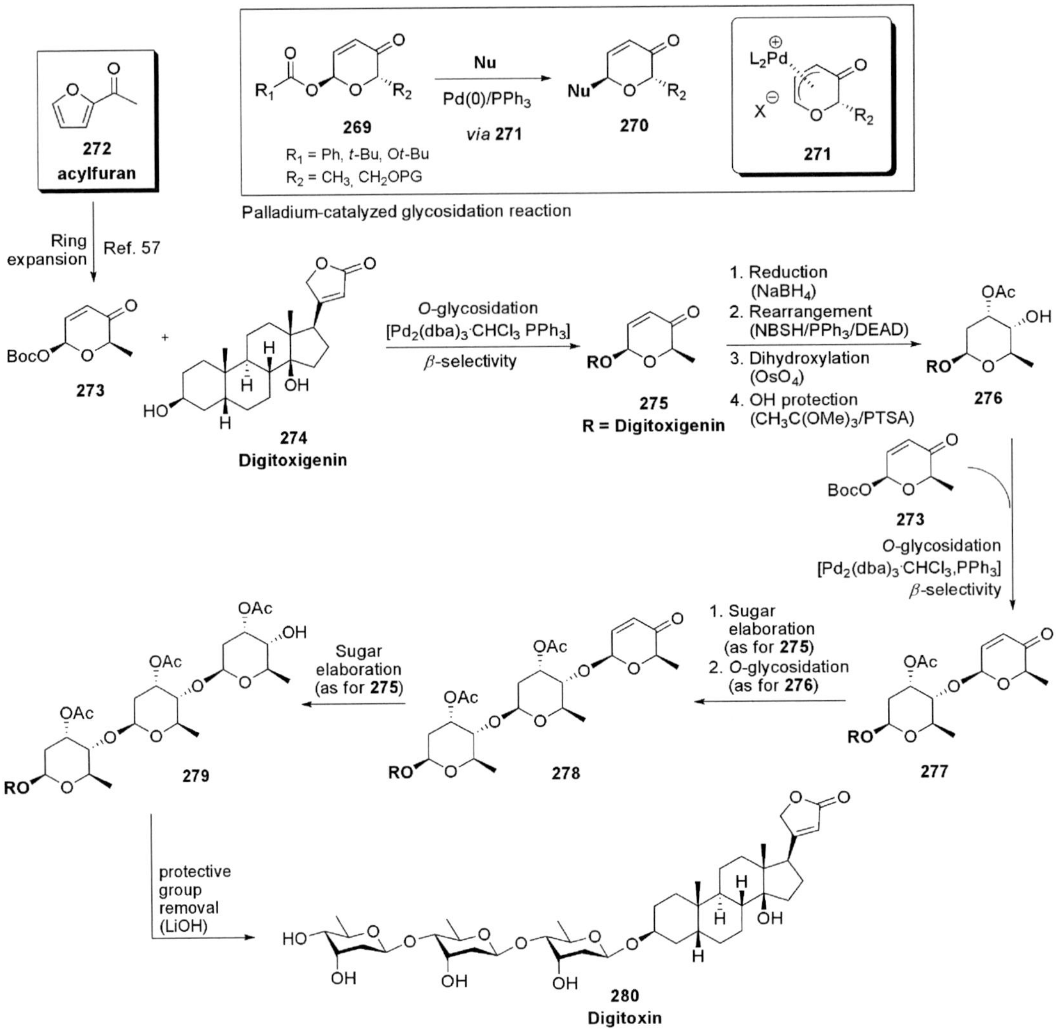


Scheme 31. Oligosaccharide assembly by application of *de novo* protocols: digitoxin (*280*) synthesis.

The Pd-catalyzed glycosylation efficiently provides challenging 2,6-dideoxy-β-glycosides. Moreover, this approach is equally amenable for the synthesis of diastereomeric L-sugar digitoxin analogues.

One of the most explanatory examples of the *de novo* protocols for oligosaccharide synthesis has been lately reported [109] (Scheme 32), in which the successful approach to the natural product Anthrax tetrasaccharide *281* has been described by a highly stereocontrolled 25 step synthesis (longest linear, 39 total steps) with a 13% overall yield from achiral acylfuran *272*. The comparison with a previous monosaccharide-based approach [110] (attained from L-rhamnose *289* and the rare D-fucose *290*) demonstrates the practicability of the *de novo* methodology: while this synthesis is longer in terms of longest linear sequence (Scheme 32), it is shorter in terms of total steps. Moreover, the *de novo* approach is preferable also in terms of overall yield (13% in *de novo* approach vs. 7% in monosaccharide-based approach).

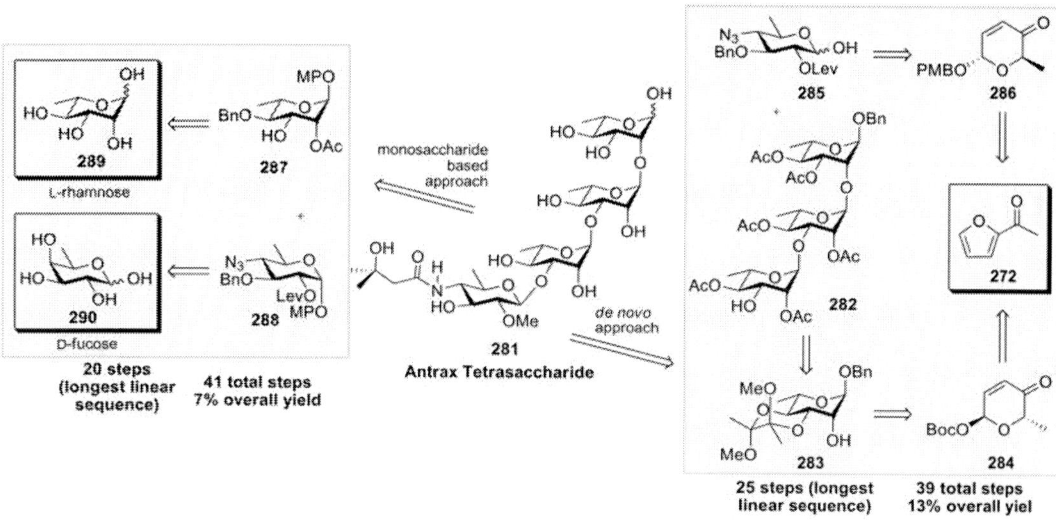


Scheme 32. Monosaccharide-based and *de novo* approaches in the synthesis of Antrax tetrasaccharide *281*.

## References


[1] Varki, A. *Glycobiology* 1993, *3*, 97-130.

[2] (a) Maeder, T. *Sci. Am.* 2002, *287*, 40-47. (b) Shriver, Z.; Raguram, S.; Sasisekharan, R. *Nat. Rev. Drug Discov.* 2004, *3*, 863-873. (c) Seeberger, P. H. *Chem. Commun.* 2003, *10*, 1115-1121.

[3] Dwek, R. A. *Chem. Rev.* 1996, *96*, 683-720.

[4] Gornik, O.; Dumić J.; Flögel M.; Lauc G. *Acta Pharm.* 2006, *56*, 19–30.

[5] (a) Hirabayashi, J.; Kasai, K.-I. *Trends Glycosci. Glycotechnol.* 2000, *12*, 1-5. (b) Feizi, T.; Mulloy. B. *Curr. Opin. Struc. Biol.* 2003, *13*, 602-604.

[6] Sacchettini, J. C.; Baum, L. G.; Brewer, C. F. *Biochemistry* 2001, *40*, 3009-3015.

[7] Simanek, E. E.; McGarvey, G. J.; Jablonowski, J. A.; Wong, C.-H. *Chem. Rev.* 1998, *98*, 833-862.

[8] Geijtenbeck, T.; Torensama, R.; van Vliet, S.; van Duijnhoven, G.; Adema, G.; van Kooyk, Y.; Figdor, C. *Cell* 2000, *100*, 575-585.

[9] Karlsson, K. A. *Biochem. Soc. Trans.* 1999, *27*, 471-474.

[10] Feizi, T. *Carbohydr. Chem. Biol.* 2000, *4*, 851-866.

[11] Tiemeyer, M.; Goodman, C. S. *Development* 1996, *122*, 925-936.

[12] Koeller, K. M.; Wong, C.-H. *Nat. Biotechnol.* 2000, *18*, 835-841.

[13] Yarema, K. J.; Bertozzi, C. R. *Curr. Opin. Chem. Biol.* 1998, *2*, 49-61.

[14] Lillehund, V. H.; Jensen, H. H.; Liang, X.; Bols, M. *Chem. Rev.* 2002, *102*, 515-553.

[15] Bach, P.; Bols, M.; Damsgård, A.; Hansen, S. U.; Lohse, A. In *Glycoscience: Chemistry and Chemical Biology*; Fraser-Reid, B.; Tatsuta, K.; Thiem, J.; Eds.; Springer: Berlin, DE, 2001; Vol. III, 2533-2540.

[16] For an example, see: Boulineau, F. P.; Wei, A. *J. Org. Chem.* 2004, *69*, 3391-3399.

[17] (a) Joyce, G. F.; Visser, G. M.; van Boeckel, C. A. A.; van Boom, J. H.; Orgel, L. E.; van Westrenen, J. *Nature* 1984, *310*, 602-604. (b) Takats, Z.; Nanita, S. C.; Cooks, R. G. *Angew. Chem., Int. Ed.* 2003, *42*, 3521-3523.

[18] Shallenberger, R. S.; Acree, T. E.; Lee, C. Y. *Nature* 1969, *221*, 555-556.

[19] Sabini, E.; Hazra, S.; Konrad, M.; Lavie, A. *J. Med. Chem.* 2007, *50*, 3004-3014.

[20] Kato, A.; Kato, N.; Kano, E.; Adachi, I.; Ikeda, K.; Yu, L.; Okamoto, T.; Banba, Y.; Ouchi, H.; Takahata, H.; Asano, N. *J. Med. Chem.* 2005, *48*, 2036-2044.

[21] Proteins and polypeptides comprised solely of D-amino acids have been synthesized and have shown to be intriguing compounds in receptor binding, as these are able to resist enzymatic degradation. For example, see: (a) Milton, R. C. D.; Milton, S. C. F.; Kent, S. B. H. *Science* 1992, *256*, 1445-1448. (b) Chalifour, R. J.;McLaughlin, R. W.; Lavoie, L.; Morissette, C.; Tremblay, N.; Boule, M.; Sarazin, P.; Stea, D.; Lacombe, D.; P., T.; Gervais, F. *J. Biol. Chem.* 2003, *278*, 34874-34881. (c) Schumacher, T. N. M.; Mayr, L. M.; Minor, D. L. J.; Milhollen, M. A.; Burgess, M. W.; Kim, P. S. *Science* 1996, *271*, 1854-1857. A similar line of thinking has been applied to nucleic acids: oligomers composed of 2-deoxy-L-ribofuranose based nucleotides have been synthesized and successfully assembled into mirror-image DNA duplexes: (a) Urata, H.; Shinohara, K.; Ogura, E.; Ueda, Y.; Akagi, M. *J. Am. Chem. Soc.* 1991, *113*, 8174-8175. (b) Williams, K. P.; Liu, X.-H.; Schumacher, T. N. M.; Lin, H. Y.; Ausiello, D. A.; Kim, P. S.; Bartel, D. P. *Proc. Natl. Acad. Sci. USA* 1997, *94*, 11285-11290.

[22] Rojo, J.; Morales, J. C.; Penadés, S. In *Host-Guest Chemistry. Mimetic Approaches to Study Carbohydrate Recognition; Topics in Currrent Chemistry*, Penadés, S.; Ed.; Springer: Berlin, DE, 2002; Vol. 218, 45-92.

[23] Liu, D.; Shriver, Z.; Venkataraman, G.; El Shabrawi, Y.; Sasisekharan, R. *Proc. Natl. Acad. Sci. USA* 2002, *99*, 568-573.

[24] (a) Smorenburg, S. M.; Van Noorden, C. J. *Pharmacol. Rev.* 2001, 53, 93-105. (b) Liu, D.; Shriver, Z.; Qi, Y.; Venkataraman, G.; Sasisekharan, R. *Semin. Thromb. Hemost.* 2002, *28*, 67-78.

[25] Shriver, Z.; Liu, D.; Sasisekharan, R. *Trends Cardiovasc. Med.* 2002, *12*, 71-77.

[26] Crocker, P. R. *Curr. Opin. Struct. Biol.* 2002, *12*, 609-615.

[27] Fu, X.; Albermann, C.; Jiang, J.; Liao, J.; Zhang, C.; Thorson, J. S. *Nat. Biotech.* 2003, *21*, 1467-1469.

[28] McNaught, A. D. *Pure Appl. Chem.* 1996, *68*, 1919-2008.

[29] (a) Collins, P. M. *Dictionary of Carbohydrates*; Collins, P.M.; Ed.; Chapman and Hall: London, EN, 1987. (b) Lee, J.-C.; Chang, S.-W.; Liao, C.-C.; Chi, F.-C.; Chen, C.-S.; Wen, Y.-S.; Wang, C.-C.; Kulkarni, S. S.; Puranik, R.; Liu, Y.-H.; Hung, S.-C. *Chem. Eur. J.* 2004, *10*, 399-415.

[30] (a) Lane, D. A.; Lindahl, U. *Heparin-Chemical and Biological Properties Clinical Applications*; Lane, D. A.; Lindahl, U.; Eds.; Edward Arnold: London, EN, 1989. (b) Lindahl, U. *Pure Appl. Chem.* 1997, *69*, 1897-1902. (c) Iozzo, R. V. *Annu. Rev. Biochem.* 1998, *67*, 609-652. (d) Rabenstein, D. L. *Nat. Prod. Rep.* 2002, *19*, 312-331.

[31] Ogita, T.; Otake, N.; Miyazaki, Y.; Yonehara, H.; Macfarlane, R. D.; McNeal, C. J. *Tetrahedron Lett.* 1980, *21*, 3203-3206.

[32] (a) Burger, R. M. *Chem. Rev.* 1998, *98*, 1153-1169. (b) Boger, D. L.; Ramsey, T. M.; Cai, H.; Hoehn, S. T.; Stubbe, J. A. *J. Am. Chem. Soc.* 1998, *120*, 9139-9148. (c)

Katano, K.; An, H.; Aoyagi, Y.; Overhand, M.; Sucheck, S. J.; Stevens, W. C.; Hess, C. D.; Zhou, X.; Hecht, S. M. *J. Am. Chem. Soc.* 1998, *120*, 11285-11296.

[33] Stack, R. J. *FEMS Microbiol. Lett.* 1987, *48*, 83-87.

[34] Kubelka, W.; Kopp, B.; Jentsch, K. *Phytochemistry* 1974, *13*, 1805-1808.

[35] Esaki, S.; Ohishi, A.; Katsumata, A.; Sugiyama, N.; Kamiya, S. *Biosci. Biotech. Biochem.* 1993, *57*, 2099-2103.

[36] Hudlicky, T.; Entwistle, D. A.; Pitzer, K. K.; Thorpe, A. J. *Chem. Rev.* 1996, *96*, 1195-1220.

[37] This route has been independently developed by two different research groups: (a) Kim, K. S.; Cho, B. H.; Shin, I. *Bull. Korean Chem. Soc.* 2002, *23*, 1193-1194. (b) Ermolenko, L.; Sasaki, N. A.; Potier, P. *Helv. Chim. Acta* 2003, *86*, 3578-3582. (c) Ermolenko, L.; Sasaki, N. A. *J. Org. Chem.* 2006, *71*, 693-703.

[38] Abushanab, E.; Vemishetti, P.; Leiby, R. W.; Singh, H. K.; Mikkilineni, A. B.; Wu, D. C.-J.; Saibaba, R.; Panzica, R. P. *J. Org. Chem.* 1988, *53*, 2598-2602.

[39] Saïah, M.; Bessodes, M.; Antonakis, K. *Tetrahedron Lett.* 1992, *33*, 4317-4320.

[40] As a matter of fact, usual Kishi's procedure for olefin dihydroxylation resulted in a low *syn/anti* selectivity, see Ref. 60.

[41] Dondoni, A.; Marra, A. *Chem. Rev.* 2004, *104*, 2557.

[42] For some recent synthetic applications, see: (a) Guaragna, A.; Palumbo, G.; Pedatella, S. *J. Org. Chem.* 2004, *69*, 7033-7037. (b) Caputo, R.; Ciriello, U.; Festa, P.; Guaragna, A.; Palumbo, G.; Pedatella, S. *Eur. J. Org. Chem.* 2003, 2617-2621. See also: Guaragna, A.; Pedatella, S.; Palumbo, G. In *e-Encyclopedia of Reagents for Organic Synthesis* (e-EROS); Paquette, L. A.; Ed.; John Wiley & Sons: New York, US, 2008, DOI: 10.1002/047084289X.rn00831.

[43] Guaragna, A.; Napolitano, C.; D'Alonzo, D.; Pedatella, S.; Palumbo, G. *Org. Lett.* 2006, *8*, 4863-4866.

[44] D'Alonzo, D.; Guaragna, A.; Napolitano, C.; Palumbo, G. *J.Org. Chem.* 2008, *73*, 5636-5639.

[45] Caputo, R.; Guaragna, A.; Palumbo, G.; Pedatella, S.; Solla, F. *J. Carbohydr. Chem.* 2000, *19*, 631-634.

[46] Guaragna, A.; D'Errico, S.; D'Alonzo, D.; Pedatella, S.; Palumbo, G. A. *Org. Lett.* 2007, *9*, 3473-3476.

[47] Guaragna, A.; D'Alonzo, D.; De Nisco, M.; Pedatella, S.; Palumbo, G. *Nucleos. Nucleot. Nucl. Acids* 2007, *26*, 959-962.

[48] Honzumi, M.; Taniguchi, T.; Ogasawara, K. *Org. Lett.* 2001, *3*, 1355-1358.

[49] (a) Takano, S.; Higashi, Y.; Kamikubo, T.; Moriya, M.; Ogasawara, K. *Synthesis* 1993, 948-950. (b) Konno, H.; Ogasawara, K. *Synthesis* 1999, 1135-1140.

[50] Hiroya, K.; Kurihara, Y.; Ogasawara, K. *Angew. Chem., Int. Ed.* 1995, *34*, 2287-2289.

[51] Ogasawara, K. *J. Synth. Org. Chem. Jpn.* 1999, *57*, 957-968.

[52] Ko, S. Y.; Lee, A. W. M.; Masamune, S.; Reed III, L. A.; Sharpless, K. B.; Walker, F. J. *Tetrahedron* 1990, *46*, 245-264.

[53] (a) Johnson, R. A.; Sharpless, K. B. In *Comprehensive Organic Synthesis*; Trost B. M.; Ed.; Pergamon Press: New York, US, 1991; Vol.7, 389-436. (b) Johnson, R. A.; Sharpless, K. B. In *Catalytic Asymmetric Synthesis*; Ojima I.; Ed.; VCH Publishers: New York, US, 1993, pp. 101-158.

[54] Ko, S. Y.; Lee, A. W. M.; Masamune, S.; Reed III, L. A.; Sharpless, K. B. *Science* 1983, *220*, 949-951.

[55] Astruc, D. *Organometallic Chemistry and Catalysis*; EDP Sciences; Ed.; Springer-Verlag: Berlin Heidelberg, DE, 2007.

[56] Rassu, G.; Zanardi, F.; Battistini, L.; Casiraghi, G. *Chem. Soc. Rev.* 2000, *29*, 109-118.

[57] (a) Harris, J. M.; Keranen, M. D.; O'Doherty, G. A. *J. Org. Chem.* 1999, *64*, 2982-2983. (b) Harris, J. M.; Keranen, M. D.; Nguyen, H.; Young, V. G.; O'Doherty, G. A. *Carbohydr. Res.* 2000, *328*, 17-36. (c) Babu, R. S.; O'Doherty, G. A. *J. Am. Chem. Soc.* 2003, *125*, 12406-12407.

[58] Such chemistry has been already developed by Ogasawara and augments the earlier work of Achmatowicz, see: (a) Achmatowicz, O.; Bukowski, P.; Szechner, B.; Zwierzchowska, Z.; Zamojski, A. *Tetrahedron* 1971, *27*, 1973-1996. (b) Takeuchi, M.; Nakamura, K.; Ogasawara, K. *Synlett* 1996, 971-972.

[59] Luche, J.-L. *J. Am. Chem. Soc.* 1978, *100*, 2226-2227.

[60] Cha, J. K.; Christ, W. J.; Kishi, Y. *Tetrahedron* 1984, *40*, 2247-2255.

[61] Donohoe, T. J.; Blades, K.; Moore, P. R.; Waring, M. J.; Winter, J. J. G.; Helliwell, M.; Newcombe, N. J.; Stemp, G. *J. Org. Chem.* 2002, *67*, 7946-7956.

[62] The opportunity to have both hexose enantiomers in high optical purity has been exploited in the *de novo* installation of both D,D- and L,L-1,4-α,α-*manno-manno*-disaccharide fragments of mannopeptimycin-E: Babu, R. S.; Guppi, S. R.; O'Doherty, G. A. *Org. Lett.* 2006, *8*, 1605-1608.

[63] Haukaas, M. H.; O'Doherty, G. A. *Org. Lett.* 2002, *4*, 1771-1774.

[64] Haukaas, M. H.; O'Doherty, G. A. *Org. Lett.* 2001, *3*, 3899-3992.

[65] Guppi, S. R.; Zhou, M.; O'Doherty, G. A. *Org. Lett.* 2006, *8*, 293-296.

[66] Haukaas, M. H.; O'Doherty, G. A. *Org. Lett.* 2001, *3*, 401-404.

[67] (a) Takeuchi, M.; Taniguchi, T.; Ogasawara, K. *Synthesis* 1999, *2*, 341-354. (b) Takeuchi, M.; Taniguchi, T.; Ogasawara, K. *Chirality* 2000, *12*, 338-341.

[68] (a) *The organic chemistry of sugars*, Levy D.E.; Fugedi, P.; Eds.; CRC Press/Taylor & Francis: Boca Raton, US, 2006. (b) Vogel P. In *Glycoscience: Chemistry and Chemical Biology*; Fraser-Reid, B.; Tatsuta, K.; Thiem, J.; Eds.; Springer: Berlin, DE, 2001; Vol. II, pp. 1023-1174. (c) Osborn, H.M.I.; Coisson, D. *Mini-Rev. Org. Chem.* 2004, *1*, 41-54.

[69] Jørgensen, K. A. *Ang. Chem., Int. Ed.* 2000, *39*, 3558-3588.

[70] Zamojski, A. In *Preparative Carbohydrate Chemistry*; Hanessian, S.; Ed.; Marcel Dekker Inc.: New York, US, 1997; pp. 615-636.

[71] Auberson, Y.; Vogel, P. *Helv. Chim. Acta* 1989, *72*, 278-286.

[72] (a) Evans, D. A.; Johnson, J. S.; Olhava, E. J. *J. Am. Chem. Soc.* 2000, *122*, 1635-1649. (b) Evans, D. A.; Johnson, J. S. *J. Am. Chem. Soc.* 1998, *120*, 4895-4896. (c) Evans, D. A.; Olhava, E. J.; Johnson, J. S.; Janey, J. M. *Angew. Chem., Int. Ed.* 1998, *37*, 3372-3375.

[73] (a) Audrain, H.; Thorhauge, J.; Hazell, R. G.; Thorhauge, J.; Jørgensen, K. A. *J. Org. Chem.* 2000, *65*, 4487-4497; Johannsen, M.; Jørgensen, K. A.; *Ang. Chem. Int. Ed.* 1998, *37*, 2404-2406.

[74] Jurczak, J.; Chmielewski, M.; Filipek, S. *Synthesis* 1979, 41-42.

[75] Malinowska, M.; Salanski, P.; Caille, J.-C.; Jurczak, J. *Synthesis* 2002, *18*, 2707-2710.

[76] Kwiatkowski, P.; Asztemborska, M.; Jurczak, J. *Synlett* 2004, *10*, 1755-1758.
[77] (a) Córdova, A. *Acc. Chem. Res.* 2004, *37*, 102-112. (b) Kazmaier, U. *Angew. Chem., Int. Ed.* 2005, *44*, 2186-2188. (c) Limbach, M. *Chem. Biodiv.* 2005, *2*, 825-836.
[78] Mangion, I. K.; Northrup, A. B.; MacMillan, D. W. C. *Angew. Chem., Int. Ed.* 2004, *43*, 6722-6724.
[79] Northrup, A. B.; MacMillan, D. W. C. *Science* 2004, *305*, 1752-1755.
[80] Córdova, A.; Ibrahem, I.; Casas, J.; Sundén, H.; Engqvist, M.; Reyes, E. *Chem. Eur. J.* 2005, *11*, 4772-4784.
[81] Córdova, A.; Engqvist, M.; Ibrahem, I.; Casas, J.; Sundèn, H. *Chem. Commun.* 2005, 2047-2049.
[82] (a) Kulkarni, S. S.; Chi, F.-C.; Hung, S.-C. *J. Chin. Chem. Soc.* 2004, *51*, 1193-1200.
[83] Hung, S.-C.; Puranik, R.; Chi, F.-C. *Tetrahedron Lett.* 2000, *41*, 77-80.
[84] Hung, S.-C.; Wang, C.-C.; Thopate, S. R. *Tetrahedron Lett.* 2000, *41*, 3119-3122.
[85] Hung, S.-C.; Wang, C.-C.; Chang, S.-W.; Chen, C.-S. *Tetrahedron Lett.* 2001, *42*, 1321-1324.
[86] (a) Hung, S.-C.; Thopate, S. R.; Chi, F.-C.; Chang, S.-W.; Lee, J.-C.; Wang, C.-C.; Wen, Y.-S. *J. Am. Chem. Soc.* 2001, *123*, 3153-3154. (b) Wang, C.-C.; Lee, J.-C.; Luo, S.-Y.; Fan, H.-F.; Pai, C.-L.; Yang, W.-C.; Lu, L.-D.; Hung, S.-C. *Angew. Chem., Int. Ed.* 2002, *41*, 2360-2362.
[87] Takahashi, H.; Shida, T.; Hitomi, Y.; Iwai, Y.; Ikegami, S. *J. Am. Chem. Soc.* 2000, *122*, 2995-3000.
[88] Takahashi, H.; Shida, T.; Hitomi, Y.; Iwai, Y.; Miyama, N.; Nishiyama, K.; Sawada, D.; Ikegami, S. *Chem. Eur. J.* 2006, *12*, 5868-5877.
[89] Boulineau, F. P.; Wei, A. *Org. Lett.* 2002, *4*, 2281-2283.
[90] Ferrier, R. J.; Hoberg, J. O. In *Advances in Carbohydrate Chemistry and Biochemistry*; Horton, D.; Ed.; Academic Press: New York, US, 2003; Vol. 58, pp 55-119.
[91] (a) Flekhter, O. B.; Baltina, L. A.; Tolstikov, G. A. *J. Nat. Prod.* 2000, *63,* 992-994. (b) Cirillo, P. F.; Danishefsky, S. J. In *Solid Support Oligosaccharide Synthesis and Combinatorial Carbohydrate Libraries*; Seeberger, P. H. Ed.; John Wiley and Sons, Inc.: New York, US, 2001, 15-40.
[92] Philips, K. D.; Zemlicka, J.; Horwitz, J. P. *Carbohydr. Res.* 1973, *30*, 281-86.
[93] Fleming, I.; Sanderson, P. E. J. *Tetrahedron Lett.* 1987, *28*, 4229-4232.
[94] Wong, C.-H.; Whitesides, G. M. *Enzymes in Synthetic Organic Chemistry*; Baldwin, J. E.; Magnus, P. D.; Eds; Tetrahedron Organic Chemistry Series Volume 12; Pergamon (Elsevier Science & Technology Books): Oxford, EN, 1994.
[95] During the last years new technologies have been further implemented to increase the scope and practicality of aldolases as tools for the synthetic chemist, including the discovery of new classes of aldolases: Dean, S. M.; Greenberg, W. A.; Wong, C.-H. *Adv. Synth. Catal.* 2007, *349*, 1308-1320.
[96] Wong, C.-H. *Pure Appl. Chem.* 1995, *67*, 1609-1616.
[97] (a) Wong, C.-H.; Halcomb, R. L.; Ichikawa, Y.; Kajimoto, T. *Angew. Chem., Int. Ed.* 1995, *34*, 412-432. (b) Gijsen, H. J. M.; Qiao, L.; Fitz, W.; Wong, C.-H. *Chem. Rev.* 1996, *96*, 443-473. (c) Takayama, S.; McGarvey, G. J.; Wong, C.-H. *Chem. Soc. Rev.* 1997, *26*, 407-415. (d) Machajewsky, T. D.; Wong, C.-H. *Angew. Chem., Int Ed.* 2000,

*39*, 1352-1374. (e) Silvestri, M. G.; Desantis, G.; Mitchell M.; Wong, C.-H. *Top. Stereochem.* 2003, *23*, 267-342.
[98] Izumori, K. *Naturwissenschaften* 2002, *89*, 120-124.
[99] Ito, H.; Okaya, H.; Khan, A. R.; Tajima, S.; Hayakawa, S.; Izumori, K. *Biosci. Biotechnol. Biochem.* 1994, 2168-2171.
[100] Takeshita, K.; Suga, A.; Takada, G.; Izumori, K. *J. Biosci. Bioeng.* 2000, 453-455.
[101] Ito, H.; Izumori, K. *J. Ferment. Bioeng.* 1996, 351-353.
[102] Shimonishi, T.; Okumura, Y.; Izumori, K. *J. Ferment. Bioeng.* 1995, 620-622.
[103] (a) Sasahara, H.; Mine, M.; Izumori, K. *J. Ferment. Bioeng.* 1998, 84-88. (b) Sasahara, H.; Izumori, K. *J. Biosci. Bioeng.* 1999, 548-550.
[104] One of the first approaches to monosaccharides can be considered the classical 1861's formose synthesis from formaldehyde: Boutlerow, M. A. *Compt. Rend.* 1861, *53*, 145-147.
[105] Boulineau, F. P.; Wei, A. *Org. Lett.* 2004, *6*, 119-121.
[106] Zhou, M.; O'Doherty, G. A. *Curr. Topics Med. Chem.* 2008, *8*, 114-125.
[107] (a) Zhou, M.; O'Doherty, G. A. *Org. Lett.* 2004, *6*, 119-121. (b) Zhou, M.; O'Doherty, G. A. *J. Org. Chem.* 2007, *72*, 2485-2493.
[108] Babu, R. S.; Zhou, M.; O'Doherty, G. A. *J. Am. Chem. Soc.* 2006, *8*, 4339-4342.
[109] Guo, H.; O'Doherty, G. A. *Angew. Chem., Int. Ed.* 2007, *46*, 5206-5208.
[110] Werz, D. B.; Seeberger, P. H. *Angew. Chem., Int. Ed.* 2005, *44*, 6315-6318.

# INDEX

## A

## D

## E



## F

## G



## H

## I



## J



## K



## L

## M



## N

## O



## P

## T